Carbonyl Compounds
Chemistry and Synthetic Applications

KURIYA MADAVU LOKANATHA RAI

M. Sc., Ph.D

Professor of Chemistry (Rtd), University of Mysore,
Visiting Scientist, Vignana Bhavan,
Manasagangotri, MYSURU-570 006 (INDIA)
kmlrai@yahoo.com

INDIA · SINGAPORE · MALAYSIA

Notion Press

Old No. 38, New No. 6
McNichols Road, Chetpet
Chennai - 600 031

First Published by Notion Press 2020
Copyright © Kuriya Madavu Lokanatha Rai 2020
All Rights Reserved.

ISBN 978-1-64760-855-2

CONTENTS

ACKNOWLEDGEMENT

I am indebted to the University of Mysore for extending co-operation for writing this book. I wish to thank my research scholars Dr. P. T. Soumya, Smt. Sumana Y. Kotian and Dr. Narayana N. Kudva for reading the entire manuscript and making helpful suggestions. I thank Dr. Manohara V. Kulkarni, Professor of Organic Chemistry (Rtd.), Karnatak University, Dharwad for providing me the Forward letter for this book. I also thank my family members for their patience and encouragement. I sincerely thank all my ex colleagues for making encouragement. I express my sincere gratitude to Notion press team for publishing this book in a very short time.

FOREWORD

It is both a great pleasure and honor for me to write a foreword to a book written by Prof. Lokanath Rai, former Professor and Chairman, Department of Chemistry, University of Mysore. He has been a true academician, a very good friend of mine and we know each other in all our academic endeavors for the last three decades. Prof. Rai has written a number of book Chapters and monographs especially on heterocyclic chemistry which has been published by reputed International book houses from Europe and USA. His depth of knowledge in chemistry, particularly in Organic Chemistry is reflected in diverse areas of geochemistry, application of thermal methods in organic synthesis, and designing innovative methods in organic synthesis. His research work has been extensively published and cited in International journals of high impact factors. All his colleagues with whom he has interacted in the state universities of Karnataka, Kerala and Tamil Nadu have unequivocally expressed their regard and appreciation for his depth of knowledge in Organic chemistry which he imparts to his students and research scholars. He has provided new experimental methods for quantitative estimations in organic chemistry which have been validated and approved by highest academic bodies in USA and have been included under the standard methods of estimations.

This book entitled "The Chemistry of carbonyl compound and its synthetic applications" is a glowing testimony of what all I have written about Prof. Rai in the above paragraphs. The book has been presented in the form of seven chapters and each chapter has been further divided in to a number of different sections. This book covers all aspects of carbonyl compounds. Beginning with systematic nomenclature, synthetic routes for carbonyl compounds and a wide variety of reactions have been extensively covered. Reactions of carbonyl compounds right from the oldest aldol condensation to Sakurai allylation have found their place in this book. Reduction of carbonyl compounds is an important aspect and has been extensively covered by Prof. Rai. It is pertinent to note that all the name reactions, photochemical reactions, have been dealt with the associated stereo chemical aspects, asymmetric synthetic strategies. Inclusion of quantitative methods and their experimental procedures, calculations has been a hall mark of this book which greatly enhances the usefulness of this book.

Prof. Rai has included has included a large number of NET-type questions with answers. The book is aptly supported by references to original research articles which is again a testimony for the author's commitment towards providing a high quality presentation of the subject.

After going through this book I feel this is a trend setting book useful for (B.Sc.) graduates, post-graduates (M.Sc.) and teachers of Organic Chemistry at both UG and PG levels. This book reminds me of the famous Saul Patai's series of books on Functional groups published in early 1970s. I can summarize my views on this book by saying that this book is a treatise with flow and flavor to all its readers.

Dr. M. V. Kulkarni,
Professor of Organic Chemistry,
and Former Dean Faculty of Science,
Karnatak University, Dharwad

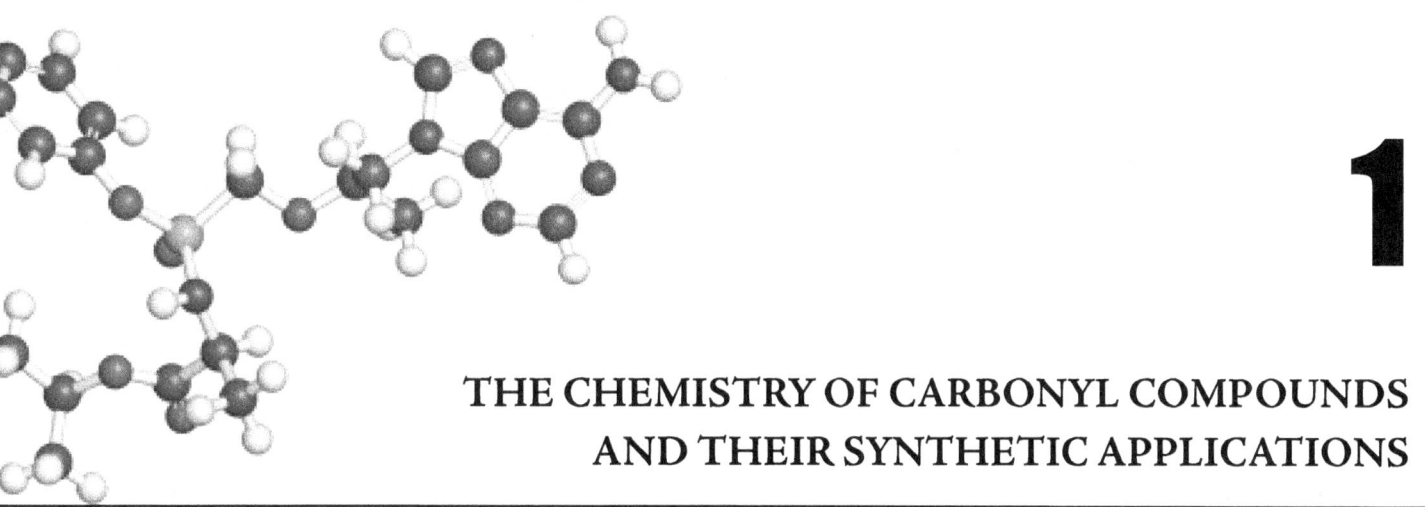

THE CHEMISTRY OF CARBONYL COMPOUNDS AND THEIR SYNTHETIC APPLICATIONS

1.1 INTRODUCTION

The chemistry of carbonyl compound is virtually the backbone of synthetic organic chemistry. Aldehydes and ketones are two types of carbonyl compounds, a more general class of compounds containing the carbonyl group that includes carboxylic acids and their derivatives. Carbonyl compound have the general formula $C_nH_{2n}O$ and contain the carbonyl functional groups (>C=O). In aldehydes, the functional group is –CHO, i.e., one of the available valences of the carbonyl group is attached to hydrogen and so the aldehyde group occurs at the end of a chain. In ketones both available valences are attached to carbon atoms and so the keto group occurs within a chain.

Aldehydes and ketones are among the most important organic compounds for several reasons. First, there are many naturally occurring aldehydes and ketones. Traces of many aldehydes are found in essential oils and often contribute to their favorable odors, e.g. cinnamaldehyde, cilantro, and vanillin. Possibly because of the high reactivity of the formyl group, aldehydes are not common in several of the natural building blocks: amino acids, nucleic acids, lipids. Most sugars, however, are derivatives of aldehydes. These aldoses exist as hemiacetals, a sort of masked form of the parent aldehyde.

Both ketones and aldehydes are found in a number of perfumes. Ketones are used to create acetophenone, which is responsible for creating almond, cherry, honeysuckle, jasmine, and strawberry fragrances. Compared to ketones, aldehydes are a more popular source for perfumes fragrances. The following chart shows common aldehydes used in perfumes and their scent:

Heptanal ($C_7H_{14}O$)	Occurs naturally in sage leaf and has an herbal odour
Octanal ($C_8H_{16}O$)	Has a citrusy scent, specifically smells like oranges
Nonanal ($C_9H_{18}O$)	Smells like roses
Decanal ($C_{10}H_{22}O$)	Has a smell strongly reminiscent of orange-rind
Undecanal ($C_{11}H_{24}O$)	Smells like lilacs and violets
Duodecanal ($C_{12}H_{24}O$)	Has a citrusy scent, specifically smells like grapefruit
Tridecanal ($C_{13}H_{26}O$)	Has a citrusy scent, specifically smells like grape fruit
Tetradecanal ($C_{14}H_{28}O$)	Famous for its peach-skin in mitsouko

Even the delicious scents of cookies baking in the oven come from aldehydes. Aldehydes are an important part of some sugars and are contained in many substances used in baking, such as cinnamon, vanilla, and more.

They also play a crucial role in the caramelization of sugars. When sugars are cooked slowly without stirring, amino acids in the sugar begin the process of turning the aldehyde group into an unsaturated aldehyde. The substance has then become caramel, which can be used to give a product a brown color, create a crust on a baked good, or be consumed plain.

Aldehydes are also contained in many herbs. The aldehyde decanal is a major component of coriander, the leaf which is often said to be the world's most widely consumed herb. Decanal, which is responsible for the coriander leaf's odor, is especially reactive. This causes coriander to quickly lose its scent when heated because the compound often reacts with other compounds. Another commonly encountered aldehyde is hexanal. Also known as the "leaf aldehyde," hexanal is responsible for the "grassy" scent of fresh leaves. This scent fades, however, when the leaves are cut or crushed because the damaged cells release enzymes that are capable of breaking up the six-carbon chain.

Aldehydes are also used in disinfectants and antiseptics. The two types of aldehydes that are most commonly used in commercial cleaners are formaldehyde and glutaraldehyde. Formalin, the aqueous form of formaldehyde, kills bacteria by dehydration. It causes the liquid inside the cells to coagulate. Bacteria can usually flush unwanted toxins from the cell. However, when they are dehydrated, the toxins remain trapped inside the cell and cause the bacteria to die. Formalin is often used to maintain aquariums. It has also been used frequently as an embalming agent, because it helped human cells to retain their form and prevented the body from decaying before the funeral. However, formalin has been discovered to be toxic, allergenic, and even carcinogenic (cancer-causing) when inhaled. It is therefore no longer used for embalmment purposes. Glutaraldehyde is another common cleaner. It kills bacteria, fungi, viruses, and more. Glutamate is able to attack the cell membrane and cell walls in bacteria and fungi, which prevents the cell from functioning. It also affects amino acids and causes proteins to denature. As proteins are responsible for many cell functions and make up cell DNA, this prevents the cell from functioning.

Aldehydes and ketones are also among the most useful compounds in organic synthesis, and the development of the reactions of carbonyl compounds has played a central role in the progress of organic synthesis.

1.2   NOMENCLATURE OF ALDEHYDES AND KETONES

Aldehydes: The lower members are commonly named after the acids that they form on oxidation. The suffix of the names of the acids is "ic"; this suffix is deleted and replaced by "aldehyde".

$$HCHO \text{ (formaldehyde)} \rightarrow HCOOH \text{ (formic acid)}$$
$$CH_3CHO \text{ (acetaldehyde)} \rightarrow CH_3COOH \text{ (acetic acid)}$$

The simplest aromatic aldehyde is benzaldehyde (C_6H_5CHO).

The positions of side chains or substituents are indicated by Greek letter, the α-carbon atoms being the one adjacent to the aldehyde groups.

$$CH_3CH(OH) CH_2CHO \qquad \beta\text{-hydroxy butyraldehyde}$$

In other cases, such as when a -CHO group is attached to a ring, the suffix -carbaldehyde may be used. Thus, $C_6H_{11}CHO$ is known as cyclohexane carbaldehyde. If the presence of another functional group demands the use of a suffix, the aldehyde group is named with the prefix formyl-.

Ketones: The lower members are commonly named accordingly to the alkyl groups attached to the keto group.

CH_3COCH_3	dimethyl ketone	simple or symmetrical
$CH_3CH_2COCH(CH_3)_2$	ethyl isopropyl ketone	mixed or unsymmetrical

Certain aromatic ketones are named by attaching the suffix –ophenone to the appropriate prefix.

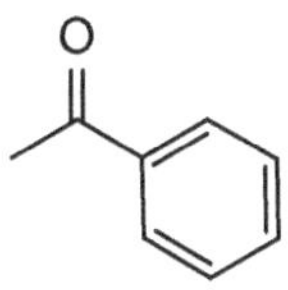

Aceto + phenone = Acetophenone 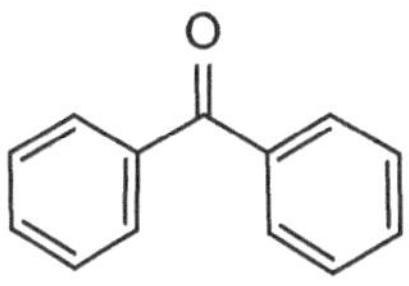   Benzophenone

Acetone (CH_3COCH_3) is nonaromatic ketones for which the common name is routinely used. It is occasionally convenient to name a ketone by citing the names of the two groups on the carbonyl carbon followed by the word ketone.

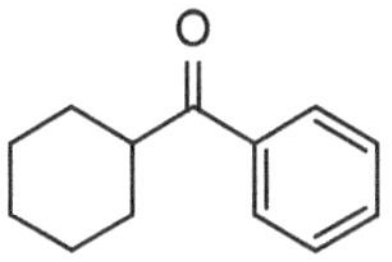

Cyclohexyl phenyl ketone 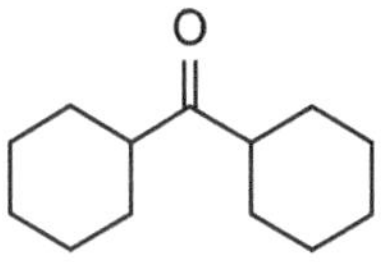   Dicyclohexyl ketone

The positions of side chains or substituents are indicated by Greek letters, the α-carbon atoms being the one adjacent to the keto groups.

$CH_3CH(Cl)$ $COCHCH_2Cl$ α, β'-dichloroethyl ketone

Many common carbonyl- containing substituent groups are named by a suffix –yl added to the appropriate prefix. The names of several groups constructed in this way are shown below:

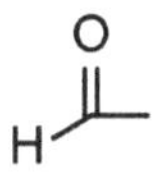

Formyl 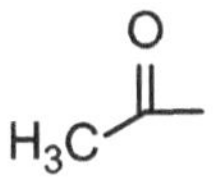   Acetyl 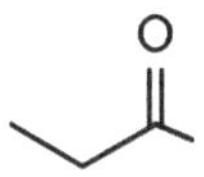   Propionyl 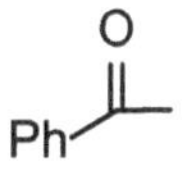   Benzoyl

Such groups are called in general acyl groups.

1.2.1 Systematic (IUPAC) Nomenclature

In the IUPAC system the name of an aldehyde is constructed from a prefix indicating the length of the carbon chain followed by the suffix –al. The prefix is the name of the corresponding hydrocarbon without the final –e. In the following compound there are four carbons in the chain and so the parent hydrocarbon is butane:

$$CH_3CH_2CH_2CHO \qquad butane + al = butanal$$

The carbonyl carbon of an aldehyde receives the number one.

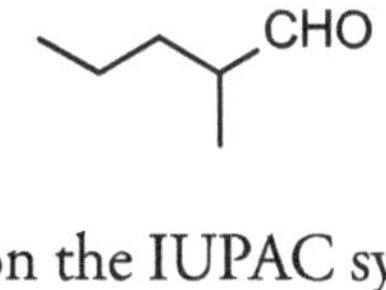

2-methylpentanal (Common name: alpha methylpentanaldehyde)

Notice that on the IUPAC system carbon-1 is the carbonyl carbon, but in the common system, the α-carbon is the carbon adjacent to the carbonyl carbon. As with diols, the final –e the hydrocarbon name is not dropped when there are more than one aldehyde group in the carbon chain.

$$OHCCH_2CH_2CH_2CH_2CHO \qquad \text{hexanedial}$$

When an aldehyde group is attached to a ring, the suffix carbaldehyde is appended to the name of the ring. Carbon-1 in such aldehyde is the ring carbon attached to the carbonyl group.

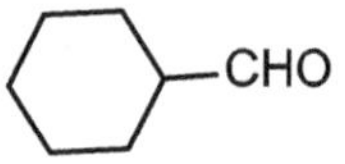

Cyclohexanecarbaldehyde 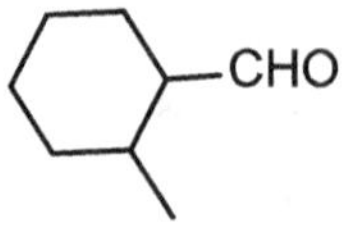   2-methylcyclohexanecarbaldehyde

A ketone is named by giving the hydrocarbon name of the longest carbon chain containing the carbonyl group, dropping the final –e, and adding the suffix –one. The position of the side chains or substituents is indicated by numbers and the keto group is given the lowest number possible.

$$CH_3COCH_2CH_2CH_3{}'$$
$$(CH_3)_2CHCOCH(CH_3)\,CH_2CH_3$$

pentan-2-one

2,4-dimethyl hexan-3-one

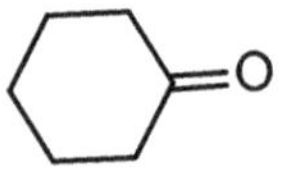

Cyclohexanone

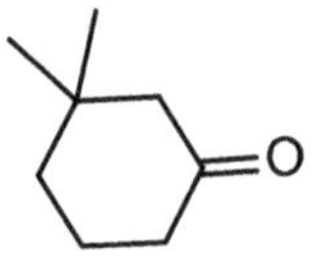

3, 3-dimethylcyclohexanone

As with diols and dialdehyde, the final -e of the hydrocarbon is not dropped in the nomenclature of diones, triones etc. The order of priority for citation of carbonyl groups in the IUPAC system is as follows: **aldehydes > ketones > alcohols > alkenes, alkynes**

Nomenclature of the compound (**1**) illustrates the use of the priority rules. Of the two functional groups present, the carbonyl group has higher priority; consequently the carbonyl group receives lowest possible number. When a ketone group is treated as a substituent, the position of the carbonyl oxygen is designated by the prefix "oxo".

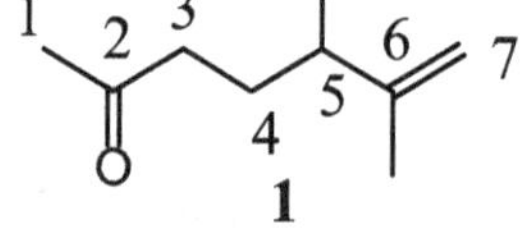

5, 6-dimethyl-6-hepten-2-one

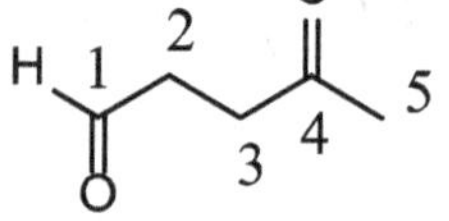

4-oxopentanal

Table 1: Structure, trivial name and IUPAC names of carbonyl compounds

Structure	Trivial name	IUPAC name
HCHO	Formaldehyde	Methanal
CH_3CHO	Acetaldehyde	Ethanal
CH_3CH_2CHO	Propanaldehyde	Propanal
$CH_2{=}CHCHO$	Acrolein	Propenal
CH_3COCH_3	Acetone (dimethyl ketone)	Propan-2-one

Structure	Trivial name	IUPAC name
$CH_3CH_2CH_2CHO$	Butyraldehyde	Butanal
$CH_3CH=CHCHO$	Crotonaldehyde	But-2-enal
$CH_3COCH_2COCH_3$	Biacetyl	Butane-2,3-dione
$CH_3CH(CH_3)\,CHO$	Isobutyraldehyde	2-Methylpropanal
$CH_3CH_2COCH_3$	Ethyl methyl ketone	Butane-2-one
$CH=CHCOCH_3$	Methyl vinyl ketone	But-3-en-2-one
$CH_3CH(OH)\,CH_2CHO$	β-hydroxy butyraldehyde	3-Hydroxybutanal
$CH_3COCH(OH)\,CH_3$	Acetyl methyl carbinol	3-Hydroxybutan-2-one
$CH_3CH_2CH_2CH_2CHO$	Valeraldehyde	Pentanal
$CH_3CH(CH_3)\,CH_2CHO$	Isovaleraldehyde	3-Methylbutanal
$CH_3CH(Cl)\,COCHCH_2Cl$	α, β'-dichloroethyl ketone	1,4-dichloro-pentan-3-one
$CH_3COCH_2COCH_3$	Acetyl acetone	Pentane-2,4-dione
$(CH_3)_2CHCOCH(CH_3)_2$	Diisopropyl ketone	24-Dimethyl-pentan-3-one
$CH_3COCH_2CH_2CHO$	3-acetylpropnaldehyde	4-oxo-pentanal
$CH_3CH_2CH_2CH_2CH_2CHO$	Hexanaldehyde	Hexanal

1.3 GENERAL PROPERTIES OF ALDEHYDES AND KETONES

Except for formaldehyde, which is a gas, the simple aldehydes and ketones are liquids at room temperature. Aldehydes and ketones are polar molecules because of the C-O bond dipole. Dipole measurements of aldehydes and ketones have shown that the values are larger than can be accounted for by the inductive effect of the oxygen atom.

However, aldehydes and ketones have low boiling points compared to with those of the corresponding alcohols. And this is probably due to their inability to form intermolecular hydrogen bonds. On the other hand, carbonyl compounds have higher boiling points than alkenes with similar molecular weights and shape, and this is due to dipole-dipole interactions in the former.

Boiling point	-47.4°	20.8°	78.3°	-6.9°	56.5°	82.3°
Dipole moment	0.4D	2.7D	1.7D	0.5D	2.7D	1.6D

The lower ketones and aldehydes are soluble in water because they can accept hydrogen bonds from water at the carbonyl oxygen. The water solubility of these compounds diminishes rapidly as molecular weight increases.

Acetone and butan-2-one especially are valued solvents because they dissolve not only in water but also in a wide variety of other organic compounds. These compounds also have low boiling points that they can be easily separated from other less volatile compounds. Acetone with a dielectric constant of 20.7D is a polar solvent and is often used as a solvent or co-solvent for nucleophilic displacement reactions.

The α-hydrogens of aldehydes and ketones are weakly acidic. Removal of α-hydrogen from aldehyde by bas gives the conjugate base anion; this anion is called an enolate ion. The pKa of the α-protons in a simple aldehyde or ketone is in the range 16-20. Thus, these compounds are about as acidic as alcohols. The name enolate comes from the fact that the same ion is also the conjugate base of an enol.

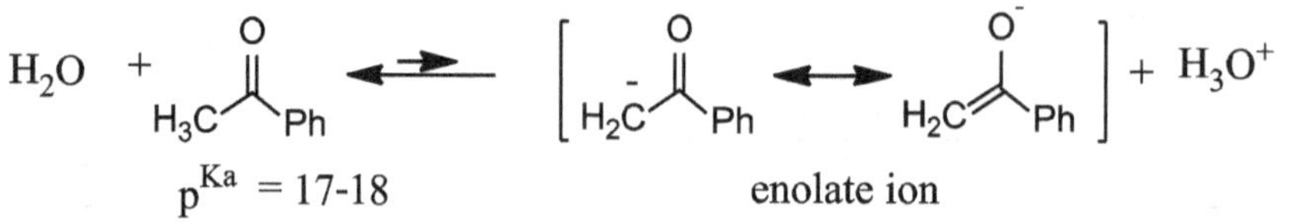

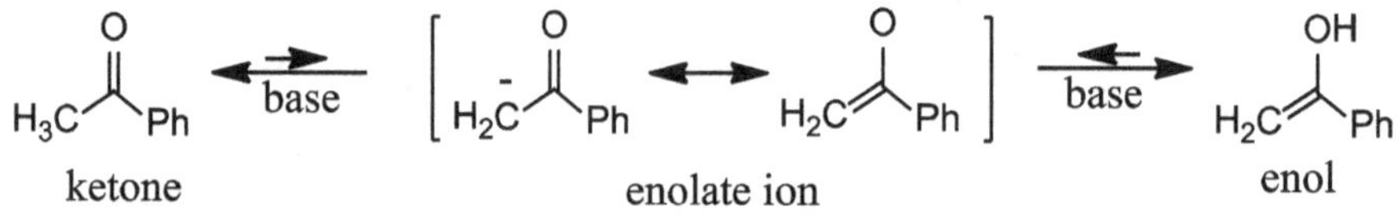

$$H_2O \; + \quad \underset{\underset{p^{Ka} = 17\text{-}18}{}}{\overset{O}{\underset{H_3C}{\|}}{\!}_{Ph}} \; \rightleftharpoons \; \left[\overset{O}{\underset{H_2C}{\|}}{\!}_{Ph} \; \longleftrightarrow \; \overset{O^-}{\underset{H_2C}{\|}}{\!}_{Ph} \right] \; + \; H_3O^+$$

enolate ion

$$\underset{ketone}{\overset{O}{\underset{H_3C}{\|}}{\!}_{Ph}} \; \underset{base}{\rightleftharpoons} \; \left[\underset{enolate\ ion}{\overset{O}{\underset{H_2C}{\|}}{\!}_{Ph} \; \longleftrightarrow \; \overset{O}{\underset{H_2C}{\|}}{\!}_{Ph}} \right] \; \underset{base}{\rightleftharpoons} \; \underset{enol}{\overset{OH}{\underset{H_2C}{\|}}{\!}_{Ph}}$$

The acidity of the aldehydes and ketones is not enough to cause a water-insoluble aldehyde or ketone to dissolve in aqueous base. However, enolate ions are important reactive intermediates in a number of reactions of aldehydes and ketones. As equation 1 indicates aldehydes and ketones are much more acidic than hydrocarbons because the enolate ion is stabilized by resonance. The α-carbon atom of the enolate ion is sp² hybridized with the unshared electron pair in a p orbital. As shown below, the enolate anion is a planar conjugated system of π-electrons much like an allylic anion.

β-Dicarbonyl compounds are unusually acidic because the enolate ion is resonance stabilized by two adjacent carbonyl groups. In fact, acetyl acetone is acidic enough (p^{Ka} =8.95) that its enolate ion can be formed quantitatively in aqueous base.

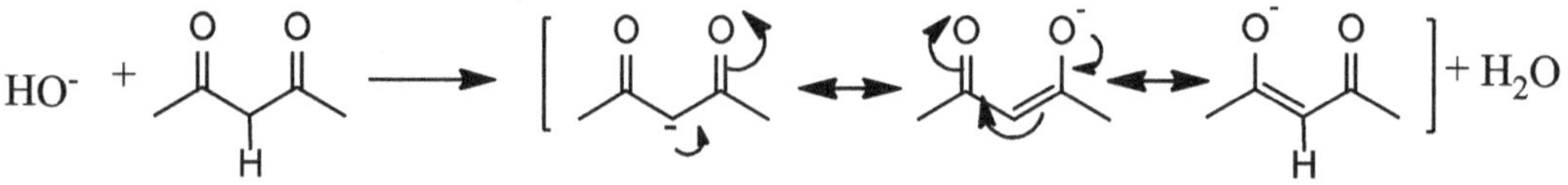

$$HO^- \; + \; \underset{H}{\overset{O \quad O}{\|\quad\|}} \; \longrightarrow \; \left[\; \overset{O \quad O}{\|\quad\|} \; \longleftrightarrow \; \overset{O \quad O^-}{\|\quad\|} \; \longleftrightarrow \; \underset{H}{\overset{O^- \quad O}{\|\quad\|}} \; \right] \; + \; H_2O$$

The carbonyl group also exhibits basic properties; it is readily protonated by strong acids to form oxonium salts. Since oxygen is more electronegative than carbon, the second resonating structures will make a larger contribution than the first. Hence protonation increases the electrophilic character of the carbonyl group and so it can be expected that nucleophilic addition will be catalyzed by acids. It should also be noted that, because of the positive charge on the carbon atom, the CO group has a strong –I effect.

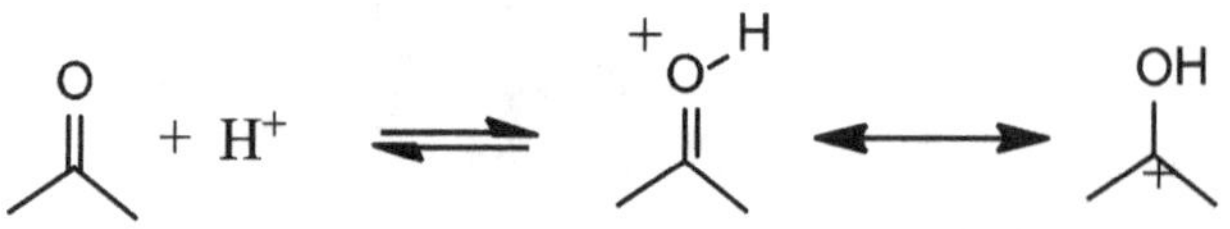

$$\overset{O}{\|} \; + \; H^+ \; \rightleftharpoons \; \overset{\overset{+}{O}\!-\!H}{\|} \; \longleftrightarrow \; \overset{OH}{\underset{+}{\wedge}}$$

Many addition reactions of carbonyl compounds may be represented by the general equation.

$$\underset{R_1}{\overset{O}{\|}}\underset{R_2}{\diagdown} + \text{H-Z} \; \rightleftharpoons \; \underset{R_1}{\overset{OH}{\diagup}}\underset{R_2}{\overset{Z}{\diagdown}}$$

Thus, the hybridization of the carbon atom changes from sp^2 to sp^3. This results in increased crowding and so reaction can be expected to be increasingly sterically hindered with increasing size of R_1 and R_2. For aldehydes, R_2 = H (for formaldehyde, R_1 = R_2 = H), and hence it can be anticipated that aldehydes will be more reactive than ketones. This is observed in practice, reaction being faster and the equilibrium is shifted more to the right.

The +I effect of an alkyl group also decreases the reactivity of carbonyl group towards nucleophilic reagents, this being due to the partial neutralization of the positive charge on the carbonyl carbon. Hence ketones will be less reactive than aldehydes due to both steric and inductive factors.

If we consider the energetics of these addition reactions:

$$\overset{O}{\|} + \text{H-Z} \; \rightleftharpoons \; \overset{OH}{\diagup}{\diagdown}Z$$

$$(+694.5)\;(+)x \qquad\qquad (-334.7\text{-}464.4\text{-}y)$$

Then $\Delta H^* = (x\text{-}y\text{-}104.6)$ kJ where x and y are the bond energies of the H-Z and C-Z bonds respectively. Suppose Z is OH, i.e., H-Z is H-OH. In this case, $\Delta H^* = (464.4\text{-}334.7\text{-}104.6) = 25.1$kj. Now suppose Z is CN, i.e., H-Z is H-CN. In this case, $\Delta H^* = (414.2\text{-}347.3\text{-}104.6) = -37.7$kj.

Since $\Delta G^* = \Delta H^* - T\Delta S^* = -RT\ln K$ and since the number of product molecules in both the cases is less than the number of reactant molecules, there will be a decrease of entropy, i.e., the term $-T\Delta S^*$ will be positive. Hence with water ΔG^* must be positive and the reaction is therefore thermodynamically unfavorable. For hydrogen cyanide, ΔG^* will be considerably less positive (it may actually be negative if $\Delta H^* > T\Delta S^*$) than for the hydration reaction and consequently, relative to the latter, will be highly thermodynamically favorable.

1.4 IDENTIFICATION OF ALDEHYDE OR KETONES

1.4.1 Bisulphite test

The majority of aldehydes and ketones may be detected by treatment with a concentrated aqueous solution of sodium bisulphite, when a crystalline solid separates out, its formation occasionally being accompanied by considerable evolution of heat. This precipitate consists of the sodium salts of α-hydroxy sulphonic acid, usually termed as 'aldehyde or ketone bisulphite compound". It is to be observed that the bisulphite compounds of many carbonyl compounds of these types are readily soluble in water, and may therefore produce no precipitate. An ethereal solution of an aldehyde or a ketone will nevertheless lose its solute on treatment with a saturated bisulphite solution, whether the resulting salt separates out or not.

The tendency of aldehydes and of ketones in which the carbonyl group is attached to methyl or exists in a ring to combine with bisulphite is so pronounced that they develop free alkali on shaking with neutral sodiumsulphite solution (Equation 1).

$$(CH_3)_2CO \; + \; Na_2SO_3 \; + \; H_2O \; \rightleftharpoons \; (CH_3)_2C(OH)SO_3Na \; + \; NaOH \qquad \text{(Equation 1)}$$

This can be observed by the formation of a red colour with phenolphthalein. On the other hand, ketones in which higher alkyl groups are attached to the carbonyl react slowly with bisulphite or not at all.

1.4.2   Iodoform test

Aldehydes and ketones containing a methyl group attached to the carbonyl readily yield iodoform on treatment with iodine and dilute alkali. This reaction is not specific for carbonyl compounds as it is given by ethyl, isopropyl and other alcohols which on oxidation yield the above type of aldehyde or ketone.

Procedure: Dissolve 50 mg (if solid) or 5 drops (if liquid) of unknown in a small vial containing water. Add 0.5 ml of 10% NaOH, and then add the KI/I_2 reagent drop wise with shaking until a definite dark color persists. Then add a few more drops of NaOH to get rid of the iodine color. Allow to stand at room temperature for 15 minutes. Formation of a yellow precipitate is a positive test. The reaction can be performed in slightly acidic or neutral medium. In the presence of large excess of alkali the colour disappears.

It was found that aldehydes give characteristic yellow colour with the solution of chlorite solution due to the formation of chlorine dioxide.

$$RCHO + 3HClO_3 \longrightarrow RCOOH + 2ClO_2 + HCl + H_2O \qquad \text{(Equation 2)}$$

Ketones, phenols, ethers, haloforms, alcohols do not give colouration with chlorite.

1.4.3   Oxime test

Aldehyde and ketones form characteristic derivatives on treatment with hydroxylamine or with hydrazines. Oximes may be prepared by warming the substance with an excess of an aqueous or alcoholic solution of hydroxylamine hydrochloride (Table 1-3) [1960Misc] (equation 3).

$$C_6H_5CHO \; + \; NH_2OH \longrightarrow C_6H_5CH=NOH \; + \; H_2O \qquad \text{(Equation 3)}$$

The carbonyl compound is dissolved in not less than three times its weight of alcohol; to the hot solution of a slight excess of solid hydroxylamine hydrochloride (1 to 1.5 molecular proportions) is added, followed at once by an excess of anhydrous sodium acetate. The mixture is then gently warmed, so as to avoid bumping until a drop of the solution gives a clear solution in 0.5 cc of 1% NaOH, without odour of aldehyde or ketone. The mixture is then concentrated on the water bath, cooled, diluted with water and extracted with ether; or should the oxime have separated out as a solid, this may be filtered off and recrystallized from alcohol.

1.4.4   Phenyl hydrazine test

Phenyhydrazones of aldehydes and ketones may be prepared by warming the substance on the water-bath for about two hours in aqueous-alcohol solution under reflux with either pure phenyl hydrazine, phenyl hydrazine acetate or phenyl hydrazine hydrochloride in presence of excess sodium acetate (equation 4).

$$C_6H_5CHO + PhNHNH_2 \longrightarrow C_6H_5CH=NNHPh + H_2O \qquad \text{(Equation 4)}$$

Procedure: A solution of phenyl hydrazine is prepared by dissolving 1g of the hydrochloride and 1.5 g of sodium acetate in 10 ml of water. The aldehyde or ketone (0.5 g) is dissolved in a little of alcohol and added to the phenyl hydrazine solution, more alcohol being added if necessary to give a clear solution. The mixture is heated on a stream bath for about 30 minutes. The phenyl hydrazones often separate first as solid on cooling. Purification is effected by recrystallization from alcohol.

1.4.5 Dinitrophenylhydrazine test

In this test an insoluble hydrazone derivative is formed. Saturated compounds usually give yellow precipitates, while aromatic or unsaturated compounds give red-orange precipitates. The test is positive for both aldehydes and ketones, but not usually for esters. Easily oxidized alcohols may also give a positive test. Procedure: Add 0.5 ml of 2,4-dinitrophenylhydrazine reagent to a small vial. Add about 10 mg of unknown dissolved in a few drops of ethanol if it is a solid, or 2 drops of unknown if it is a liquid. Allow to stand up to 15 minutes. If a red, orange, or yellow precipitate forms, the test is positive. The reaction between 2,4-dinitrophenylhydrazine and a generic ketone to form a hydrazone is as shown below:

$$RR'C = O + C_6H_3(NO_2)_2NHNH_2 \rightarrow C_6H_3(NO_2)_2NHN = CRR' + H_2O \qquad \text{(Equation 5)}$$

1.4.6 Semicarbazone test

Aldehyde or ketone (1g) is dissolved in alcohol (10ml) and water added until a faint turbidity appears, which is removed by the addition of a few drops of alcohol. Then semicarbazide hydrochloride (1g) and sodium acetate (1.5g) are added. The mixture is vigorously shaken and the test tube is placed in boiling water for half to one hr. The cooled solution is filtered, the residue washed with water and recrystallized from water or aqueous alcohol.

1.4.7 Tollen's test

In this test the aldehyde is oxidized to a carboxylic acid, reducing silver ions to silver metal. Ketones cannot be oxidized, so this is a good way to distinguish ketones from aldehydes.

Procedure: Mix 1 ml of 10% silver nitrate and 0.5 ml of 10% sodium hydroxide in a small test tube. Add 10% ammonium hydroxide dropwise under the dark silver oxide precipitate just dissolves. Then add about 10 mg (if solid) or 2 drops (if liquid) of the unknown and shake well. Allow to stand at room temperature for ten minutes. Formation of a silver coating on the outside of the test tube is a positive test.

1.4.8 Fehling's test

In this test the aldehyde is oxidized to a carboxylic acid, reducing cupric oxide to cuprous oxide. Ketones cannot be oxidized, so this is a good way to distinguish ketones from aldehydes.

Procedure: Mix 2 ml of Fehling's A and 2 ml of Fehling's B in a small test tube. Then add about 10 mg (if solid) or 2 drops (if liquid) of the unknown, shake well and boil the solution for a minute. Appearance of red precipitate indicates the presence of aldehyde group.

1.4.9   The Chromic Acid Tests

The Chromic Acid Tests, sometimes known as the Bordwell-Wellman Test, uses chromic acid to oxidize the aldehydes to carboxylic acids. Ketones do not react. When oxidized, the color changes from orange to blue-green.

1.4.10   A colorimetric method

A colorimetric method for the quantitative determination of acetaldehyde is described which is sensitive to 0.5 microgram. The test uses 2-(4′-phenylazo) phenylhydrazine sulfonic acid for the development of a lavender color which is proportional in intensity to the amount of acetaldehyde present.

1.4.11   The Schiff test

The Schiff test is an reaction developed by Hugo Schiff, and is a relatively general chemical test for detection of many organic aldehydes that has also found use in the staining of biological tissues. The Schiff reagent is the reaction product of a dye formulation such as fuchsine and sodium bisulfite; pararosaniline. In its use as a qualitative test for aldehydes, the unknown sample is added to the decolorized Schiff reagent; when aldehyde is present a characteristic magenta color develops.

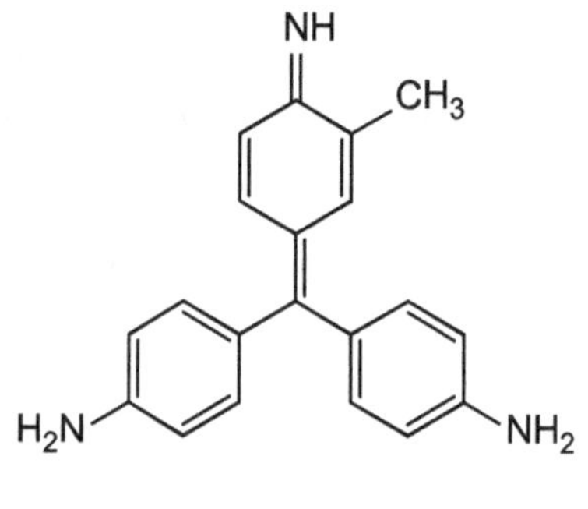

Fuchsine solutions appear colored due to the visible wavelength absorbance of its central quinonoid structure but are "decolorized" upon sulfonation of the dye at its central carbon atom by sulfurous acid or its conjugate base, bisulfite. This reaction disrupts the otherwise favored delocalized extended pi-electron system and resonance in the parent molecule.

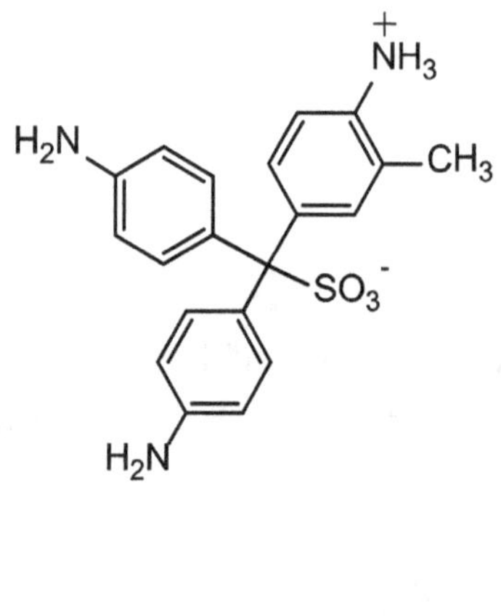

The further reaction of the Schiff reagent with aldehydes is complex with several research groups reporting multiple reaction products with model compounds. In the currently accepted mechanism, the pararosaniline and bisulfite combine to yield the "decolorized" adduct with sulfonation at the central carbon as described and shown. The free, uncharged aromatic amine groups then react with the aldehyde being tested to form two aldimine groups; these groups have also been named for their discoverer as Schiff bases (azomethines), with the usual carbinolamine (hemiaminal) intermediate being formed and dehydrated en route to the Schiff base. These electrophilic aldimine groups then react with further bisulfite, and the $Ar-NH-CH(R)-SO_3^-$ product give rise to the magenta color of a positive test.

1.4.12   Miscellaneous tests

Solution of many aldehydes and ketones develop characteristic red colours with **sodium nitroprusside** and alkali. If instead of alkali, piperidine be employed, blue colors result.

On adding a dilute solution of **m-phenyldiamine** hydrochloride to aqueous or alcoholic solutions of aldehyde or ketones, a green fluorescence is developed. Alkaline solutions of some aldehydes and ketones give red colourations on addition of m-dinitrobenzene. These colour reactions are given only by ketones in which at least one methyl or ethyl radicals is attached to the carbonyl group.

1.4.13   Spot tests

A new spot test on silica gel thin-layers of some carbonyl compounds was described, which was based on their fluorigenic reactions with o-aminodiphenyl dissolved in diluted sulfuric acid. Higher fatty aldehydes, glycol aldehyde, glyoxylic acid and 2,3- pentanedione gave brilliantly fluorescent spots in UV light by heating with the reagent sprayed. Some other non- or sparingly volatile carbonyls also gave positive results.

A method involving the use of ammonium molybdate and excess 70% perchloric acid is described as a detection test for certain classes of organic reducing agents. Mo(VI) is reduced by the organic reducing agents in the presence of concentrated perchloric acid to molybdenum blue. The procedure distinguishes aliphatic aldehydes from aromatic aldehydes, alicyclic ketones from acyclic and aromatic ketones, unsaturated alicyclic hydrocarbons and unsaturated acyclic hydrocarbons possessing conjugate double bonds from monoolefinic acyclic hydrocarbons and aromatic hydrocarbons, phenols, and saturated alicyclic alcohols from saturated acyclic alcohols.

Ammonium molybdate solution (0.1 ml) was placed on a white porcelain spot plate and 0.4 ml of 70% perchloric acid was added to it. After the mixture was stirred, 10 mg of the organic reducing agent were added directly to the molybdic-perchloric acid mixture and the contents were stirred again. The appearance of a blue color or precipitate, either immediately, or within seconds of addition of the reducing agent, was considered a positive test. Where the organic reducing agents were insoluble in the molybdie-perchloric acid mixture, the reducing agents, liquid or solid, were first spotted on the porcelain plate and dissolved in one drop of absolute ethanol before the test was applied.

Table 2: metlting points (in °C) of derivatives for aldehydes - Liquids

Compound	m.p	Oxime	Phenyl hydrazone	Semi- carbazones	2,4-dinitrophenyl hydrazone
Formaldehyde	-21	-	-	169	
Acetaldehyde	21	47	128	163	162
Propionaldehyde	49	40	205	89	155
Glyoxal	50	178		270	
Acrolein	52	-	51	171	
Isobuteraldehyde	73	139		125	182
Buteraldehyde	75	152		126	122
2-Methyl propenal	73			198	
Isovaleraldehyde	92	161		107	123
Chloral	98				
Crotonaldehyde	102	119	56	144	
Valeraldehyde	104	52			111
Paraldehyde	124				
Hexanal	131			106	
Heptanal	154	57		106	106
Furfural	161			202	202
Hexahydrobenzaldehyde	162	91		173	
2-Ethylhexanal	163			254	
Octanal	170	60			101
Benzaldehyde	179	35	158	214	235
Nonanal	190				
Phenylacetaldehyde	194	99	63	156	
Salicylaldehyde	196	57	142	231	248
m-Tolualdehyde	199	60	143	223	
o-Tolualdehyde	200	49	106	212	
p-Tolualdehyde	204	80	112	234	
(+(-citronellal	207			84	171
Decanal	208	69		102	
Citral	208			84	
Glyceraldehyde		118			
Anisaldehyde	248	92	120	203	
o-Chlorobenzaldehyde	213		86	146	221
m-Chlorobenzaldehyde	214		134	229	
o-Bromobenzaldehyde	230	102		214	
m-Bromobenzaldehyde	234		141	205	

Table 3: Melting (in °C) of derivatives for aldehydes - solids

Compound	m.p	Oxime	Phenyl hydrazone	Semi- carbazones	2,4-dinitrophenyl hydrazone
o-Iodobenzaldehyde	37	108	79	206	
Piperonal	37	92	106	215	136

Compound	m.p	Oxime	Phenyl hydrazone	Semi-carbazones	2,4-dinitrophenyl hydrazone
o-methoxybenzaldehyde	38	99	138	231	
o-Aminobenzaldehyde	40	135	221	247	
o-Nitrobenzaldehyde	44	103	156	256	
3,4-Dichlorobenzaldehyde	44		127	232	
p-Chlorobenzaldehyde	47	62	155	226	
2,3-di methoxybenzaldehyde	54	95	121	177	
Veratraldehyde	58	156	206	245	
m-NItrobenzaldehyde	58	122	121	246	
2-Naphthaldehyde	61	106			
p-Bromobenzaldehyde	67	111	113	228	
2,4-di methoxybenzaldehyde	69	117	105	239	
2,6-dichlorobenzaldehyde	71	150			
2,4-dichlorobenzaldehyde	72				
p-Dimethylminobenzaldehyde	74	185	148	222	
p-Iodobenzaldehyde	78		121	224	
Vanilin	81	90	130	198	
p-NItrobenzaldehyde	106	133	159	221	
m-hydroxybenzaldehyde	108	72	178	224	
p-hydroxybenzaldehyde	116	72	177	224	157
Terphthalaldehyde	116	200			
B-Resorcyladehyde	136	192	160	260	
protocatechuicaldehyde	153	157	176	230	
m-Aminobenzaldehyde	amorph	195	162		

Table 4: metlting points (in °C) of derivatives for ketones

Compound	b.p	Oxime	Phenyl hydrazone	Semi-carbazones	2,4-dinitrophenyl hydrazone
Acetone	56	59	42	187	128
Ethyl methyl ketone	80		190	135	115
2,3-Butane dione (Biacetyl)	88	234 -di	243 –di		
Isopropyl methyl ketone	90				120
Diethyl ketone	102			139	
Methyl propyl ketone	102	58			144
Diethyl ketone	102	69			132
Pinacolone	106	78			124
Isobutyl methyl ketone	117	58			91
Chloroacetone	119				125
Diisopropyl ketone	124	34			86
Ethyl propyl ketone	124				110
Butyl methyl ketone	128	49			107
4-methylpent-en-2-one	130	49			203

contd...

Compound	b.p	Oxime	Phenyl hydrazone	Semi-carbazones	2,4-dinitrophenyl hydrazone
Mesityl oxide	130	49		156	
Cyclopentanone	131		142		146
Acetyl acetone	139	149-di			209
2-Methylcyclopentanone	139				
Dipropyl ketone	144				75
3-Hydroxybutan-2-one	145				118
Hydroxyacetone (acetol)	146	71			129
Heptan-2-one	151				89
Cyclohexanone	156	88	74-77	166	162
Diacetone alcohol	164	57			
2-Methylcyclohexanone	165	43			137
4-hydroxy-4-methyl pentan-2-one	168	58			92
Methyl acetoacetate	170				
3-Methylcyclohexanone	170		94		155
4-Methylcyclohexanone	171	39	110		134
Hexyl methyl ketone	173				58
1,3-dichoroacetone	173				133
Cycloheptanone	180				148
Cyclohexyl methyl ketone	180				140
Ethyl acetoacetate	181				93
Dibutyl ketone	188				
(+)-Fenchone	193	167			140
Hexan-2,5-dione	194	137-di	120-d		257-di
Methyl levulinate	196		96		142
Acetophenone	202	59	105	162	
Ethyl levulinate	206		104		102
(-)-Mentone	209	59	53	184	146
(+)-Campour	209	118	233	237	177
Benzalacetone	262	115	156		
Benzophenone	305	141	137	164	229
Benzil	347	237-di	225 -di		
Benzoin	343	151	106		

1.5   SPECTRAL ANALYSIS

1.5.1   UV spectroscopy

The $\pi \text{---}{>} \pi^*$ absorption of unconjugated carbonyl compounds occurs at about 150nm, below the wavelength range of common UV spectrometers. Simple carbonyl compounds, however, have another, much weaker absorption at higher wavelength, about 260-290nm. This absorption is caused by excitation of the unshared electrons on oxygen. This high-wavelength absorption is usually referred to as an $n \text{---}{>} \pi^*$ absorption. [$(CH_3)_2 C{=}O$ $n \text{---}{>} \pi^*$ 271 nm ($\varepsilon = 16$) (in ethanol)]

This absorption is easily distinguished from a π---$>\pi^*$ absorption because it is only 10^{-2}-10^{-3} times as strong. However, the n---$>\pi^*$absorption is strong enough that carbonyl compounds cannot be used as solvents for UV spectroscopy.

The π-electrons of conjugated carbonyl compounds like those of dienes, have strong absorption in the UV spectrum. For instance in the UV spectrum of 1-acetyl-1-cyclohexene, the 232 nm peak is caused by the absorption of light by the conjugated π-electron system and is thus a π---$>\pi^*$ absorption. It has a very large extinction coefficient, much like that of a conjugated diene. The weak absorption at 308 nm is the n--$>\pi^*$absorption.

The λ_{max} of an α,β-unsaturated aldehyde or ketone is governed by the same variables that affect the λ_{max} of conjugated dienes, number of double bonds, type of substitution and the presence of exocyclic double bonds.

Woodward's rule for predicting λ_{max} in the UV spectra of α,β-unsaturated aldehydes and ketones.

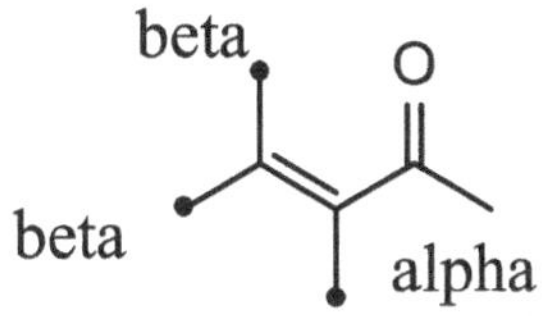

Base values:	nm
Acyclic α,β-unsaturated ketone	215
Acyclic α,β-unsaturated aldehydes	207
6-membered cyclic α,β-unsaturated ketone	215
5-membered cyclic α,β-unsaturated ketone	202

Increments for:	
Double bond extending conjugation	+30
Alkyl group α- substituent	+10
β- substituent	+12
γ- substituent	+18
δ- substituent	+18
-OH, OR α- substituent	+35
(R = alkyl β- substituent	+30
Exocyclic double bond	+05
Sum	λ_{max} (ethanol)

For instance,

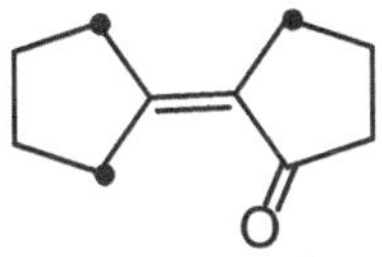

base	215
alpha-substituent	+10
two beta-substituent	+24
exocyclic double bonds (2)	+10
λ_{max} (predicted)	259nm

Note in this example that the base absorption is that for an acyclic ketone. The base for a cyclic ketone would be used if the double bond were endocyclic (within the ring).

For

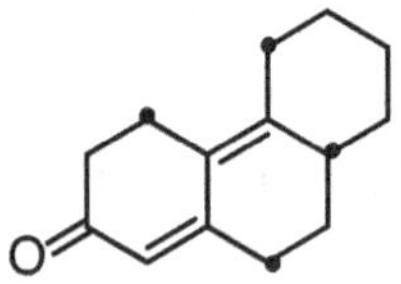

base	215
one beta-substituent	+24
one gama-substituent	+18
two delta-substituent	+36
exocyclic double bonds (3)	+15
Extended conjugation (1)	+30
λ_{max} (predicted)	338nm

For benzene, λ_{max} 204nm (ε = 7900) : for acetophenone, λ_{max} 240nm (ε = 13,000)
204nm (ε = 212)   278nm (ε = 1100)
319nm (ε = 50) (n--->π^*)

Acyclic a-diketones exist in s-trans conformation and show the normal weak R band and a weak K-band arising from the conjugation between carbonyl groups, e.g. biacetyl shows Amax 275 nm (R-band) and -450 nm (K-band). Cyclic α-diketones with α -hydrogen atom(s) exist almost exclusively in the enolic form.

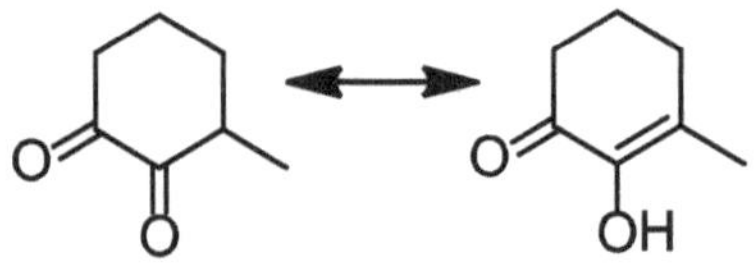

base	215
two beta-substituent	+24
one alpha OH	+35
λ_{max} (predicted)	274nm

Similar to the cyclic a-diketones, cyclic ß-diketones, like 1 ,3-cyclohexanedione exist almost exclusively in the enolic form even in polar solvents.

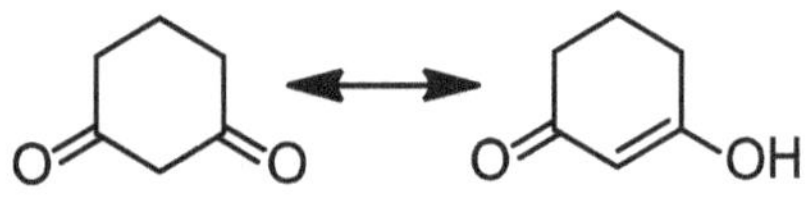

base	215
one beta-substituent	+12
one beta OH	+35
λ_{max} (predicted)	262nm

1.5.2 IR spectroscopy

A carbonyl compound may be considered as a resonance hybrid of the following structures (i) and (ii).

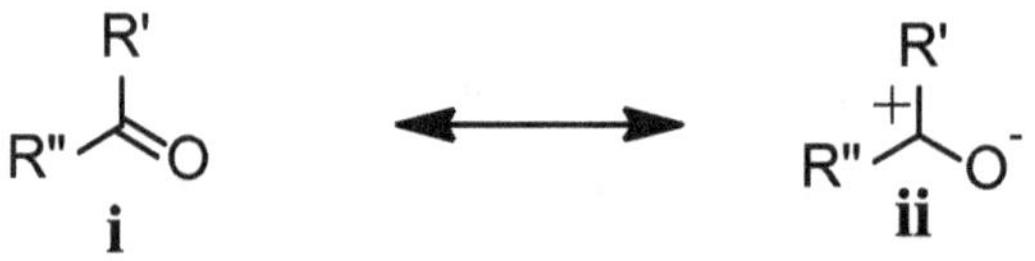

The Stretching frequency of a carbonyl group decreases with increasing number of alkyl groups attached to it. This is due to +I effect of alkyl groups which favours structure (ii) and lengthens (weakens) the carbon-oxygen double bond, and hence its force constant is decreased resulting in the lowering of the C=O Stretching frequency. For example, HCHO, CH_3CHO and CH_3COCH_3 show $\gamma_{c=o}$ absorption at 1750, 1730 and 1720 cm^{-1}, respectively. It should be noted why aldehydes absorb at higher frequency than ketones. Similarly, when a group with -1 effect is attached to a C=O group, it favours structure (i) and its Stretching frequency is increased due to increase in the bond order (force constant) of the carbon-oxygen double bond. For example, CH_3COCH_3, CH_3COCF_3 and CF_3COCF_3 show $\gamma_{c=o}$ bands at 1720, 1769 and 1801 cm^{-1}, respectively. In addition, the aldehyde carbon-hydrogen stretch is an important absorption that occurs at about 2720 cm^{-1}(3.68μ) and typically stands out from other C-H absorptions. Intermolecular hydrogen bonding between a C=O group and a hydroxylic solvent causes a slight decrease in the absorption frequency of the carbonylgroup.

Conjugation of a carbonyl group with an olefinic double bond or an aromatic ring lowers the stretching frequency of the C=O groups by about 30cm^{-1}. This is because the double bond character of the C=O group is reduced by mesomeric effect.

	1715cm^{-1}	1685cm^{-1}	1670cm^{-1}
	(5.83)	(5.93)	(5.99)
		1600cm^{-1}	1613cm^{-1}
	-	(6.25)	(6.20)

This effect can be explained by the resonance structures for α, β-unsaturated carbonyl compounds, which shows that the C=O and C=C bonds have some single bond character. Since these bonds are more like single bonds, they are somewhat weaker than ordinary double bonds and therefore absorb in the IR at lower frequency. The lowering of the frequency of both peaks is consistent with a weakening of both π-bonds (notice that the polarized structure has only single bonds where the C=O and C=C double bonds were). The increase in the intensity of the C=C absorption is consistent with polarization brought about by conjugation with C=O a conjugated C=C bond has a significantly larger dipole moment than its unconjugated cousins.

The carbonyl stretching frequency in cyclic ketones having ring strain is shifted to a higher value. The C-CO-C bond angle in strained rings is reduced below the normal value of 120° (acyclic and six-membered cyclic ketones have the normal C-CO-C angle of 120°). This Ieads to an increase in s character in the sp2 orbital of carbon involved in the C=O bond. Hence, the C=O bond is shortened (strengthened) resulting in an increse in the $g_{c=o}$ frequency. This increase in the s character of the outside sp2 orbital is there because it gives more p character to the sp2 orbitals of the ring bonds which relieves some of the strain, as the preferred bondangle of p orbitals is 90°. In ketones where C-CO-C angle is greater than the normal angle (120°), an opposite effect operates and they have lower $g_{c=o}$ frequency.

| C=O | $1715 cm^{-1}$ (5.83) | $1745 cm^{-1}$ (5.73) | $1780 cm^{-1}$ (5.62) | $1850 cm^{-1}$ (5.42) | $H_2C=C=O$ (a two membered ring) $2150 cm^{-1}$ (4.65) |

1.5.3 NMR spectroscopy

The NMR spectra of carbonyl compounds are affected by the strong inductive effect (-I) of the Carbonyl group and also by the magnetic anisotropy of the carbon-oxygen double bond. Both effects deshield an aldehydic proton, the result being a very low τ value 0.1-0.7 ppm. On the other hand, protons in ketones are deshielded mainly by the –I effect and consequently the shift downfield is much less, e.g., aliphatic ketones containing the MeCO group have τ value 7.8-8.1 ppm for the methyl protons. In the case of α,β-unsaturated carbonyl compounds, the polarization of the C=C bond is also evident in the ^{13}C NMR spectrum, with the signal for the sp^2 carbon atom furthest from the carbonyl group moving downfield relative to an unconjugated alkene to about 140ppm, and the signal for the other double bond carbon atom staying at about 120ppm.

1.5.4 Mass spectra

Aldehydes give molecular ions of low intensity and readily undergo α-cleavage to produce acylium ions. The presence of ions M-1 and m/z 29 (M-R) are usually characteristic of aldehydes (R+ is also formed). It should be noted that the ion m/z 29 could also be $C_2H_5{}^+$, which is given by the higher aldehydes (the two ions may be distinguished by high resolution).

In the C4 and higher aldehydes, McLafferty cleavage of the α,β- C-C bond occurs to give a major peak at m/z44, 58, or 72, depending on the α-substituents.

Aromatic aldehydes are characterized by a large molecular ion peak and by an M-1 peak (Ar-C=O+) that is always large and may be larger than the molecular ion peak. The M-1 ion eliminates CO to give the phenyl ion (m/z 77), which in turn eliminates HC=CH to give C_4H_3+ ion (m/z 51).

Ketones undergo fragmentation patterns similar to those of aldehydes, but the intensity of the molecular ion is very strong and loss of the group with the heavier mass occurs predominantly. Hence, for methyl ketones, the acylium ion CH_3CO^+ (m/z 43) are often the base peak. Alkyl ions are also produced, as well as alkenes and CH_2=CR-OH by the McLafferty rearrangement.

The molecular ion peak in cyclic ketones is prominent. As with aliphatic ketones, the primary cleavage of cyclic ketones is adjacent to the C=O group, but the ion thus formed must undergo further cleavage in order to produce a fragment. The base peak in the spectrum of cyclopentanone and cyclohexanone is m/z 55. The mechanism is similar in both cases: hydrogen shift to convert a primary radical to a conjugated secondary radical followed by formation of the resonance-stable ion, m/z 55.

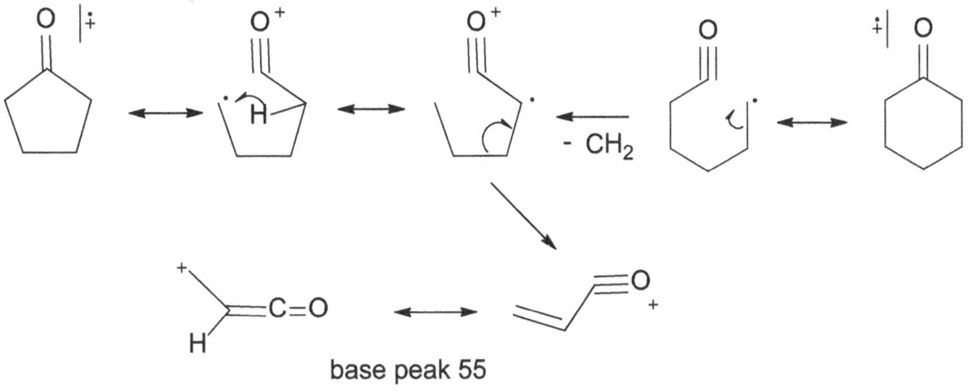

base peak 55

1.6   PREPARATION OF ALDEHYDES AND KETONES

1.6.1   From alcohols

Primary and secondary alcohols, both saturated and unsaturated may be oxidized to the corresponding carbonyl compound by means of manganese dioxide in acetone solution (equation 6). The configuration of the double bond is conserved in the reaction. The correspondingacetylenic alcohols are also suitable substrates, although the resulting propargylic aldehydes can be quite reactive.

$$\text{RCH}_2\text{OH} \xrightarrow[\text{acetone}]{\text{MnO}_2} \text{RCHO} \qquad \text{(Equation 6)}$$

Another method of oxidizing primary and secondary alcohols to the corresponding carbonyl compound is that of Barton et al. (1964). The alcohol is converted into its alkyl chloroformate by reacting it with phosgene, which when dissolved in DMSO underwent substitution followed by the removal of carbon dioxide to form the intermediate alkoxysulfonium ion. This ion on treatment with triethylamine loses dimethyl sulfide to yield the desired aldehyde or ketone via five membered transition states (Scheme 1).

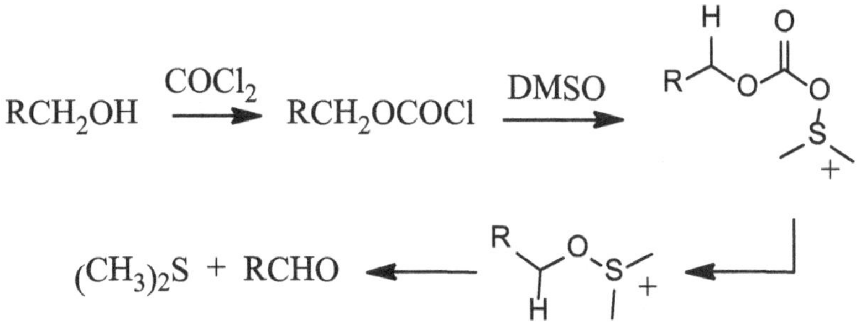

Scheme 1

The **Kornblum oxidation**, named after Nathan Kornblum, is a chemical reaction of a alkyl halide or alkyl tosylates with dimethyl sulfoxide (DMSO) creates an alkoxysulphonium ion, which, in the presence of a base, such as triethylamine (Et$_3$N), will eliminate to form the desired aldehyde (Scheme 2) [1959JACS4113, 1957JACS6562, 1986SC1343].

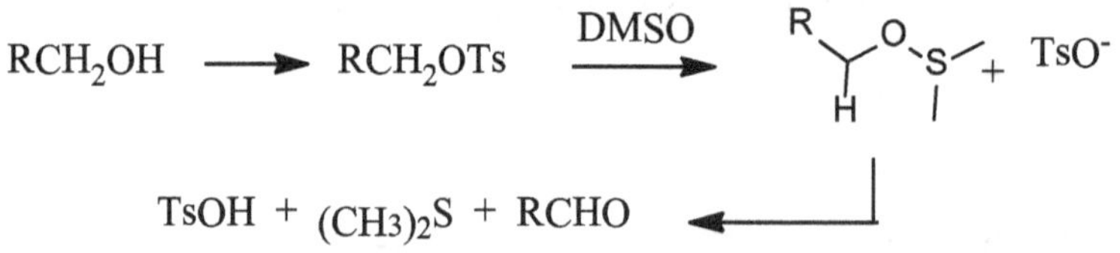

Scheme 2

The **Swern oxidation**, named after Daniel Swern, is a chemical reaction whereby a primary or secondary alcohol is oxidized to an aldehyde or ketone using oxalyl chloride, dimethyl sulfoxide (DMSO) and an organic base, such as triethylamine. The reaction is known for its mild character and wide tolerance of functional groups (Equation 7) [1979JOC4148, 1992JOC5979].

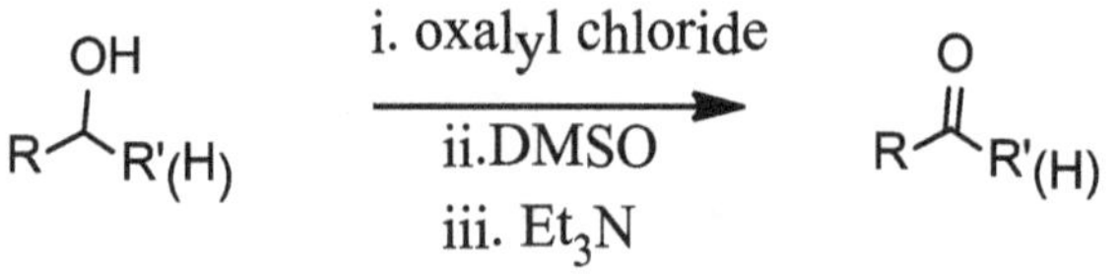

(Equation 7)

The by-products are dimethyl sulfide (Me$_2$S), carbon monoxide (CO), carbon dioxide (CO$_2$) and when triethylamine is used as base triethylammonium chloride (Et$_3$NHCl). Two of the by-products, dimethyl sulfide and carbon monoxide, are very toxic volatile compounds, so the reaction and the work-up needs to be performed in a fume hood.

The first step of the **Swern oxidation** involves the formation of chloro(dimethyl) sulfonium chloride by the reaction of DMSO with oxalyl chloride at low temperature via the elimination of carbon dioxide and carbon monoxide (Scheme 3) [1981Syn145, 1990Syn857].

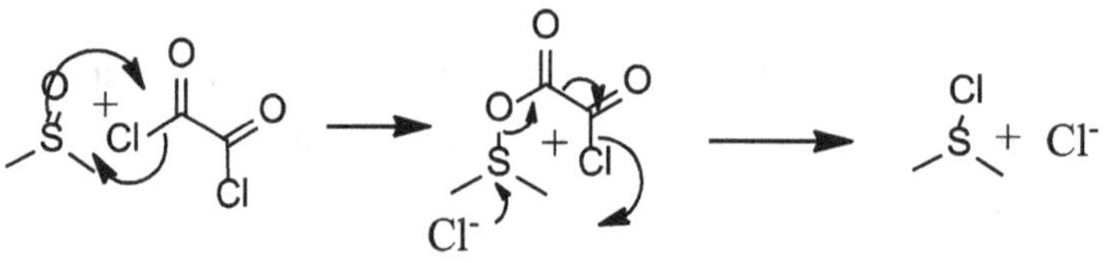

Scheme 3

The resulted chloro(dimethyl)sulfonium chloride reacts with the alcohol to give the intermediate alkoxysulfonium ion, which on treatment with triethylamine loses dimethyl sulfide to yield the desired aldehyde or ketone via five membered transition states (Scheme 4). When using oxalyl chloride as the dehydration agent, the reaction must be kept colder than -60°C to avoid side reactions.

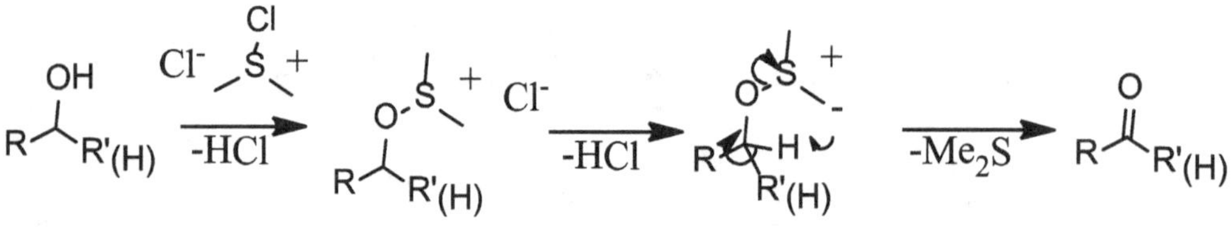

Scheme 4

The reaction conditions allow oxidation of acid-sensitive compounds, which might decompose under the acidic conditions of a traditional method such as Jones oxidation. For example, in Thompson & Heathcock's synthesis of isovelleral (an sesquiterpene), the final step uses the Swern protocol (Equation 8), avoiding rearrangement of the acid-sensitive cyclopropanemethanol moiety.

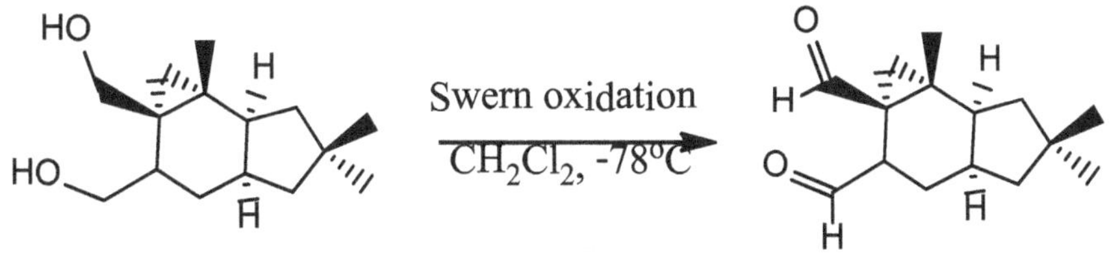

(Equation 8)

The **Parikh–Doering oxidation** is an oxidation reaction that involves the transformation of primary and secondary alcohols into aldehydes and ketones, respectively by the use of dimethyl sulfoxide (DMSO) as the oxidant, activated by sulfur trioxide-pyridine complex in the presence of triethylamine base (Equation 9) [1967JACS5505]. The reaction can be run at mild temperatures, often between 0°C and room temperature, without formation of significant amounts of methylthiomethyl ether side product.

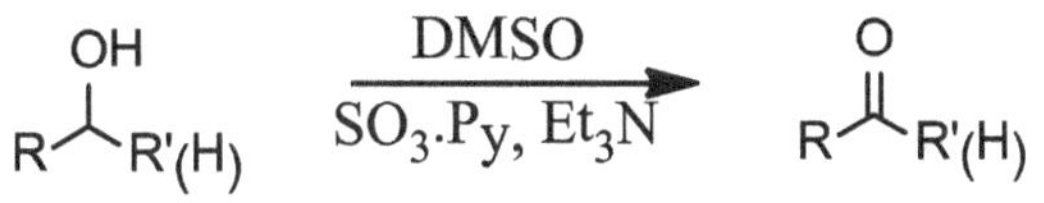

(Equation 9)

The first step of the **Parikh–Doering oxidation** involves the formation of alkoxysulfonium ion associated with the anionic pyridinium sulfate complex by the reaction of DMSO with sulphur trioxide in the presence of pyridine at low temperature (Scheme 5). The resulted anionic pyridinium sulfate complex on treatment with triethylamine loses dimethyl sulfide to yield the desired aldehyde or ketone via five membered transition states.

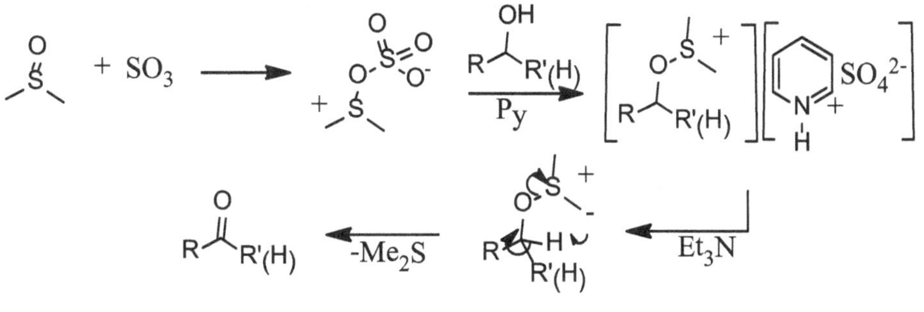

Scheme 5

Reagent **Dess–Martin periodinane (DMP)** is prepared by heating a solution of potassium bromate, sulfuric acid, 2-iodobenzoic acid followed by acetylation with acetic acid and acetic anhydride. DMP is used to oxidize alcohols to ketones (Equation 10). This periodinane has several advantages over chromium- and DMSO-based oxidants that include milder conditions (room temperature, neutral pH), shorter reaction times, higher yields, simplified workups, high chemoselectivity, tolerance of sensitive functional groups, and a long shelf life. It is named after the American chemists Daniel Benjamin Dess and James Cullen Martin who developed the reagent in 1983 [1991JACS7277].

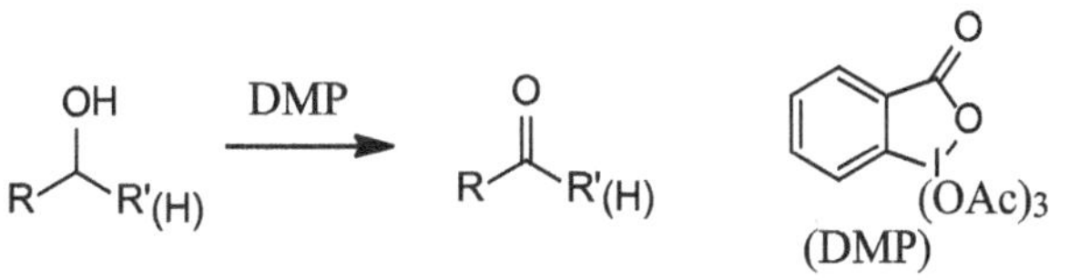

(Equation 10)

Dess-Martin periodinane is mainly used as an oxidant for the oxidation of complex, sensitive and multifunctional alcohols. One of the reasons for its effectiveness is its high selectivity towards complexation of the hydroxylic group, which allows alcohols to rapidly perform ligand exchange; the first step in the oxidation reaction. Proton NMR has indicated that using one equivalent of alcohol forms the intermediate diacetoxy alkoxyperiodinane. The acetate then acts as a base to deprotonate the α-H from the alcohol to afford the carbonyl compound, iodinane, and acetic acid (Scheme 6).

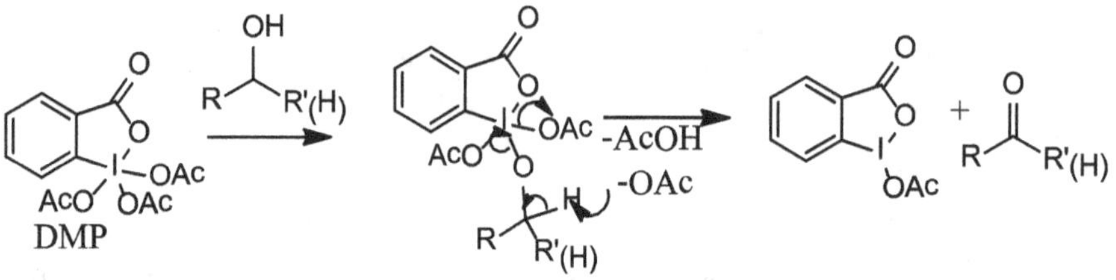

Scheme 6

Schreiber and coworkers have shown that water increases the rate of the oxidation reaction. Dess and Martin had originally observed that the oxidation of ethanol was increased when there was an extra equivalent of ethanol. It is believed that the rate of dissociation of the final acetate ligand from the iodine is increased, because of the electron-donating ability of the hydroxyl group (thus weakening the I-OAc bond).

Corey–Kim oxidation:

In 1972, Corey and Kim developed a new process for the efficient conversion of alcohols to aldehydes and ketones using N-chlorosuccinimide (NCS), dimethyl sulfide (Me_2S) and triethylamine (TEA). The oxidation of primary and secondary alcohols with NCS/DMS is known as Corey–Kim oxidation. The active reagent, S,S-dimethylsuccinimidosulphonium chloride is formed in situ when NCS and DMS are reacted and is called the Core-Kim reagent. The sulphonium salt is then attacked by the nucleophilic alcohol to afford alkoxy sulphonium salt. This alkoxy sulphonium salt is deprotonated by triethylamine and the desired carbonyl compound is formed (Scheme 7) [1972JACS7586]. The dimethyl sulfide is regenerated and it is easily removed from the reaction mixture in vacuum. In the odourless Corey-Kim oxidation instead of dimethyl sulfide, dodecylmethyl sulfide is used. This sulfide lacks the unpleasant odour of DMs due to its low volatility.

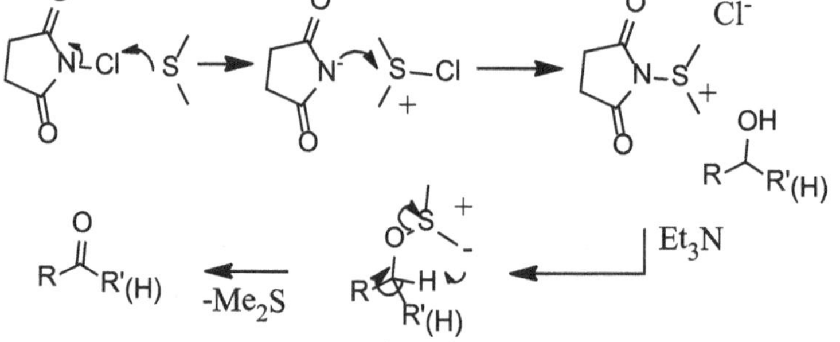

Scheme 7

Pfitzner–Moffatt oxidation

In 1963, Moffatt and Pfitzner observed that primary and secondary alcohols were efficiently oxidized to the corresponding aldehydes and ketones in a solution of dimethyl sulfoxide (DMSO) upon the addition of dicyclohexyl carbodiimide (DCC) and catalytic amount of anhydrous phosphoric acid (H_3PO_4). This transformation is known as the **Pfitzner–Moffatt oxidation (Moffatt oxidation)** and falls into the general category of activated dimethyl sulfoxide mediated oxidation.

Probable mechanism for the oxidation of alcohols by this method involves the initial activation of DMSO by reacting with DCC. Activated DMSO complex reacts with alcohols resulted in the formation of alkoxysulfonium ylide. The decomposition of alkoxysulfonium ylide yields ketone with the elimination of dimethyl sulfide (Scheme 8) [1963JACS3027].

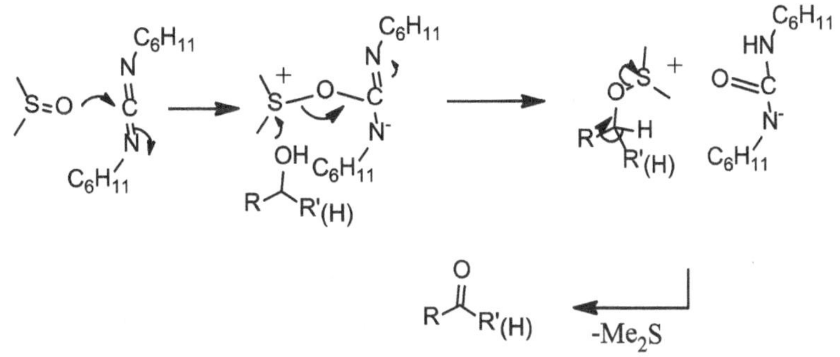

Scheme 8

Pyridinium chlorochromate (PCC) is a yellow-orange salt with the formula $[C_5H_5NH][CrO_3Cl]$ and was prepared via addition of pyridine into a cold solution of chromium trioxide in concentrated hydrochloric acid: PCC is used as an oxidant. In particular, it has proven to be highly effective in oxidizing primary and secondary alcohols to aldehydes and ketones, respectively. A typical PCC oxidation involves addition of an alcohol to a suspension of PCC in dichloromethane [2007Misc].

Oppenauer oxidation, named after Rupert Viktor Oppenauer, is a gentle method for selectively oxidizing secondary alcohols to ketones. Aluminium t-butoxide is specific reagent for oxidizing secondary alcohols to ketones. The secondary alcohol is refluxed with the reagent and then acetone is added. The mechanism of the reaction is the reverse of that of the Meerwein–Pondorf–Verley reduction (Scheme 9) [1937RTCPB137]. The oxidation is highly selective for secondary alcohols and does not oxidize other sensitive functional groups such as amines and sulfides.

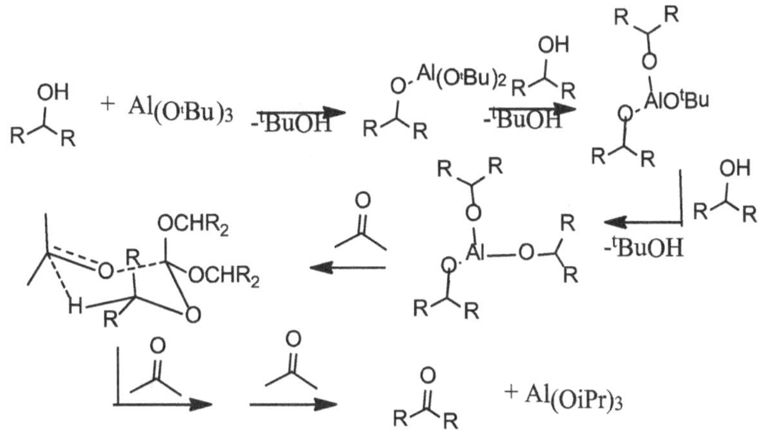

Scheme 9

In the first step of the mechanism, the alcohol coordinates to the aluminium to form a complex, which then, in the second step, gets deprotonated by an alkoxide ion to generate an alkoxide intermediate. In the third step, both the oxidant acetone and the substrate alcohol are bound to the aluminium. The acetone is coordinated to the aluminium which activates it for the hydride transfer from the alkoxide. The aluminium-catalyzed hydride shift from the α-carbon of the alcohol to the carbonyl carbon of acetone proceeds over a six-membered transition state. The desired ketone is formed after the hydride transfer. An advantage of the Oppenauer oxidation is its use of relatively inexpensive and non-toxic reagents. Reaction conditions are mild and gentle since the substrates are generally heated in acetone/benzene mixtures. Another advantage of the Oppenauer oxidation which makes it unique to other oxidation methods such as pyridinium chlorochromate (PCC) or Dess-Martin periodinane is that secondary alcohols are oxidized much faster than primary alcohols, thus chemoselectivity can be achieved. Furthermore, there is no over oxidation of aldehydes to carboxylic acids as opposed to another oxidation methods such as Jones oxidation.

This reagent is particularly useful for oxidizing unsaturated secondary alcohols because it does not affect the double bond. On the other hand, primary alcohols particularly unsaturated may also be oxidized to aldehydes if acetone is replaced by p-benzoquinones.

In the **Wettstein-Oppenauer reaction**, discovered by Wettstein in 1945, Δ 5–3β-hydroxy steroids are oxidized to Δ 4,6-3-ketosteroids with benzoquinone as the hydrogen acceptor. This reaction is useful in that it affords a one-step preparation of Δ 4,6-3-ketosteroids (Scheme 10) [1955JACS3199].

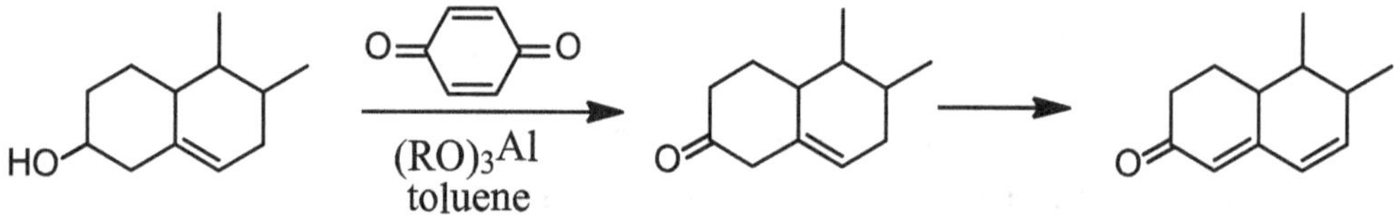

Scheme 10

Fétizon oxidation

Silver(I) carbonate absorbed onto the surface of celite (**Fétizon's reagent,** is prepared by adding silver nitrate to an aqueous solution of a carbonate, such as sodium carbonate or potassium bicarbonate, while being vigorously stirred in the presence of purified celite) is a mild oxidizing agent which will oxidize primary and secondary alcohols to the corresponding aldehyde and ketones respectively. This reaction is called **Fétizon oxidation** first employed by Marcel Fétizon in 1968. This reagent is suitable for both acid and base sensitive compounds [1968ACSP900, 1971JOC1339].

A proposed mechanism for the oxidation of an alcohol by Fétizon's reagent involves single electron oxidation of both the alcoholic oxygen and the hydrogen alpha to the alcohol by two atoms of silver(I) within the celite surface (Scheme 11). The carbonate ion then proceeds to deprotonate the resulting carbonyl generating bicarbonate which is further protonated by the additionally generated hydrogen cation to cause elimination of water and generation of carbon dioxide.

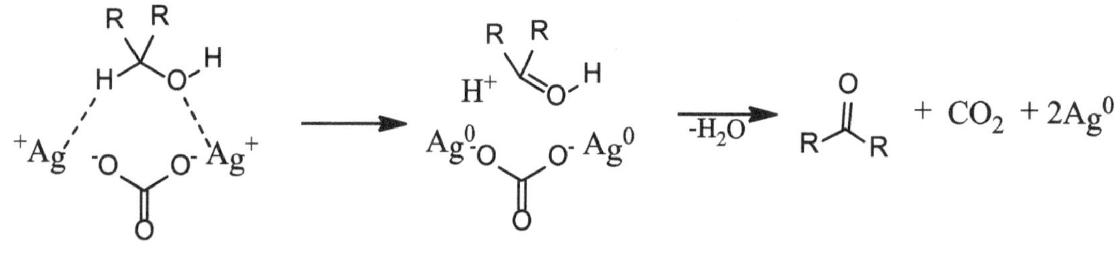

Scheme 11

The rate limiting step of this reaction is proposed to be the initial association of the alcohol to the silver ions. As a result, the presence of even weakly associating ligands to the silver can inhibit the reaction greatly. As a result, even slightly polar solvents of any variety, such as ethyl acetate or methyl ethyl ketone, are avoided when using this reagent as they competitively associate with the reagent. Additional polar functionalities of the reactant should also be avoided whenever possible as even the presence of an alkene can sometimes reduce the reactivity of a substrate 50 fold. Commonly employed solvents such as benzene and xylene are extremely non-polar and further acceleration of the reaction can be achieved through the use of the more non-polar heptane. The solvent is also typically refluxed to drive the reaction with heat and remove the water generated by the reaction through azeotropic distillation. Steric hindrance of the hydrogen alpha to the alcohol is a major determination of the rate of oxidation as it affects the rate of association. Tertiary alcohols lacking an alpha hydrogen are selected against and generally do not oxidize in the presence of Fétizon's reagent. The inability of Fétizon's reagent to oxidize tertiary alcohols makes it extremely useful in the monooxidation of a [1,2] diol in which one of the alcohols is tertiary while avoiding cleavage of the carbon-carbon bond.

Fetizon's reagent oxidizes secondary alcohols selectively in the presence of tertiary alcohols. The mildness and structural sensitivity of the reagent also makes this reagent ideal for the monooxidation of a symmetric diol (Equation 11) [1969JCSCS1102].

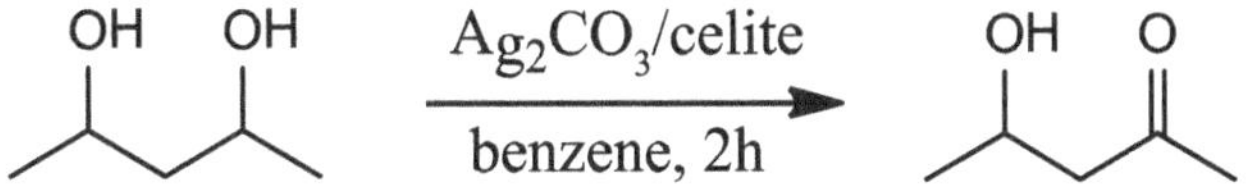

(Equation 11)

Lactols are extremely sensitive to Fétizon's reagent, being oxidized very quickly to lactone functionality. This allows for the selective oxidation of lactols in the presence of other alcohols. This also allows for a classic use of Fétizon's reagent to form lactones from a primary diol. By oxidizing one of the alcohols to an aldehyde, the second alcohol equilibrates with the aldehyde to form a lactol which is reacted quickly with more Fétizon's reagent to trap the cyclic intermediate as a lactone (Scheme 12) [1969JCS1118]. This method allows for the synthesis of seven-member lactones which are traditionally more challenging to synthesize. Treatment of a terminal diol with Fetizon's reagent can result in lactone formation

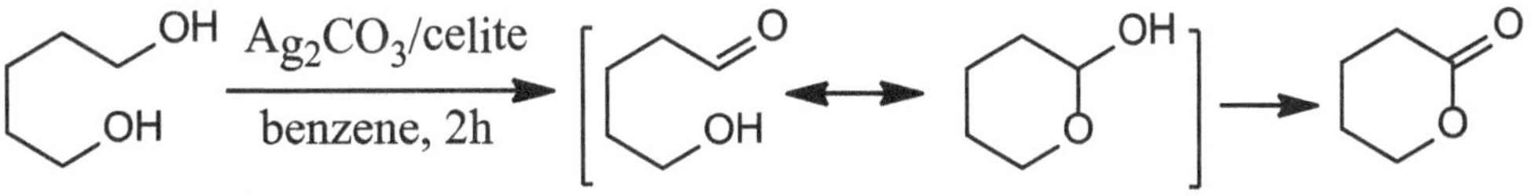

Scheme 12

Phenol functional groups can be oxidized by Fetizon's reagent to their respective quinone forms. These quinones can further couple within solution producing numerous dimerizations depending upon their substituents.

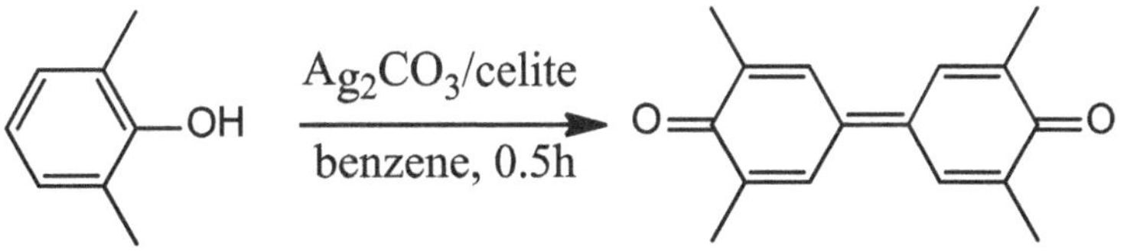

(Equation 12)

Rai et al [1983CS62] developed chloramine-T as effective oxidizing agent for the oxidation of alcohols to ketone under reflux condition in ethanol as solvent. In a typical experiment, they used podophyllotoxin, an anticancer agent as alcohol to form picropodophyllone. Probable mechanism for the reaction is as shown in the scheme 13.

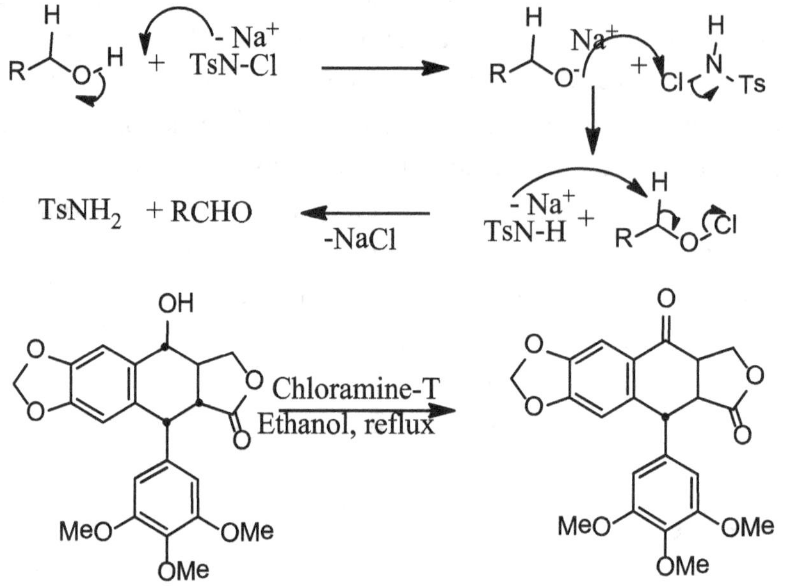

Podophyllotoxin   Picropodophyllone

Scheme 13

Oxidation of primary and secondary alcohols, using catalytic amounts of TEMPO and tetra-n-butylammonium bromide in combination with periodic acid and wet alumina in dichloromethane is compatible with a broad range of functional groups and acid-sensitive protecting groups. The system also enables a chemoselective oxidation of secondary alcohols in the presence of primary alcohols [2014Synl2923].

1.6.2   Cleavage of glycols

Cleavage of glycols with lead tetraacetate:

The carbon-carbon bond between the OH groups of a vicinal glycol can be cleaved with lead tetraacetate to give two carbonyl compounds. The cleavage of glycols with lead tetraacetate takes place through a cyclic intermediate which then breaks down by a concerted electronic rearrangement (Scheme 14).

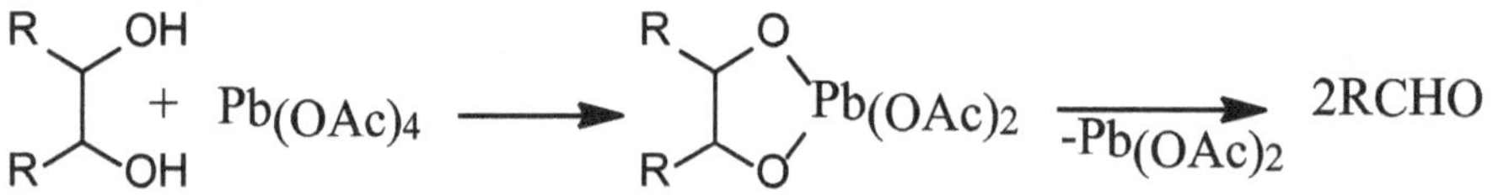

Scheme 14

Cleavage of glycols with periodic acid:

The carbon-carbon bond between the OH groups of a vicinal glycol can be cleaved with periodic acid to give two carbonyl compounds. The cleavage of glycols with periodic acid takes place through a cyclic ester intermediate which forms when glycol displaces water from H_5IO_6. The cyclic ester then breaks down by a concerted electronic rearrangement (Scheme 15).

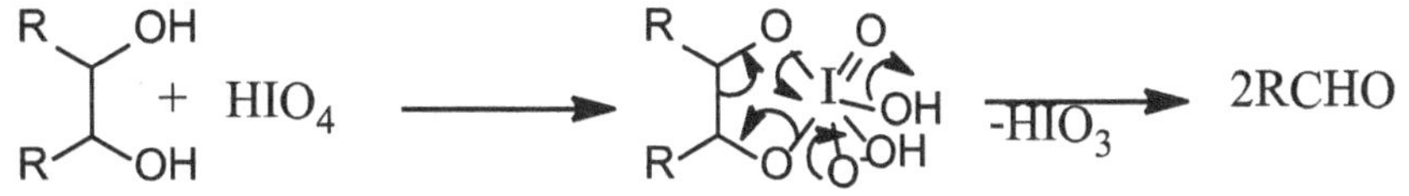

Scheme 15

Since alkenes can be oxidized to glycols with osmium tetroxide (OsO_4) and the resulting glycols cleaved with periodic acid, it follows that the successive application of these two reagents will affect the net cleavage of an alkene across the carbon-carbon double bond. These reactions can be carried out separately, but they can also be carried out in the same reaction vessel. Oxidation of alkene with OsO_4 followed by treatment with two equivalents of periodic acid yield the corresponding aldehydes in almost quantitative yield (Scheme 16). This reaction is called the Lemiux-Johnson oxidation, after its discoverers. The OsO_4 of course forms the cyclic osmate ester, which breaks down to the glycol. One equivalent of periodate reoxidizes the osmium back to Os(VIII) for another cycle of oxidation, the other equivalent of periodate brings about the glycol cleavage.

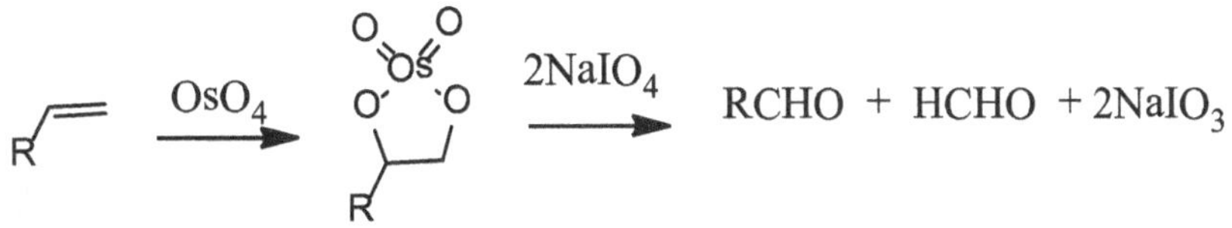

Scheme 16

1.6.3   From aromatic compounds

1.6.3.1   Vilsmeir-Haack formylation:

The Vilsmeir-Haack formylation is a method used for synthesizing aromatic aldehyde from aromatic compounds that are highly activated towards substitution. The method is frequently used with phenol and phenol ethers such as anisole. The source of the formyl group is formamide. The reaction is an electrophilic aromatic substitution related to Friedel-Crafts acylation. The electrophile is a complex formed between phosphorous oxychloride ($POCl_3$) and the formamide (Scheme 17) [1991Mis777].

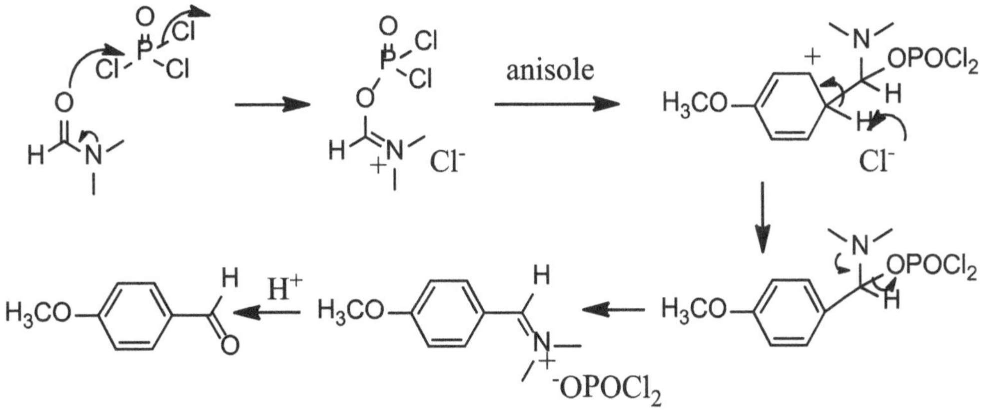

Scheme 17

1.6.3.2   Gattermann-Koch formylation:

In 1897, Gattermann and Koch successfully introduced a formyl group (CHO) on toluene by using formyl chloride (HCOCl) as the alkylating agent under Friedel-Crafts acylation conditions. Although the researchers were not able to prepare the acid chloride, they assumed that by reacting carbon monoxide (CO) with hydrogen chloride, formyl chloride would be formed in situ and in the presence of catalytic amounts of $AlCl_3$-Cu_2Cl_2 formylation of the aromatic ring would occur. The introduction of a formyl group into electron rich aromatic rings by applying CO/HCl/Lewis acid catalyst to prepare aromatic aldehydes is known as the Gatterman-Koch formylation [1897Ber1622, 1981SynMeth26].

The mechanism of the Gattermann-Koch formylation belong to the category of electrophilic aromatic substitution but are not in detail, since they have a tendency to vary from one substrate to another and the reaction conditions may also play a role. When CO is used, the electrophilic species is believed to be the formyl cation, which is attacked by the aromatic ring to form a α-complex. This α-complex is the converted to the aromatic aldehyde upon losing a proton. When HCN is used, the initial product after the reaction is an imine hydrochloride, which is subsequently hydrolyzed to the product aldehyde (Scheme 18).

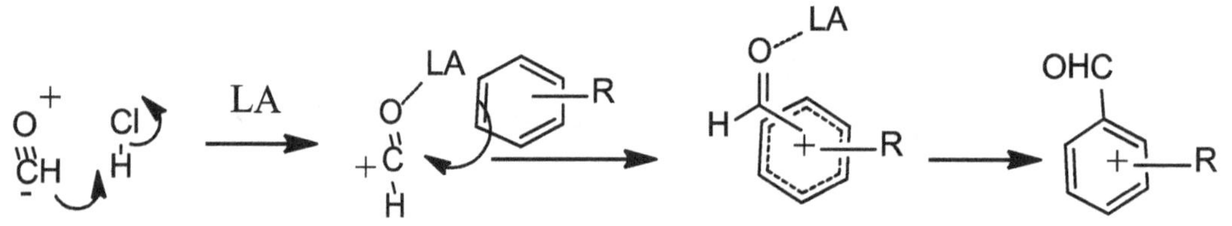

Scheme 18

1.6.3.3   Freidel-Craft's Acylation

Friedel–Crafts acylation is the acylation of aromatic rings with an acyl chloride using a strong Lewis acid catalyst. Friedel–Crafts acylation is also possible with acid anhydrides. This reaction has several advantages over the alkylation reaction. Due to the electron-withdrawing effect of the carbonyl group, the ketone product is always less reactive than the original molecule, so multiple acylations do not occur. Also, there are no carbocation rearrangements, as the carbocation is stabilized by a resonance structure in which the positive charge is on the oxygen (Scheme 19) [1948OR229].

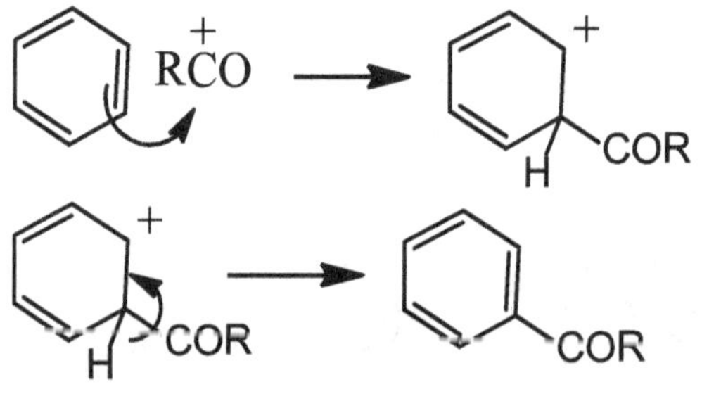

Scheme 19

The reagents of the reaction are usually acid halides, acid anhydrides and carboxylic acids. The viability of the Friedel–Crafts acylation depends on the stability of the acyl chloride reagent. The usual solvents are nitrobenzene, carbon disulfide and tetra chloroethylene. However, it has been found in some cases that orientation of acyl group changes with the change of solvent. For instance, when naphthalene is treated with acetyl chloride-aluminium chloride in nitrobenzene gives predominantly β-acetyl naphthalene while in tetrachloroethylene gives predominantly α-acetyl naphthalene. The reaction has good synthetic and preparative value.

1.6.4 From carboxylic acid

The **Rosenmund reduction** is a hydrogenation process in which an acyl chloride is selectively reduced to an aldehyde by passing hydrogen gas in boiling xylene using palladium catalyst supported on barium sulphate (Scheme 20), which is sometimes called the **Rosenmund catalyst**. The reaction was named after Karl Wilhelm Rosenmund who first reported it in 1918 [1948Misc362].

$$RCOOH \xrightarrow{PCl_5} RCOCl \xrightarrow[Pd\text{-}BaSO_4]{H_2} RCHO$$

Scheme 20

Aldehydes are more readily reduced than are acid chlorides and therefore one would expect to obtain the alcohol as the final product. It is the barium sulphate that prevents the aldehyde from being reduced, acting as a poison (to the palladium catalyst) in this reaction since it has a low surface area. Generally when the Rosenmund reduction is carried out, a small amount of quinolone and sulphur is added; these are very effective in poisoning the catalyst in the aldehyde reduction.

Reduction of esters either by DIBAL-H or sodium aluminium hydride quantitatively yield aldehydes directly (Equation 13). Aldehydes can also be synthesized by the reduction of alkyl nitrile or aryl nitrile with lithium aluminium hydride followed by acid hydrolysis (Scheme 21).

$$RCOOEt \xrightarrow[\text{or DIBAL-H}]{NaAlH_4} RCHO \qquad \text{(Equation 13)}$$

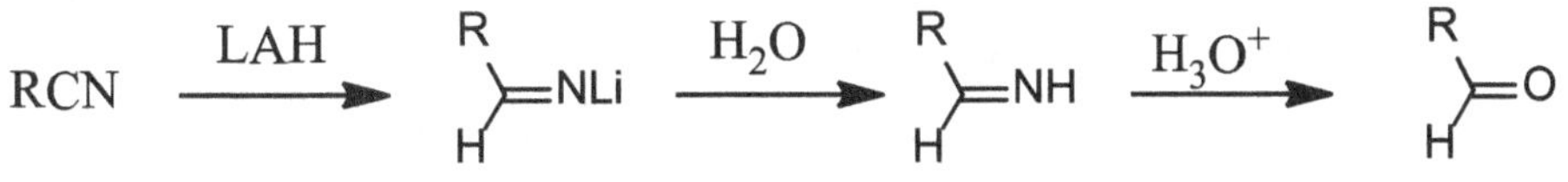

Scheme 21

If a carboxylic acid is treated with an organolithium compound, an acid-base reaction first takes place. In such a reaction, the acidic proton is abstracted by the organolithium compound's alkyl or aryl anion, as alkyl and aryl anions are extremely strong bases. As a result, the carbonyl carbon of the carboxylate anion which is formed in the first reaction step is nucleophilically attacked by an additional alkyl or aryl anion. The result of a subsequent hydrolysis is the protonation of the dianion. This yields a geminal diol and lithium hydroxide. The geminal diol represents a ketone's hydrate. Thus, it spontaneously eliminates water to yield the ketone (Scheme 22). The reaction may be carried out with primary, secondary, and tertiary alkyllithium compounds, as well as with aryllithium compounds. In order to obtain a ketone in this reaction, two equivalents of the organolithium compound to one equivalent of carboxylic acid must be applied, as the first equivalent is consumed by the acid-base reaction which cannot be prevented.

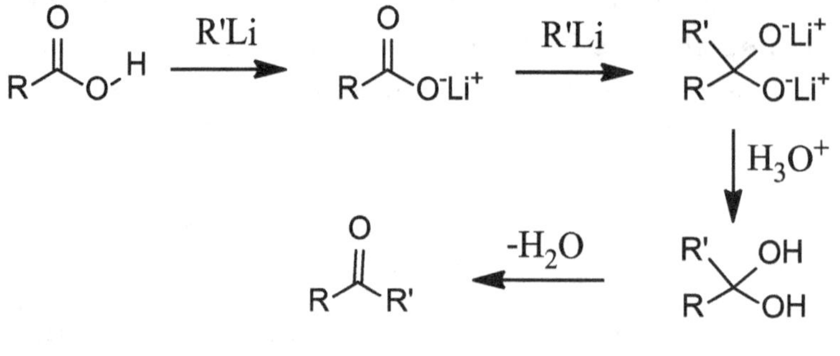
R'Li
R'Li
H_3O^+
$-H_2O$

Scheme 22

Aldehydes can be prepared by heating a mixture of the calcium salt of formic acid and homologous acid (Equation 14) while ketones can be prepared by heating the calcium salt of monocarboxylic acid (Equation 15). If a mixture of calcium salts is used, mixed ketones are obtained.

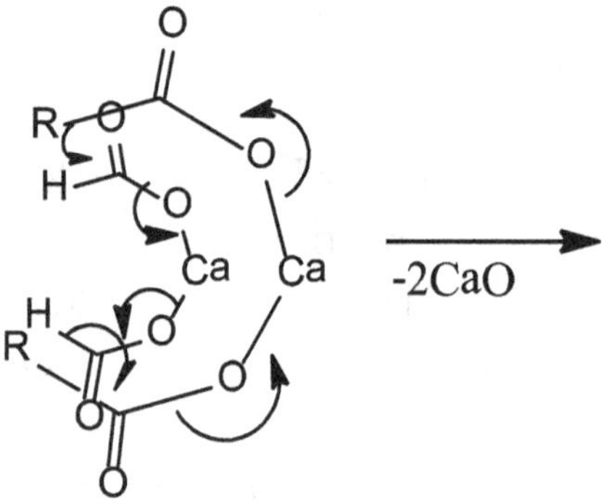
-2CaO

$2RCHO + 2CO_2$ (Equation 14)

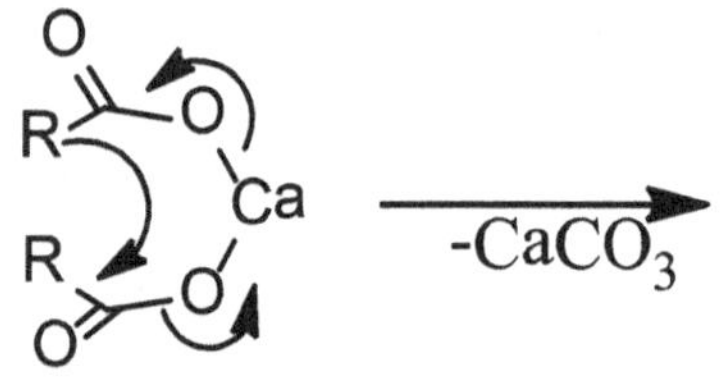
$-CaCO_3$

RCOR (Equation 15)

The yields are usually low but are high when the iron salts of n-acids are heated. The mechanism of the reaction is uncertain, but a possibility is one via the formation of an aldol and then a β-keto acid (Scheme 23).

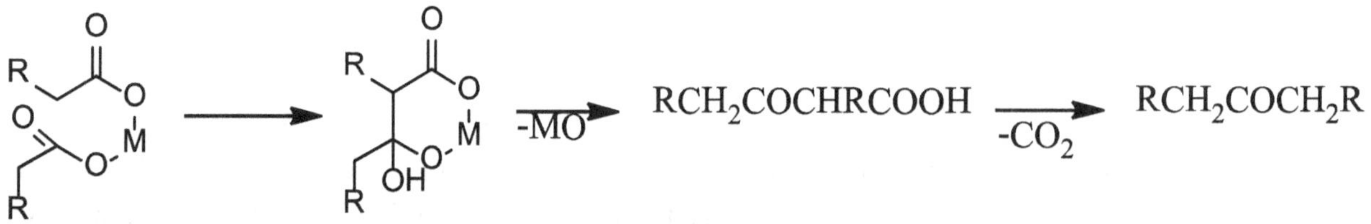
-MO
$RCH_2COCHRCOOH$
$-CO_2$
RCH_2COCH_2R

Scheme 23

Rai et al [2004TL7969] described the high yield conversion of carboxylic acids to the corresponding aldehydes. The highlight of this methodology involves the reduction of 2-oxazoline derived from the reaction of carboxylic acid with ethanolamine using $NiCl_2/NaBH_4$ leads to the formation of dihydrooxazoline derivatives. Finally hydrolysis of the resulted dihydrooxazoline derivatives leads to the formation of aldehyde in almost 90% yield (Scheme 24).

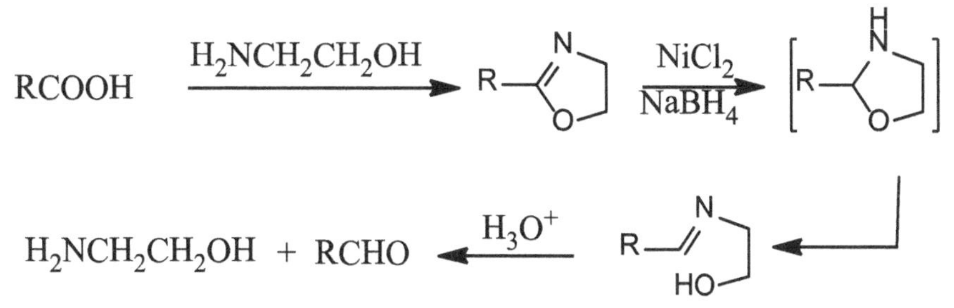

Scheme 24

Cyrous O.Kangani et al [2006TL6289] described the one-pot, high yield conversion of carboxylic acids to the corresponding aldehydes and ketones. The highlight of this methodology is the insitu generation of Weinreb amides with the Deoxo-Fluor reagent, which undergo nucleophilic reaction with DIBAL-H and Grignard reagents leads to the formation of aldehydes or ketones respectively (Scheme 25).

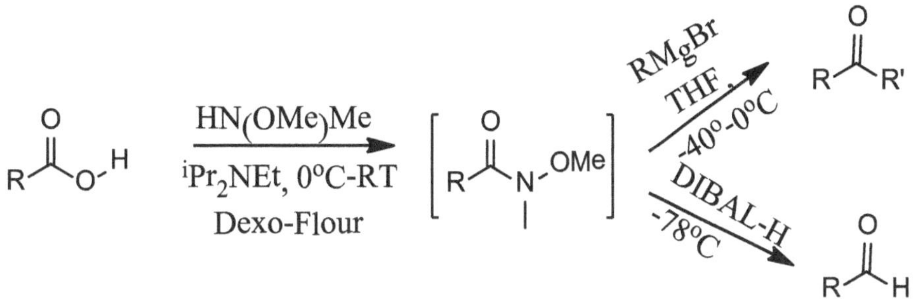

Scheme 25

Stephen aldehyde synthesis: In Stephen reported that when aromatic or aliphatic nitriles were added to a solution of stannous chloride ($SnCl_2$) in diethyl ether with anhydrous hydrogen chloride gas, imine hydrochlorides ($[R\text{-}CH\text{=}NH_2]^+Cl^-$) were obtained that readily underwent hydrolysis in warm water to give the corresponding aldehyde in good yield [1925JCST1874]. Overall, the reaction scheme is as follows (Scheme 26):

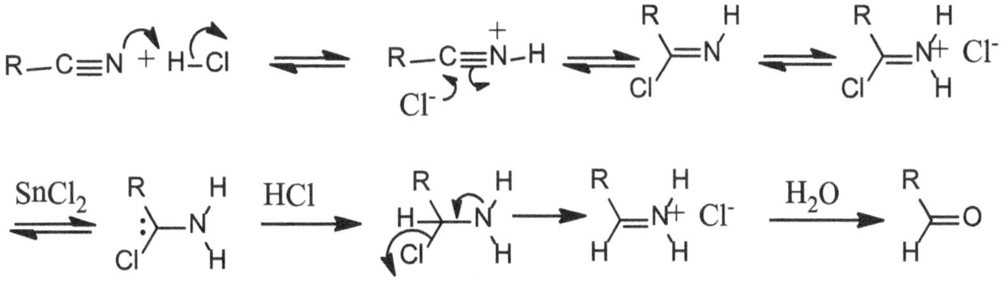

Scheme 26

1.6.5　Form alkyl halides

The **Sommelet reaction** is an organic reaction in which a benzyl halide is converted to an aldehyde by the action of hexamine and water [1985Misc] (Equation 16).

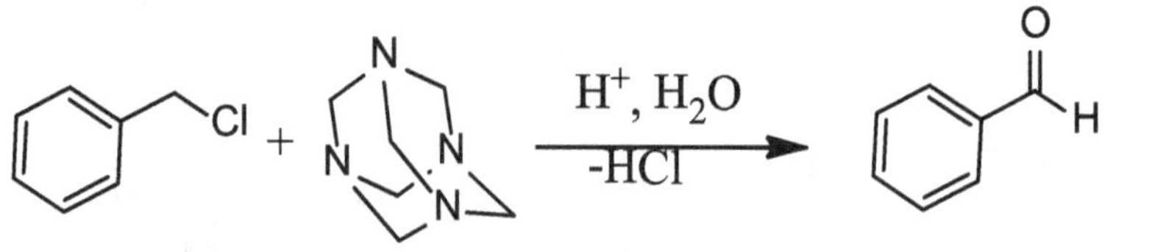

(Equation 16)

Chakraborthy et al synthesized carbonyl compounds from alkyl halides using α-chloronitrone as oxidizing agent in good yield. The mechanism for the oxidation of alkyl halide is as shown in the scheme 27 [2012SC1804].

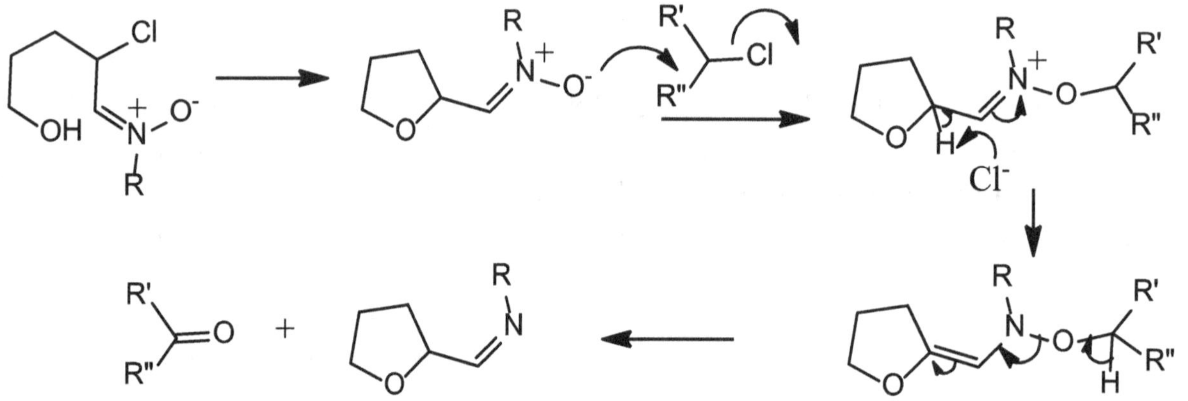

Scheme 27

Bouveault aldehyde synthesis

The Bouveault aldehyde synthesis (also known as the Bouveault reaction) is a one-pot substitution reaction that replaces an alkyl or aryl halide with a formyl group using a N,N-disubstituted formamide [1904BSC1306]. The first step of the Bouveault aldehyde synthesis is the formation of the Grignard reagent. Upon addition of a N,N-disubstituted formamide (such as DMF) a hemiaminal is formed, which can easily be hydrolyzed into the desired aldehyde (Scheme 28).

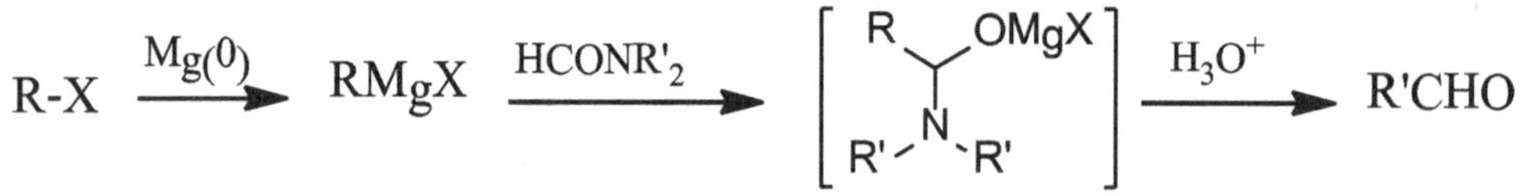

Scheme 28

1.6.6　From alkenes

The addition of ozone to alkenes in chloroform or carbon tetrachloride to give the primary ozonide is the first addition reaction in which a ring is formed. Because the formation of the primary ozonide occurs in one step, it is said to be a concerted cycloaddition reaction. This compound is unstable and is rapidly converted into a second adduct, called simply an ozonide. Reduction of the ozonide with zinc-acid, H_2-Raney nickel, triphenyl phosphine etc., gives aldehyde or ketones. The complete process of preparing the ozonide and decomposing it is known as ozonolysis (Scheme 29).

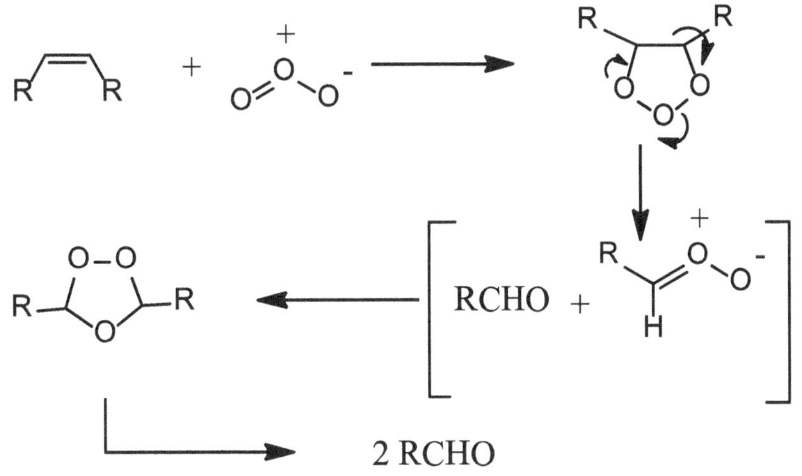

Scheme 29

Tetrasubstituted alkenes on oxidation with oxidants like potassium permanganate or chromic acid yields ketones in good yield (Equation 17).

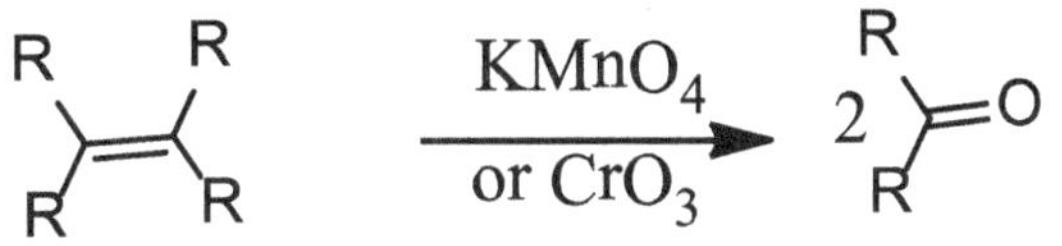

(Equation 17)

1.6.6.1   Wacker oxidation

The one pot oxidation of terminal or or 1,2-disubstituted olefins to a ketones through the action of catalytic palladium(II) salts, water, and a co-oxidant Pd(II) is known as **Wacker oxidation**. [84S369, 98OrgCat513] (Equation 18).

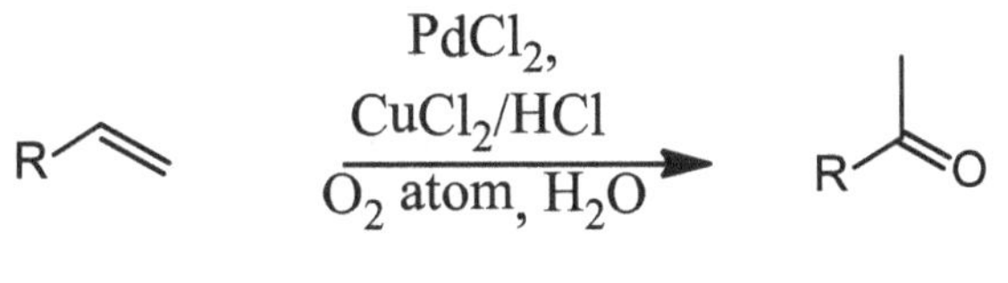

(Equation 18)

The first step of the Wacker oxidation involves coordination of the alkene to the palladium center to form π-complex **3**. Hydroxypalladation of this complex then occurs to yield zwitterionic complex **4** which later looses hydrogen chloride leading to neutral complex **5**. Next step involves the β-hydride elimination to afford enol complex **6**, which re-inserts into the Pd-H bond to afford complex **7**. Final step involves the reductive elimination of hydrogen chloride leads to the formation of respective ketone and palladium(0). Oxidation of palladium(0) by copper(II) then occurs, regenerating palladium(II) species **2** (Scheme 30).

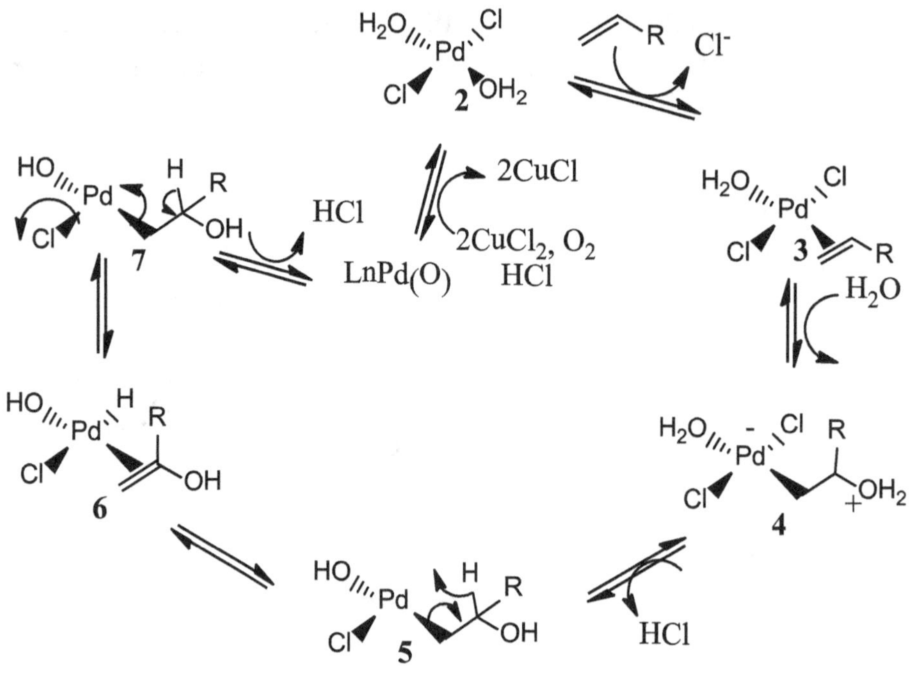

Scheme 30

The mode of hydroxypalladation is an important issue for the Wacker oxidation. Hydroxypalladation may occur either in a syn fashion through an inner-sphere mechanism or in an anti fashion via nucleophilic attack on the coordinated alkene. Although the stereocenter in **4** is ultimately destroyed upon elimination, the mode of hydroxypalladation can influence the site selectivity of the reaction. Markovnikov-type addition of water to the more substituted carbon of the alkene forms a methyl ketone, while attack of water at the less substituted position ultimately yields an aldehyde.

1.6.7   From alkyl benzene

Oxidation of toluene and propene can be oxidized by dichlorodicyanoquinone (DDQ) to benzaldehyde. A variety of mechanisms have been proposed for this reaction, but the generally accepted mechanism of this involves a hydride transfer to DDQ molecule from benzyllic carbon to form phenoxide ion which then react with resulted benzyl carbocation to form the benzyl acetate. This further react with second molecule of DDQ to produce acetal derivative, which is hydrolysed under the reaction conditions (Scheme 31).

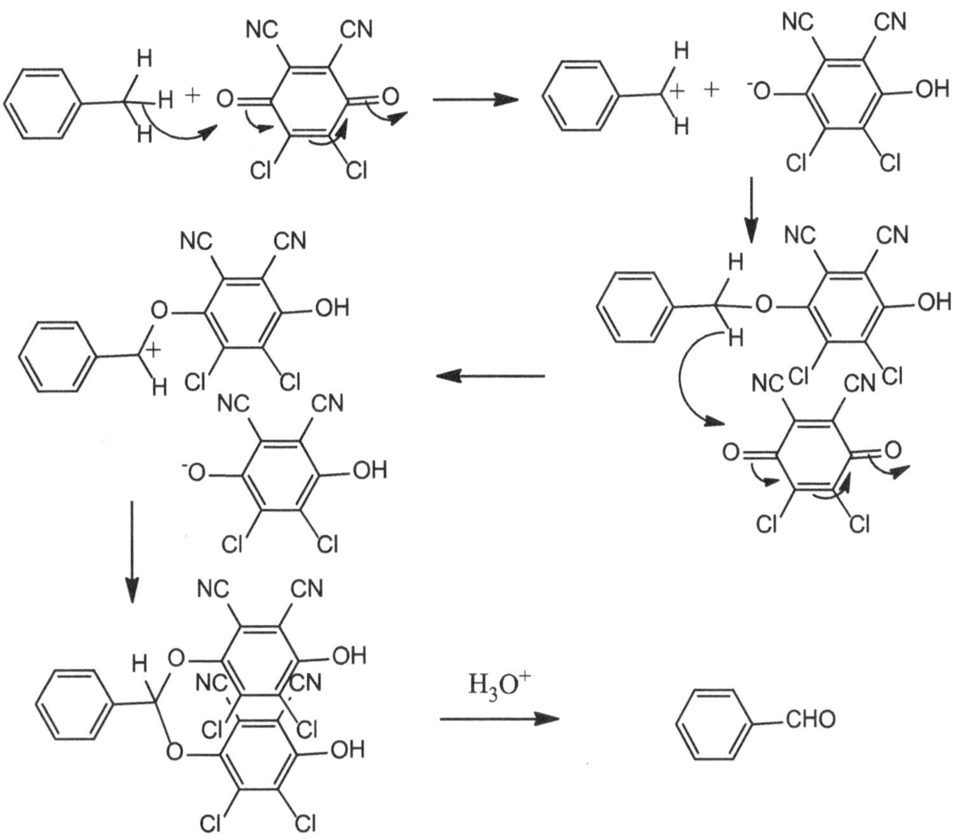

Scheme 31

1.7 STRUCTURE AND REACTIVITY

The carbonyl group consists of two atoms carbon and oxygen sharing four electrons while the four electrons are shared more or less equally by the two carbon atoms in an alkene, the four electrons in the carbonyl systems are more likely to be found near the oxygen atom, the electron density due to these four electrons is greater near the oxygen atom.

The electronic distribution of the carbonyl system can be described using either the concepts of molecular orbital theory or those of the valance band theory. The carbonyl carbon of a typical aldehyde and ketone has sp^2 hybridization with bond angles approximating 120°. The carbon-oxygen double bond consists of a σ-bond and a π-bond, much like the corresponding double bond of an alkene. Just as C-O single bonds are shorter than C-C single bonds, C=O bonds are shorter than C=C bonds.

Some carbonyl compounds have a carbon-carbon double bond or triple bond in conjugation with the carbonyl double bond. Since this additional bond connects the α- and β-carbons, such compounds are called α,β-unsaturated carbonyl compounds. Acrolein is an example of such a compound. Like conjugated dienes, α,β-unsaturated carbonyl compounds possess a bonding π-molecular orbital resulting from the overlap of the p-orbitals on the carbons and the oxygen. The delocalization of electrons throughout this π-electron system is sometimes indicated with resonance structure.

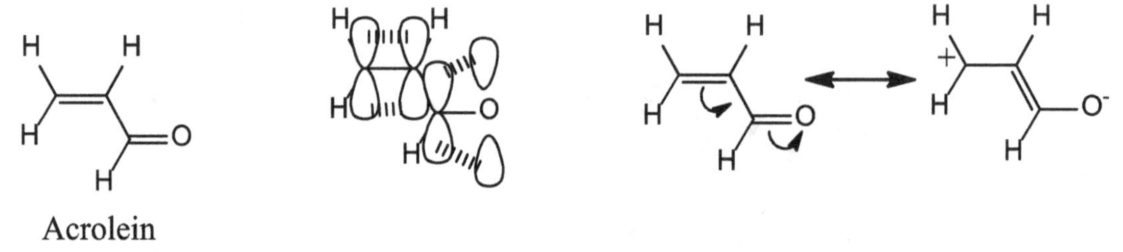

Acrolein

In the ground state of the carbonyl group, the electron density due to the four shared electrons is greater near oxygen than near carbon. In addition, oxygen has four nonbonded electron occupying sp^2 orbitals. In accordance with these ideas carbonyl systems display dipole moments and relatively high boiling points. Formaldehyde has a dipole moment equal to 2.3D while that of acetone is 2.8D. This corresponds to 50% ionic character for the carbonyl bond. Larger dipole moment for these can be accounted for the inductive effect of the oxygen atom, but can be accounted for if carbonyl compounds are resonance hybrids.

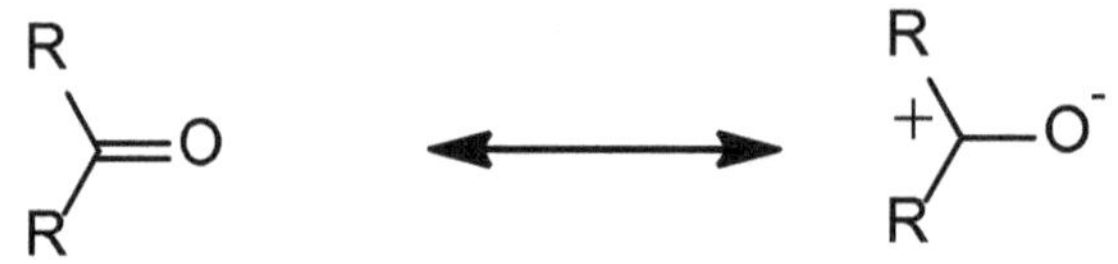

2.3D   2.8D

Thus the carbon atom has a positive charge and consequently can be attacked by nucleophilic reagents. The carbonyl group also exhibits basic properties; it is readily protonated by strong acids to form oxonium salts. Since oxygen is more electronegative than carbon, the second resonating structure will make a larger contribution than the first.

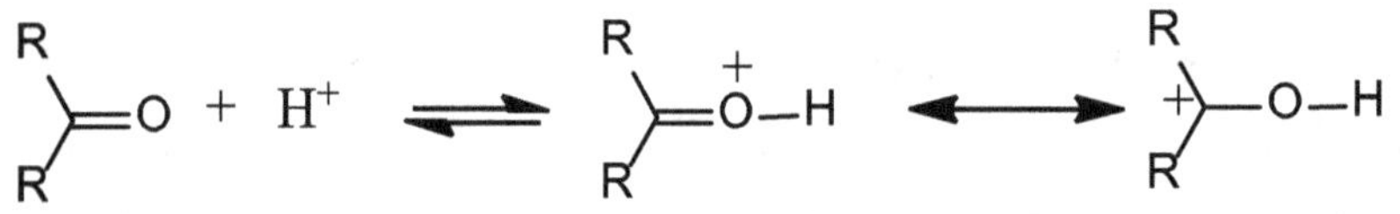

Hence protonation increases the electrophilic character of the carbonyl group and so it can be expected that electrophilic additions will be catalyzed by acids. It should be noted that because of the positive charge on the carbon atoms, the CO group has a strong –I effect. Many addition reactions of carbonyl compounds may be represented by the general equation:

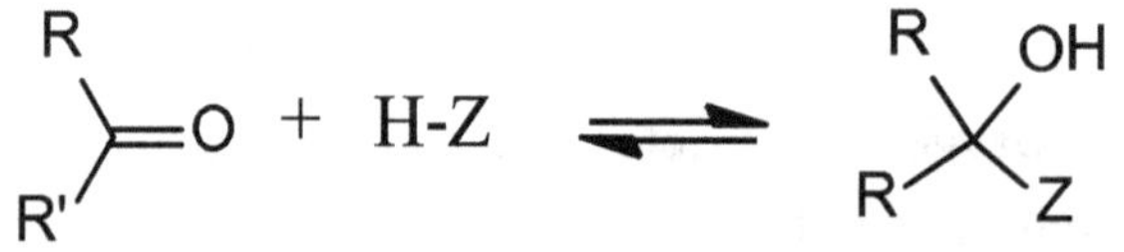

Thus, the hybridization of the carbon atom changes from sp^2 to sp^3. These results in increased crowding and so reaction can be expected to be increasingly sterically hindered with increasing size of R and R'. For aldehydes, R'=H and hence it can be anticipated that aldehydes will be more reactive than ketones. This is observed in practice, reaction being faster and the equilibrium shifted more to right. The +I effect of an alkyl group also decreases the reactivity of the carbonyl group towards nuceophilic reagents, this being due to the partial neutralization of the positive charge on the carbonyl carbon atom. Hence ketones will be less reactive than aldehydes due to both steric and inductive factors.

Cyclic ketones almost always react more rapidly in addition process than the open chain analogues. This is because the alkyl group of the open-chain compounds has considerably more freedom of motion and produce greater steric hindrance in transition state for addition.

Electrical effects are also important in influencing the ease of addition to carbonyl groups. Electron attracting groups facilitate the addition of nucleophilic reagents to carbon by increasing its positive character. Thus compound such as chloral (Cl_3CCHO) add nucleophilic reagents readily.

Because of their polarity, aldehydes and ketones have higher boiling points than the corresponding alkanes with similar molecular weights and shapes. However, aldehydes and ketones are not hydrogen bond donors; so their boiling points are considerably lower than those of corresponding alcohols.

	$CH_3CH=CH_2$	$CH_3CH=O$	CH_3CH_2OH
Boiling Point	-47.4°	20.8°	78.3°
Dipole moment	0.4D	2.7D	1.7D
Boiling Point	-6.9°	56.5°	82.3°
Dipole moment	0.5D	2.7D	1.6D

The lower ketones and aldehydes are soluble in water because they can accept hydrogen bonds from water at the carbonyl oxygen. As the molecular weight increases, the solubility in water goes on decreasing.

Acetone and 2-buatnone especially are valued as solvents because they dissolve not only in water but also in a wide variety of other organic compounds. These compounds also have low boiling points that they can be easily separated from other less volatile compounds. Acetone, with a dielectric constant 2.7D, is a polar solvent and is often used as a solvent or co-solvent for nucleophilic displacement reactions.

The carbon-oxygen bond is both a strong double bond and a reactive bond. Its bond energy (179 Kcals) is rather more than that of two carbon-oxygen single bonds (2 × 85 Kcals) in contrast to the carbon-carbon double bond (145.8 Kcals), which is weaker than two carbon-carbon single bonds (2 × 82.6 Kcals). A possible explanation of the greater strength of a >C=O bond is that repulsion between the unshared electron is greater for single bonded oxygen than double bonded oxygen.

EXERCISE

1. Among aldehydes and ketones, which one is more reactive and why?
2. Cyclic ketones are more reactive than alicyclic ketones. Justify this statement.
3. Mention the different methods available for the synthesis of aldehydes from primary alcohols.
4. Explain the mechanism for ozonolysis of alkenes.
5. Write systematic name for the following compounds:

6. Mention the different methods available for the synthesis of aldehydes from carboxylic acids.
7. With suitable mechanism explain Kornblum reaction.
8. Using IR spectral data, how you differentiate the isomers:

$$CH_3COOH \text{ and } HOCH_2CHO$$

9. Write the reaction sequence for the conversion toluene to benzaldehyde employing DDQ as oxidant.
10. How Fehling solution helps in detecting the presence of aldehyde functional group in given compound?
11. Mention the different derivatives for identifying the carbonyl compounds and how are prepared?
12. Among CH_3CHO and CH_3CH_2OH, which one has higher boiling point and why?
13. Explain Vilsmeir-Haack formylation reaction with suitable example.
14. How Tollen's reagent helps in detecting the presence of aldehyde functional group in the given compound?
15. Most of the aliphatic aldehydes possessing C4 and more carbon atoms predominantly give an ion at m/z 44. Justify this statement.
16. How chloramine-T is useful in oxidizing alcohols to ketones with suitable example.
17. Key step for the Oppenauer oxidation of alcohols to ketone involve hydride ion shift. Justify this statement.
18. In Moffatt oxidation reaction how DCC will activate DMSO? Explain with mechanism.
19. Predict the products for the following and explain with suitable mechanism.

$$RCH=CHR + O_3 \text{ ------?}$$

20. Mention the different electronic transitions occur in formaldehyde.
21. List the Woodward's rules for predicting λ_{max} for α,β-unsaturated aldehydes and ketones.
22. 3-Buten-2-one has absorption bands at 219 and 324 nm. Assign electron transitions to those absorptions and account for the differences from acetone (λ_{max} 189 and 270 nm)
23. Predict the wavelengths of maximum absorption for the following:

24. Give the probable chemical shift and splitting pattern for $CH_3CH_2COCH_3$

25. Account for the base peak observed at m/z 55 in the E1 mass spectrum of cyclopentanone and cyclohexanone.

26. Account for the peak observed at m/z 43 in the E1 mass spectrum of $CH_3CH_2CH_2COCH_2CH_2CH_2CH_3$

27. During mass spectral fragmentation, which carbonyl compounds undergo β-cleavage? Explain with example

28. Discuss the influence of ring size on carbonyl IR absorption in cyclic ketones.

29. How is benzaldehyde distinguished from acetophenone using IR?

30. Which of the following isomers (a or b) show higher wavelength and why?

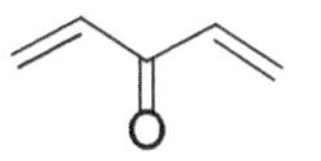

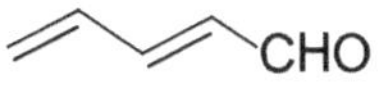

31. Using IR spectral data, how you differentiate:

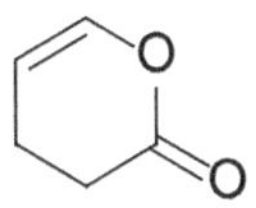

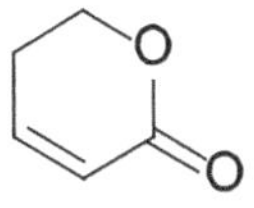

32. Predict the splitting pattern for the ^1H NMR of $CH_3COCH_2COOC_2H_5$ at -78°C and at 25°C.

33. Discuss the mechanism for the oxidation of toluene to benzaldehyde by DDQ.

ADDITION REACTIONS OF ALDEHYDE AND KETONES

Nucleophiles add to carbonyl groups to give compounds in which the trigonal carbon atom of the carbonyl group becomes tetrahedral.

2.1 ADDITION OF HYDROGEN AS NUCLEOPHILE

Nucleophilic attack by the hydride ion [H⁻] is an almost unknown reaction. This species which is present in the salt sodium hydride, NaH has such a high charge density that it only ever reacts as a base. The reason is that its filled "1s" orbital is of an ideal size to interact with the hydrogen atom's contribution to the σ^* orbital of an H-X bond, but much too small to interact easily with carbon's more diffuse "2p" orbital contribution to the LUMO (π^*) of the C=O group.

Nevertheless, adding H⁺ to the carbon atom of a C=O group would be a very useful reaction as the result would be the formation of an alcohol. This process would involve going down from the aldehyde or ketone oxidation level to the alcohol oxidation level and would therefore be a reduction. It cannot be done with NaH but it can be done with some other compounds containing nucleophilic hydrogen atoms.

There are reactions in which one or more hydride ions (H⁻) get transferred from one species to the other directly. A hydride ion being a good electron donor, the reactions involving hydride ion transfer are redox reactions and the hydride donors are good reducing agents. Metallic hydrides such as lithium aluminium hydride ($LiAlH_4$), sodium borohydride ($NaBH_4$), lithium (triethoxy)aluminium hydride ($Li(OEt_3) AlH$), lithium (tri-t-butoxy) aluminium hydride ($Li(t-BuO)_3AlH_4$) etc. reduce compounds containing >=O moiety such as RCHO, RCOR', RCOCl, RCOOH, RCOOR', $RCONH_2$. Besides metallic hydrides, aluminium isopropoxide also reduces carbonyl compounds by hydride transfer. The Cannizzaro reaction, a self-redox reaction also involves hydride transfer.

2.1.1 Cannizzaro reaction

When aldehydes, which do not possess α-H atom, are treated with strong bases (usually 50% ethanolic NaOH) (50%) or other strong bases (e.g., alkoxides), undergo self – oxidation reaction in which one molecule of the aldehyde acts as an oxidizing agent and gets reduced to the corresponding alcohol with another molecule gets oxidized to the corresponding carboxylate ion and the reaction and the intermolecular hydride transfer reaction known as Cannizzaro reaction. Alternatively, high yields of alcohol can be obtained from any aldehyde when the reaction is performed in presence of an excess of formaldehyde. This process is called the crossed Cannizzaro reaction.

A variety of mechanisms have been proposed for this reaction, but the generally accepted mechanism of the Cannizzaro reaction involves a hydride transfer (Scheme 32). First hydroxyl anion adds across the carbonyl group and the resulting species is deprotonated under the applied basic conditions to give the corresponding dianion. This dianion facilitates the ability of the aldehydic hydrogen to leave as a hydride ion. This leaving hydride ion attacks another aldehyde molecule in the rate determining step to afford the alkoxide of a primary alcohol, which gets protonated by the solvent water. By running the reaction in the presence of D_2O, it was shown that the reducing hydride ion came from another aldehyde and not the reaction medium, since the resulting primary did not contain a deuterium.

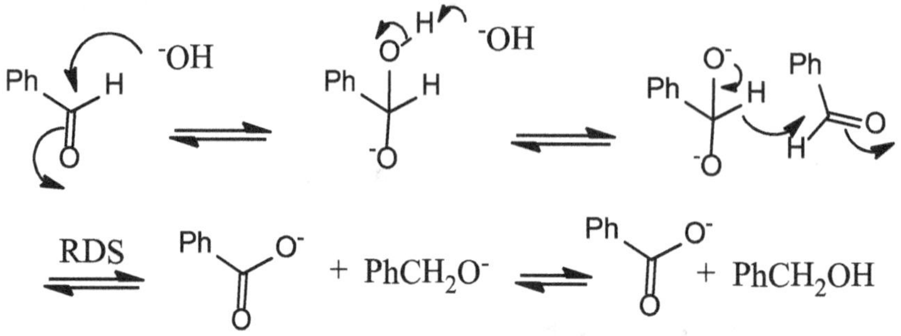

Scheme 32

When a mixture of formadehyde and benzaldehyde is treated with caustic soda solution, a crossed Cannizzaro reaction takes place and benzyl alcohol forms along with sodium formate (Equation 19). In this connection it is important to note that in a crossed Cannizzaro reaction formaldehyde gets oxidized to formate and the other gets reduced to the corresponding alcohol; this is because the C of formaldehyde is more electrophilic and less sterically congested than that of the other and –OH attacks the former C readily to give $HOCH_2O^-$ which transfers H^- directly.

$$PhCHO \ + \ HCHO \ \xrightarrow{\text{NaOH}} \ PhCH_2OH \ + HCOO^-Na^+ \quad \text{(Equation 19)}$$

2,2-bis(hydroxymethyl) propane-1,3-diol can be easily synthesized via crossed Cannizzaro reaction of acetaldehyde with excess of formaldehyde (Equation 20).

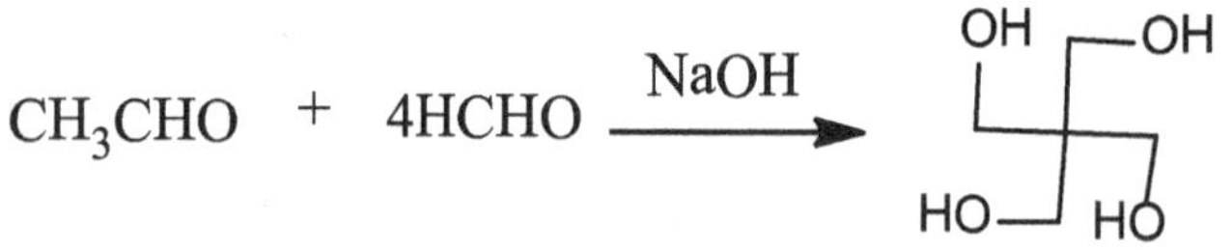

$$CH_3CHO \ + \ 4HCHO \ \xrightarrow{\text{NaOH}} \qquad \text{(Equation 20)}$$

2. 1.2 Tischenko Reaction

Tischenko found that both enolizable and non-enolizable aldehydes can be converted to the corresponding esters in the presence of magnesium or aluminium alkoxide. The reaction involves a hydride shift from one aldehyde to another that leads to the formation of the ester product. This transformation is known as Tischenko Reaction [1906RPCS355; 2003COC1713].

The mechanism of the Tischenko reaction was extensively studied and there three different mechanisms proposed. The most accepted mechanism is depicted in the Scheme 33. According to this proposal, first the aluminium alkoxide coordinates to the aldehyde. This is followed by the attack of a second molecule of aldehyde. Subsequent hydride shift leads to the regeneration of the catalyst and formation of the product.

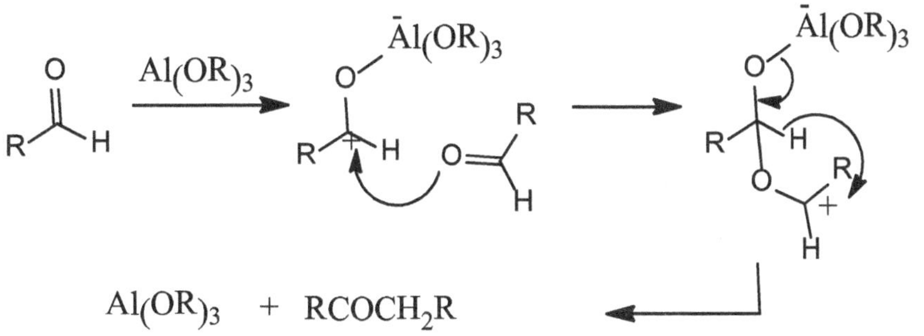

Scheme 33

Disodium tetracarbonylferrate ($Na_2Fe(CO)_4$, **8**) is an efficient catalyst for the dismutation of aromatic aldehydes to esters while aliphatic aldehydes undergo aldol condensation. For instance, reaction of benzaldehyde was treated with (**8**) in tetrahydrofuran at 25° for 40 hr. yielded benzyl benzoate in 95% yield [1976BCS3597]. In the reactions of p-substituted benzaldehydes with (**8**), there is a remarkable influence of the substitution on the reaction rate. The apparent order of their reactivities was as follows; p-ClC_6H_4CHO > C_6H_5CHO > p-$CH_3C_6H_4CHO$ > p-$CH_3OC_6H_4CHO$. The reactivity decreases with an increase in the electron-releasing property of the substituents. This suggests that the reaction is influenced by the electron density on the carbonyl carbon of the aldehydes which the ferrate (**8**) attacks nucleophilically.

Dismutation of aldehydes by the initial coordination of the base with the aldehyde carbonyl group is supported by the investigations of the effect of substituents upon the rate of reaction. The formation of p-methoxybenzoate implies that p-anisaldehyde is able to react with adduct, as a second aldehyde, though it does not form adduct upon treatment with (**8**) at 25°C.

2.2 ADDITION OF OXYGEN NUCLEOPHILE

2.2.1 Hydration (addition of water)

Most aldehydes and ketones in aqueous solution undergo rapid and reversible addition of water to the C=O bond. This reaction called hydration is similar to the hydration of a carbon-carbon double bond: hydrogen of the aldehydes and ketones is, as a rule, considerably faster than hydration of alkene. The product of the hydration reaction is a geminal diol, a compound with two hydroxyl groups on one carbon atom. We have learned that gem-diols are unstable and break down to the corresponding carbonyl compounds.

Formaldehyde is an extremely reactive aldehyde as it has no substituent to hinder attack; it is so reactive that it is rather prone to polymerization. And it is quite happy to move from sp² to sp³ hybridization because there is very little increased steric hindrance between the two hydrogen atoms as the bond angle changes from 120° to 109°. This is why our aqueous solution of formaldehyde contains essentially no CH_2O, it is completely hydrated.

A mechanism for the hydration reaction is shown below (Scheme 34). The hydration of an aldehyde or ketone for example occurs by both acid and base catalyzed pathways. The reaction also occurs in the absence of a catalyst. Acid catalyzed hydration of an aldehydes and ketones is closely analogous to acid-catalyzed hydration of alkenes. In this mechanism the protonated carbonyl compound, which is really a carbocation is attacked by water.

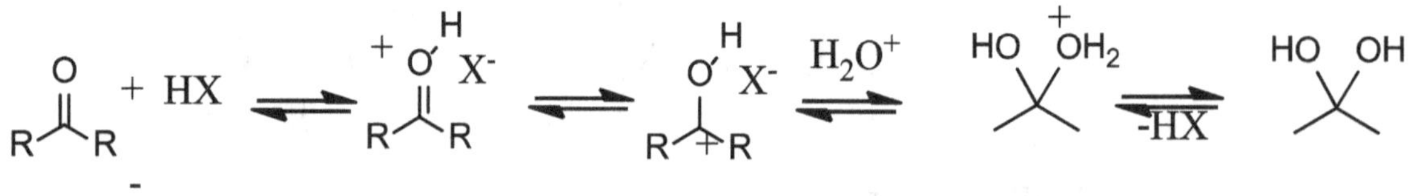

Scheme 34

Base catalyzed hydration of aldehydes and ketones occur by direct nucleophilic attack of ⁻OH on the carbonyl carbon (Scheme 35).

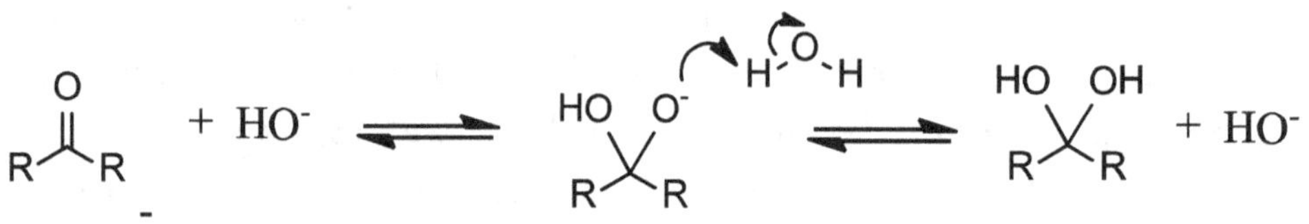

Scheme 35

In a carbonyl compound the carbonyl carbon is trigonal with bond angles of about 120°; in an addition product such as a gem-diol it is tetrahedral, forcing the groups around this carbon closer together. Thus there is less room for large groups in the addition compound than in the carbonyl compound. It follows that the increasing the size of groups at the carbonyl carbon increases the congestion in the addition compound and therefore makes carbonyl ion more difficult. Electronic effects can also favour the reaction with nuceophiles – electronegative atoms such as halogens attached to the carbon atoms next to carbonyl group can increase the extent of hydration by the inductive effect according to the number of halogen substituents and their electron - withdrawing power. They increase the polarization of the carbonyl group, which already has positively polarized carbonyl carbon and make it even more prone to attack by water. Trichloroacetaldehyde (chloral CCl_3CHO) is hydrated completely in water and the product chloral hydrate can be isolated as crystals.

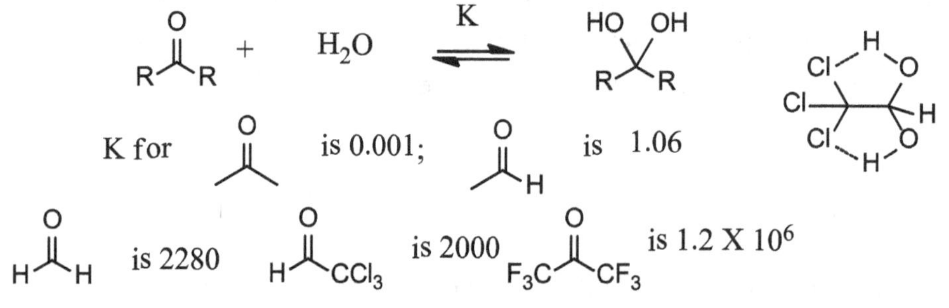

The unusual stability of chloral hydrate has been attributed to the –I effect of chlorine and to the formation of intramolecuar hydrogen bonds, the presence of which has been shown by Davies (1940) from infra-red studies. Resonance in the carbonyl compound is another factor that affects addition to the carbonyl group.

For example, benzaldehyde has important resonance structures that stabilize the aldehyde; this resonance is absent in the hydrate.

To the extent that resonance stabilizes the carbonyl compound, it is more difficult to form the hydrate ($K_{eq}u = 8.3 \times 10^{-3}$).

Cyclopropanones, three membered ring ketones are also hydrated some extent but for a different reason. Cyclopropanones conversely prefer the small bond angle because their substituents are already confined within a ring. A three membered ring is really strained with bond angles forced to be 60°. For the sp² hybridized ketone this means bending the bands 60° away from natural 120°. But for the sp3 hybridized hydrate the bonds have to be distorted by only 49° (=109°-60°). So the addition to the C=O group allows some of the strain inherent in the small ring to released hydration is favoured and indeed cyclopropanone and cyclobutanone are very reactive electrophiles.

sp² wants 120° but gets 60°   sp³ wants 109° but gets 60°

Steric factors can also influence the position of equilibrium, for the alkyl groups in a ketone are farther apart in the reactant than in the product. Steric interactions are more pronounced in the product and with large alkyl groups thus has an unfavorable influence upon the equilibrium.

2.2.2 Addition of alcohols

When an aldehyde or ketone reacts with a large excess of alcohol in the presence of a trace of mineral acid, acetal (from an aldehyde) or a ketal (from ketone) is formed. Acetals and ketals differ in the same sense as aldehydes and ketones; acetals have a proton at the central carbon; ketals do not. The formation of acetals and ketals is reversible. The reaction is driven to the right by the use of an excess of alcohol as the solvent or by the removal of the water by-product.

The first step in the mechanism of acetal and ketal formation is an acid-catlyzed carbonyl addition, completely analogous to acid-catalyzed hydration. The product of this first step is called hemiacetal (if derived from an aldehyde) or hemiketal (from ketone) [hemi = half].

The hemiactal or hemiketal reacts further by protonation of the OH group and loss of water to give a relatively stable carbocation (**11**). The loss of a water from the hemiacetal or hemiketal is essentially like the dehydration of an alcohol to a carbocation intermediate but it occurs more easily in this case because of the greater stability of the carbocation. Attack of an alcohol molecule on the cation and deprotonation completes the mechanism (Scheme 36).

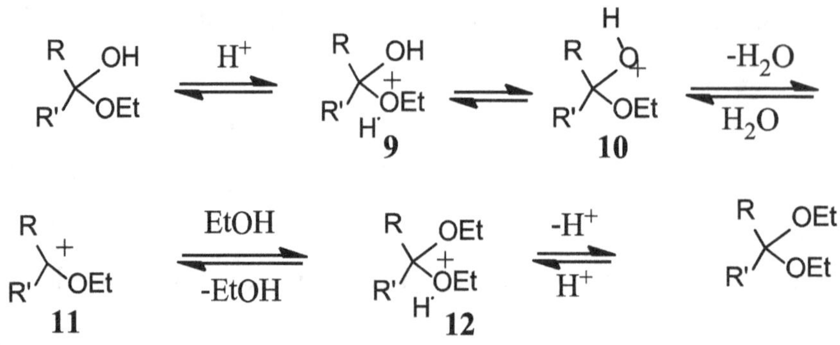

Scheme 36

The hemiacetal is rarely isolated since it readily forms the acetal. Acetals are diethers of the unstable 1,1,-dihydroxy alcohols and may be named as 1,1,-dialkoxy alkanes. Until the parent dihydroxy alcohols these acetal are stable. They are also stable in the presence of alkali but are converted into aldehydes by acid. Thus acetal formation may be used to protect the aldehyde group against alkaline oxidizing agents. The stability of acetals to alkaline media may be attributed to the fact that the alkoxy group is a poor leaving group, whereas the protonated group is a very good leaving group.

Acid hydrolysis of hemiacetal formation might involve activation of either the acetal or the carbonyl compound. However the only simple reaction one would expect between various species of alcohols and proton donors is oxonium salt formation, while hardly seems the proper way to activate an alcohol for nucleophilic attack at the carbonyl group of an aldehyde. On the other hand, formation of oxonium salt (or conjugate acid) of the carbonyl compound is expected to provide activation for hemiacetal formation by increasing the positive charge of the carbonyl cation.

In contrast to hemiacetal formation, acetal formation is catalyzed only by acids. Addition of a proton to a hemiacetal can occur in two ways to give "9" or "10" (Scheme 33). The first of these "9" can lose EtOH and yield the conjugate acid of acetaldehyde. This is the reverse of acid catalyzed hemiacetal formation. The second of these "10" can lose water and give a new entity "9", the methyloxonium cation of the aldehyde. The reaction "11" with water gives back "10", but reaction with alcohol leads to the formation of conjugate acid "12" of the alcohol. Loss of proton from "12" gives the acetal.

The fact that acetals are formed only in an acid catalyzed reaction has the chorology that acetal groups are stable to base. This can be synthetically very useful as illustrated by the following synthesis of glyceraldehyde from readily available acrolein (Scheme 37). Hydrogen chloride in ethanol adds in the anti-Markownikoff's manner to acrolein, which then react with ethanol to give the acetal. It was undergo cis hydroxylation followed by acidic hydrolysis leads to the formation of glyceraldehyde.

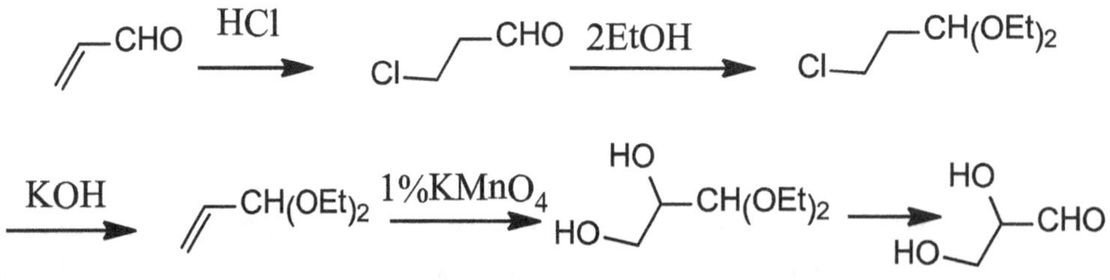

Scheme 37

The position of equilibrium in acetal and hemiacetal formation is rather sensitive to steric hindrance. Large groups in either the aldehyde or the alcohol tend to make the reaction less favorable. Ketals cannot usually be made in practical yields by the direct reaction of alcohols with ketones because of unfavorable equilibria. Laboratory preparations are possible through the reaction of ketones with ethyl formate (Equation 21). This process requires an acid catalyst.

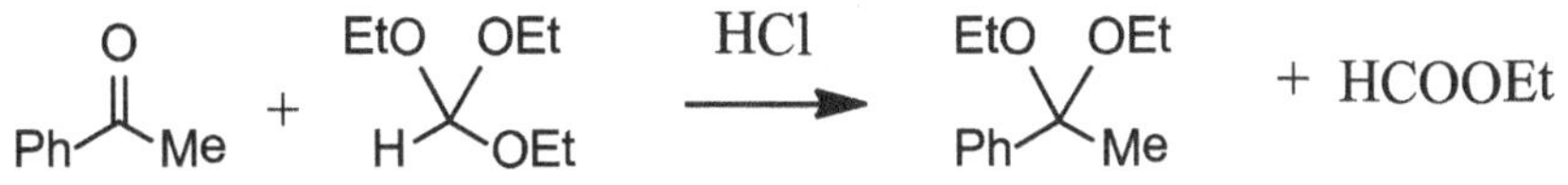

Hemiacetal can also gain stability by being cyclic, when the carbonyl group and the attacking hydroxyl group are part of the same molecule (Scheme 38). The reaction is now an intramolecular addition as opposed to the intermolecular ones we have considered so far. Notice again the preference for an intramolecular reaction when a five or six-membered ring can form.

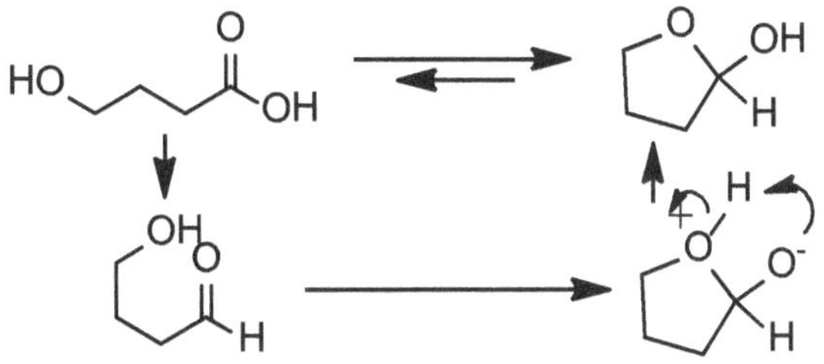

Scheme 38

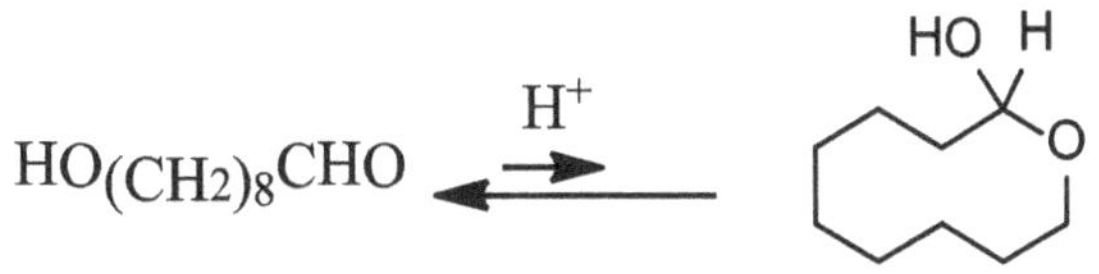

Although the cyclic hemiacetal (also called lactol) product is more stable, it is still in equilibrium with some of the open-chain hydroxyl aldehyde form. Its stability and how easily forms are, depends upon on the size of the ring: five or six membered rings are free from strain (their bonds are free to adopt 109° or 120° angles compare the three membered rings) and five or six membered hemiacetals are common. Among the most important examples are many sugars. Glucose for example, is a hydroxyl aldehyde that exists mainly as a six membered cyclic hemiacetal (>99% of glucose is cyclic in solution) while ribose exists as a five membered hemiacetal.

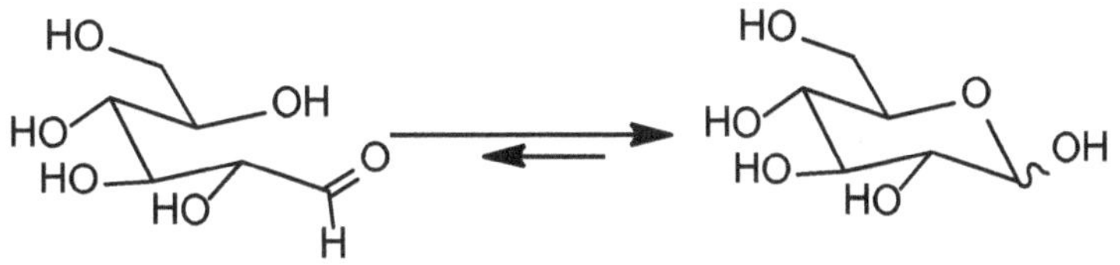

Acetaldehyde, when treated with a trace of acid, readily forms a cyclic trimer called paraldhyde. This compound is really a cyclic acetal. Paraldehyde, with a boiling point of 125°C, is a particularly convenient way to store acetaldehyde, which itself boils near room temperature. On heating with a trace of acid, acetaldehyde can be distilled from a sample of paraldehyde.

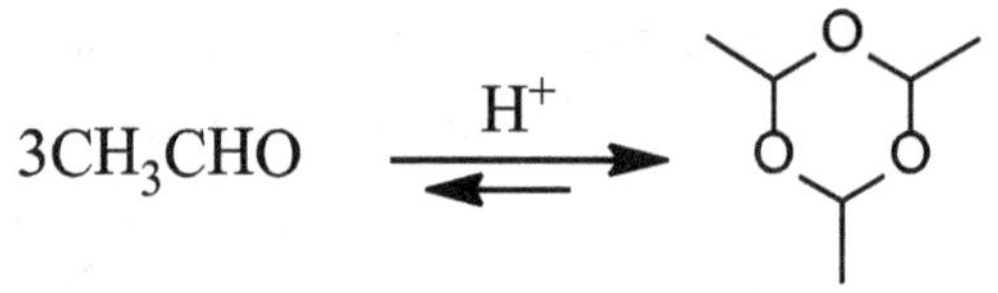

$$3CH_3CHO \quad \underset{\longleftarrow}{\overset{H^+}{\longrightarrow}}$$

Formaldehyde can also exist in a variety of acetal forms. One common form is called paraformaldehyde, which is a linear polymer of formaldehyde [H-(CH$_2$-O)$_n$H]. This material precipitates from concentrated formaldehyde solution. Because it is a solid, it is a convenient way of storing formaldehyde, itself a gas. Formaldehyde is liberated from paraformaldehyde by heating.

Cyclic acetals formed by the reaction of single molecule of a diol, a compound containing two hydroxyl groups are more stable than the acyclic acetals. When the diol is ethylene glycol, the five membered cyclic acetal formed is known as dioxolane (Scheme 39).

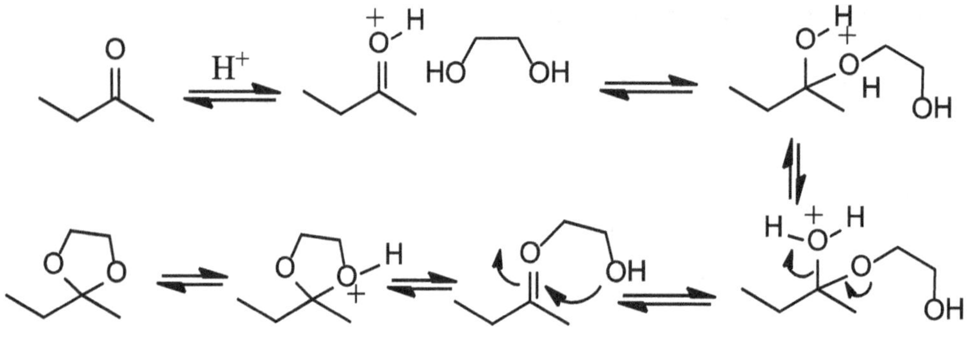

Scheme 39

Cyclic acetals like this are more resistant to hydrolysis than acyclic one, and easier to make – they form quite readily even from ketones. One explanation for this is that whenever the second oxonium forms, the hydroxyl group is always held close by, ready to snap shut and give back the dioxolane; water gets less of a chance to attack it and hydrolyze the acetal. For the formation of acyclic acetal, three molecules go in and two come out, but for a cyclic one, two molecules go in and to molecules come out so the usually unfavorable entropy factor is no longer against. A mixture has more entropy than a pure substance because they are many more ways of arranging a mixture. Imagine lining up every molecule in a mole of substance and a mole of 1:1 mixture.

An acetal of polyvinyl alcohol and buteraldehyde, called polyvinyl butyral, is used in safety glass; this is prepared by the acetal formation reaction.

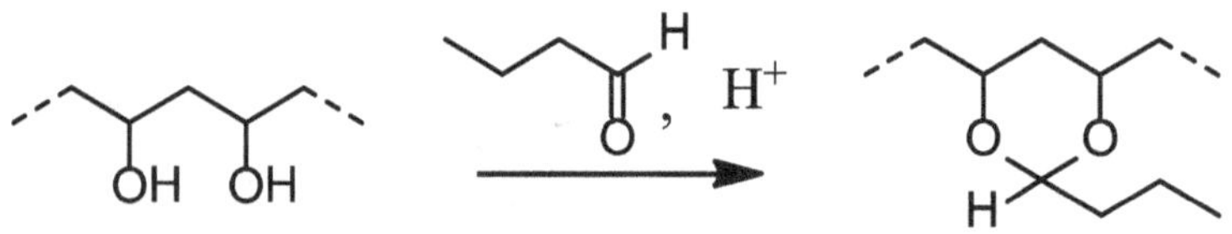

(Equation 22)

Vinyl ethers:

Vinyl or enol ether is the ethers of vinyl alcohols or enols. Vinyl ethers are considered as derivatives of aldehydes and ketones because, like acetals and ketals, they hydrolyze to aldehydes and ketones in very dilute aqueous acid. Enol ethers and enamines are so-called activated alkenes or electron rich alkenes because the oxygen atom donates electrons to the double bond by forming a resonance structure with the corresponding oxonium ion. This property makes them reactive substrates in certain organic reactions such as the Diels-Alder reaction. An enol ether (e.g., methyl vinyl ether and 2,3-dihydrofuran) can be considered the ether of the corresponding enolate, hence the name.

A usefull route to preparing enol ethers is to convert an aldehyde into an acetal, then to add a catalytic amount of acid and to remove the excess alcohol (EtOH) by distillation (Scheme 40).

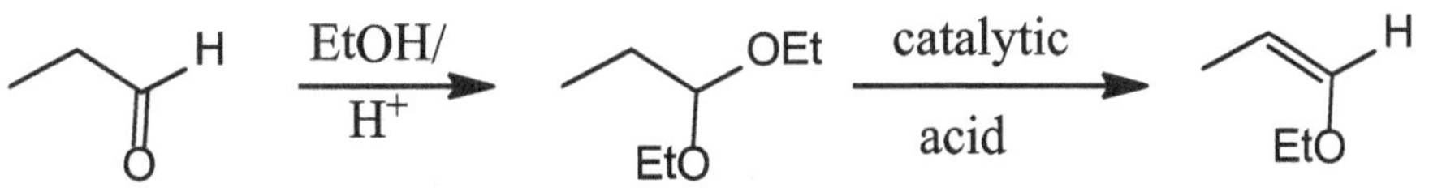

Scheme 40

Vinyl ethers, like acetals and ketals stable toward base. Thus, they can be used in the presence of aqueous NaOH or other bases but reacts immediately in dilute aqueous acid to yield the corresponding carbonyl compounds.

The ease of hydrolysis of vinyl ethers contrasts markedly with the vigorous conditions (high acid concentration, heat) required for hydrolysis of simple dialkyl ethers. In fact, it can be estimated that the hydrolysis of vinyl ether takes place at a rate which is about 10^{13} times that for the hydrolysis of diethyl ether. This large difference suggests a hydrolysis mechanism for vinyl ethers that is not available to dialkyl ethers. The first step in the hydrolysis of vinyl ethers is protonation of the double bond to give a resonance-stabilized cation (Scheme 41).

Scheme 41

Vinyl ethers and alcohols also react in the presence of anhydrous acid catalysts, such as p-toluenesulphonic acid to give acetals or ketals (Equation 23).

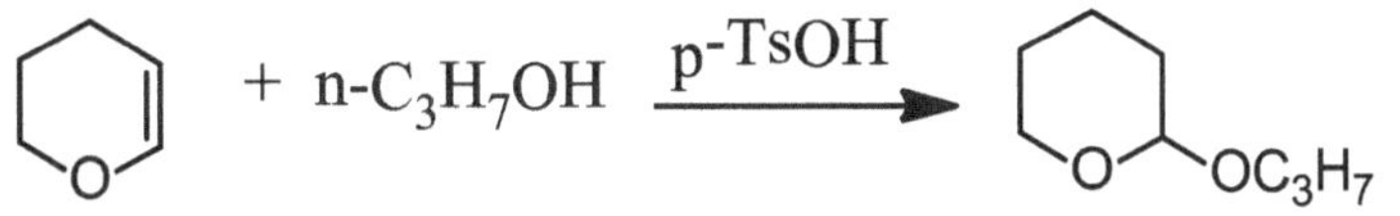

(Equation 23)

Vinyl ethers, like acetals and ketals, are stable toward base. Thus they can be used in the presence of aqueous NaOH or other bases but reacts immediately in dilute aqueous acid to yield the corresponding carbonyl compounds.

2.3   ADDITION OF SULPHUR NUCLEOPHILES

2.3.1   Addition of sodium bisulfite

Sodium bisulphite ($Na^+HSO_3^-$) reacts with most aldehydes and some ketones to give bisulfite addition products (Scheme 42). The bisulfite addition reaction is also reversible. Depending on reaction conditions, some bisulfite addition products precipitate. This can be used as a test for carbonyl compounds.

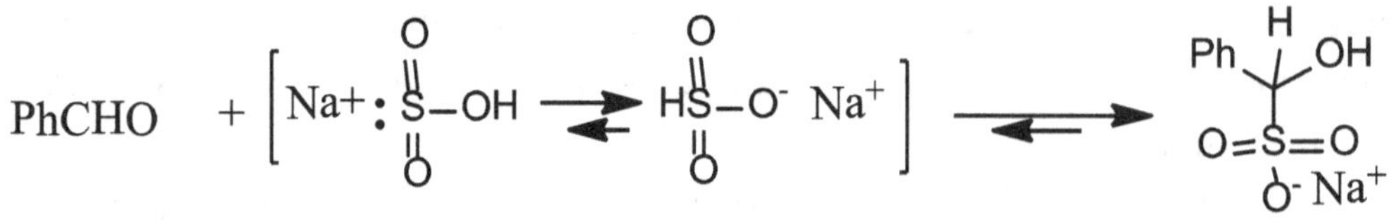

Scheme 42

The position of equilibrium lies largely to the right for most aldehydes and to the left for most ketones, e.g., Hooper (1967) showed from NMR studies of solutions of ketones in aqueous sodium hydrogen sulfite that the amounts of unreacted ketone were Me_2CO, 24.6; MeCOEt, 71.8; MeCOisoPr, 100. These results may be partly explained in terms of steric effects.

Here it is important in RCOR' that if R and R' groups are larger than methyl, the addition compound does not form unless the groups are held out of the way of the carbonyl group. If one group is methyl and the other group is branched, e.g., t-butyl group, the reaction will be vey slow and the position of the equilibrium will also be unfavourable.

The main importance of the bisulfite addition compounds is their use to separate carbonyl compounds from their mixtures with other organic compounds [2007Misc712]. Bisulfite compounds are usually crystalline solids, insoluble in sodium hydrogen sulfite solution. Since bisulfite addition compounds are slats, they are insoluble in organic solvents. So they are freed from other organic compounds by filtering and washing the residue with ether. The bisulfite adduct formation being a reversible reaction, the carbonyl compound is regenerated by any reagent that reacts irreversibly with the bisulfite adduct. Usually, alkalis are used to regenerates the ketones and acids to regenerate the aldehydes. However, since the position of equilibrium for the reaction of formaldehyde with bisulfite is more right towards the right than with any aldehyde or ketone, the carbonyl compound can be regenerated by heating a bisulfite adduct with a slight excess of aqueous solution of formaldehyde through an exchange reaction as shown below.

Another important application of bisulfite addition is the synthesis of cyanohydrin derivatives. Here the bisulfite compound forms first but only as an intermediate on the route to the cyanohydrin. When the cyanide ion is added, reversing the formation of the bisulfite compound provides the singe proton necessary to give back the hydroxyl group at the end of the reaction (Scheme 43). No dangerous HCN is released.

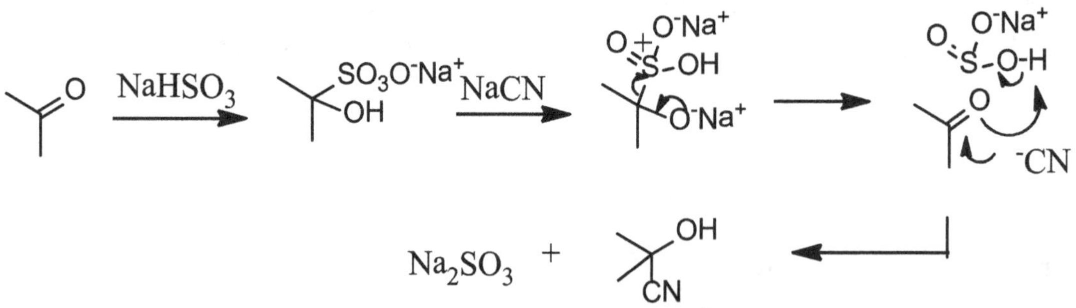

Scheme 43

Desponse is an antileprocy drug which is insoluble in water. That can be made soluble in water by mixing it with formaldehyde bisulfite addition product where it exchanges the OH group for one of the amino groups in desponse (Equation 24).

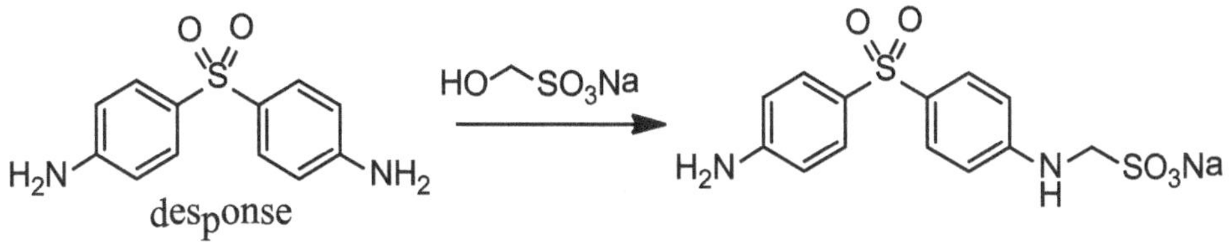

(Equation 24)

2.3.2 Addition of thioalcohol

Dithioacetals, 1,3-dithianes or 1,3-dithiolanes are prepared by reaction of the corresponding carbonyl compounds with thioalcohol or dithioethylne glycol in the presence of an acid catalyst as $ZnCl_2$ or BF_3 [1981Misc129, 1991Misc541] (Equation 25).

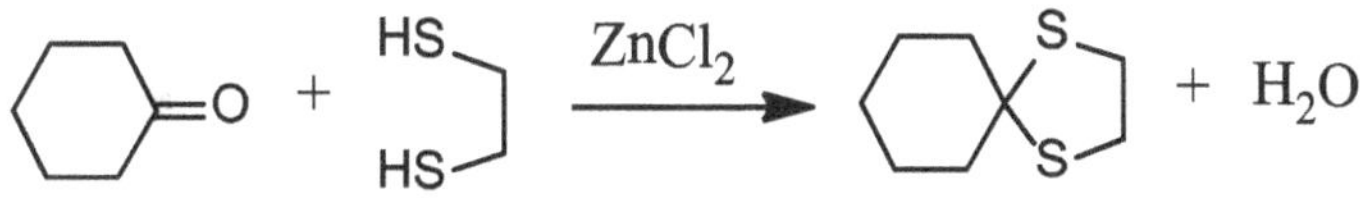

(Equation 25)

The nucleophilicity of sulphur being greater than that of oxygen, thiols undergo the acetal formation reaction more readily than alcohols. The thioacetal are relatively stable to dilute acids and bases. Regeneration of the carbonyl group from the dithioacetal sometimes present difficulties but can be carried out by hydrolysis in polar solvents in the presence of metallic ions such as Hg^{2+}, Ag^+, Cu^{2+} or Ti^{4+} (Equation 26).

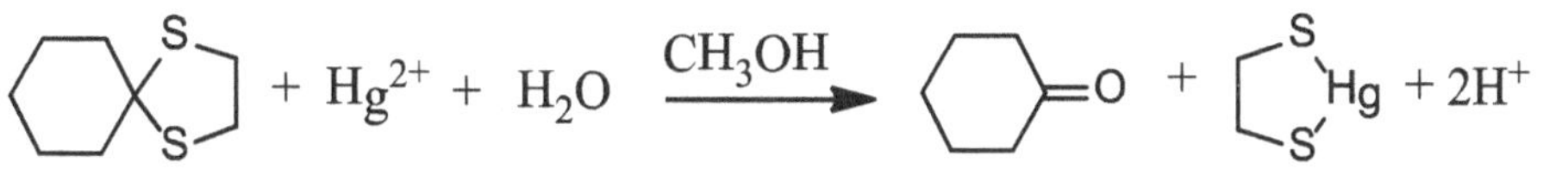

(Equation 26)

Dithioacetals and dithioketals like other sulfides are desulphorized with Raney nickel. This reaction provides a particularly mild method for reducing the carbonyl groups of aldehydes and ketones to methylene groups (Equation 27). Raney nickel is converted irreversibly to nickel sulfide. This method is milder than either the Clemmensen or Wolff-Kishner reductions, which employ strongly acidic or basic conditions, respectively, that might interfere with other functional groups. This reaction is called Mozingo reduction.

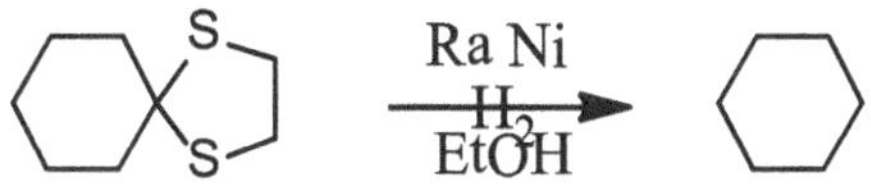

(Equation 27)

Dithioacetal derived from an aldehyde can be further functionalized at the aldehyde carbon with an alky halide, followed by thioacetal cleavage to produce a ketone. 1,2-Addition of lithiodithanes to carbonyl compounds followed by hydrolysis gives α-hydroxy ketones in good yield (Scheme 44).

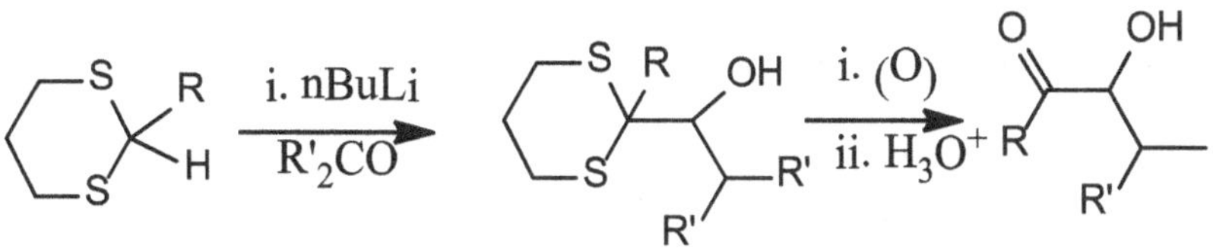

Scheme 44

The benzylic alcohols readily prepared by the conjugate addition of appropriate sulfur stabilized carbanions, either as aryldithianes or as arylbis(phenylthiomethanes) to a butenolide, followed by trapping of the general enolate with an aromatic aldehyde. Cyclization of the derived ketone provides a short efficient synthesis of clinically important podophyllotoxin derivatives (Scheme 45) [1985TL6377, 1988JCSP1603].

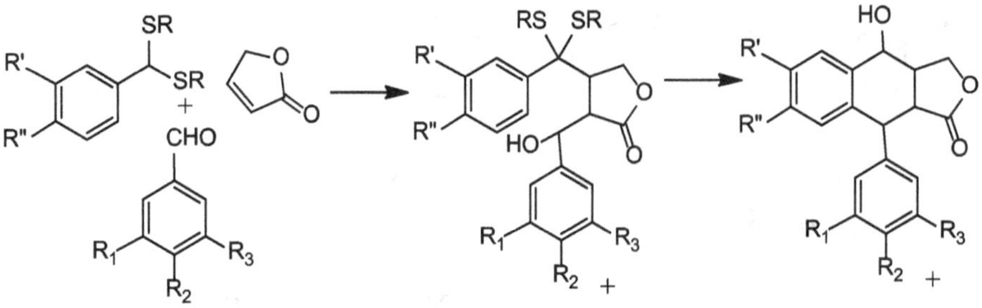

Scheme 45

Acetals/thioacetals and ketals/thioketals as protecting groups:

Acetals and ketals are useful protecting groups for alcohols as well as for aldehydes and ketones. They are easily introduced are stable to base and are easily removed under mild acidic conditions when no longer needed. These points can be illustrated with the following synthesis.

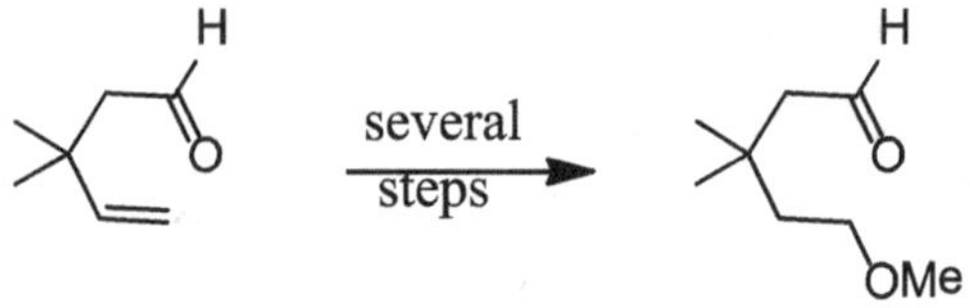

The obvious way to effect this conversion is to use B_2H_6, then $H_2O_2/^-OH$ to make the alcohol, followed by a Williamson synthesis of ether. However, because diborane reacts with aldehydes and ketones, the aldehyde group would not survive this sequence of reactions. The aldehyde can, however, be protected as an acetal, which does not react with diborane. The following synthesis incorporates this strategy (Scheme 46).

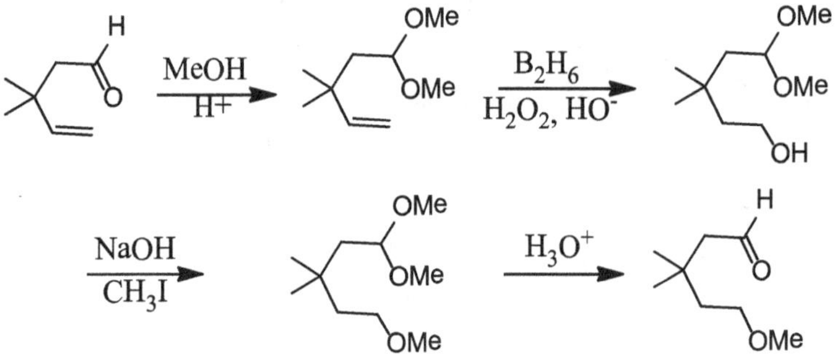

Scheme 46

2.4   ADDITION OF SELENIUM NUCLEOPHILES

Addition of selenols: Diselenoacetals are also useful for the protection of aldehydes or ketones. They are prepared by the reaction of carbonyl compounds with selenols in the presence of acid catalysis. The selenium-stabilized carbanions derived by deprotonation of selenoacetals by strong bases react readily with a variety of electrophiles including primary halides, epoxides, ketones, aldehydes and enones followed by deprotection to give ketones, β-hydroxy ketones, α-hydroxy ketones and 1,4-dicarbonyl compounds respectively [1991Mis541]. Regeneration of the carbonyl group from a diselenoacetal or ketal can be carried out by oxidative hydrolysis using clay supported Iron(III) or copper(II) nitrate Hg(OAc)$_2$ etc.

2.5   ADDITION OF NITROGEN NUCLEOPHILES

Certain derivatives of ammonia undergo addition reactions with aldehyde and ketones to yield α-hydroxy derivatives. These in turn, undergo water elimination reaction to give compounds containing >=N- (Scheme 47).

Scheme 47

The addition-elimination reactions involves derivatives of ammonia which may be represented in general as GNH$_2$

when G is – H/R, the reagent is NH$_3$(ammonia)/ RNH$_2$ (amine) and the product is imine

G is – OH, the reagent is NH$_2$OH (hydroxylamine) and the product is oxime

G is – NH$_2$, the reagent is NH$_2$NH$_2$ (hydraizine) and the product is hydrazone

G is – NHPh, the reagent is PhNHNH$_2$ (phenyl hydrazine) and the product is phenylhydrazone

G is – NHCONH$_2$, the reagent is NH$_2$NHCONH$_2$ (semicarbazide) and the product is semicarbazone

G is – NHCSNH$_2$, the reagent is NH$_2$NHCSNH$_2$ (thiosemicarbazide) and the product is thiosemicarbazone

2.5.1   Imines

Since each of these derivatives of ammonia possesses a lone pair of donatable electrons like ammonia, all of them are basic in nature and get oxidized by the oxygen of air. Being basic, they form salts with acids e.g., with hydrochloric acid they form hydrochloride salts. Hydrochlorides are more stable to oxygen than the bases. For this reason these reagents are kept in the laboratory as hydrochlorides. At the time of reaction between a carbonyl compound and any of the reagents they are taken together in the aqueous medium and a base, sodium acetate, is added to the mixture to liberate the base. Sodium acetate is a salt of weak acid, acetic acid and a strong base, sodium hydroxide. Its addition raises the pH of the solution and the reaction takes place in an optimum pH which is essential for the reaction to occur with an appreciable rate. The optimum pH for a reaction depends on the basicity of the reagent and on the reactivity of the carbonyl compound.

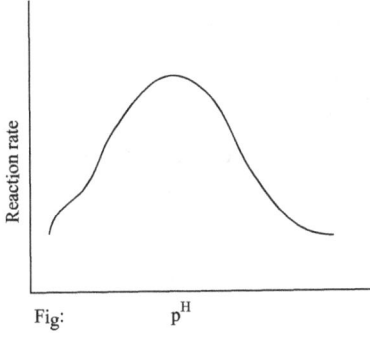

Fig: p^H

Figure shows schematically the type of behavior observed. We can understand the maximum rate by consideration of the possible equilibrium involving RNH_2 and the carbonyl compound.

$$RNH_2 \quad + \quad H^+ \quad \rightleftharpoons \quad RNH_3$$

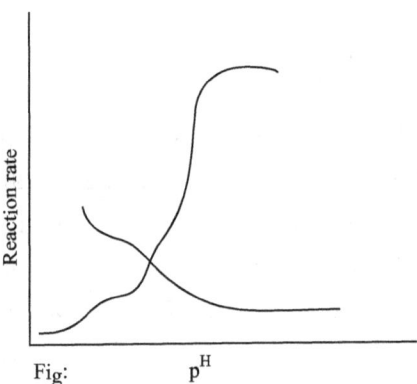

If the unpaired on the nitrogen of RNH_2 is protonated, it cannot then attack the carbon of the carbonyl group. On the other hand, protonation of the carbonyl group should enhance its reactivity toward nucleophilic agent. The favorable combination of reactant is then shown as below.

If this is the slow step in the reaction, we can see that maximum rate will be found when the product of the concentration $[(CH_3)_2C=O^+H][RNH_2]$ is a maximum. As the figure shows these concentrations are appositively affected by P^H. The optimum P^H falls in the region where not all of the RNH_2 is converted to R^+NH_3 and there is a sufficient concentration of the conjugate acid of the carbonyl compound to give a reasonable reaction rate. The rest of the steps for formation of $>=NR$ are usually faster.

Scheme 48

In these cases the reaction products are white or coloured solids having definite crystalline structure and sharp melting point. Thus these derivatives help identify the parent carbonyl compound and that is why the reactions have some analytical importance. In fact, phenylhydrazones and 2,4-dinitrophenylhydrazones are prepared and their m.p.s are determined to identify the carbonyl compounds very often.

When treated with ammonia, formaldehyde does not form an aldehyde-ammonia, but gives instead hexamethylenetetramine (m.p. 260°C) (Equation 28). The structure of hexamethylene-tetramine consists of three fused chair configurations.

$$6HCHO + 4NH_3 \longrightarrow H_2O + (CH_2)_4N_4 \qquad \text{(Equation 28)}$$

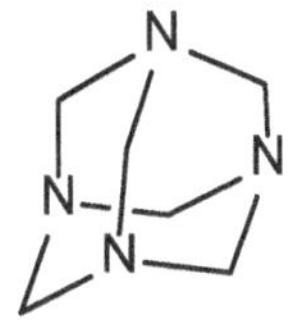

Use of imines as synthetic intermediates has been limited to mainly two processes: reduction to amines and precursors to azaallyl anions for reaction with a variety of electrophiles (Scheme 49). The former transformations can often provide the best access to highly substituted amines and the latter represents one of the highest yield methods for carbon-carbon bond formation α to the carbonyl group of an aldehyde or ketone [1982T1975].

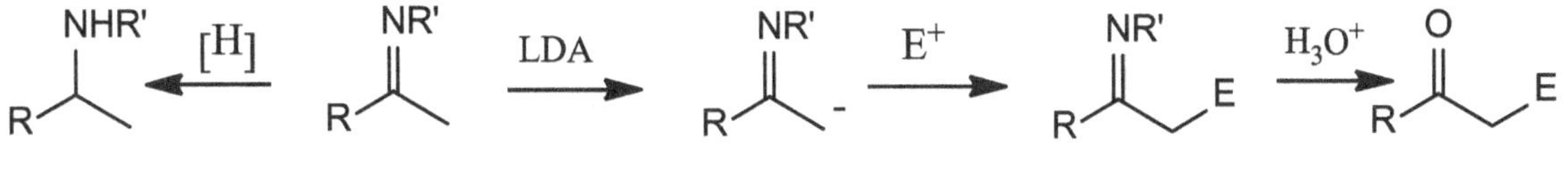

Scheme 49

The regiochemistry of deprotection of imines derived from unsymmetrical ketones is of special significance for the synthetic applications of these anions for carbon-carbon bond formation. This selectivity is sensitive to both the amine moiety and the base used (Equation 29) [1975Syn236].

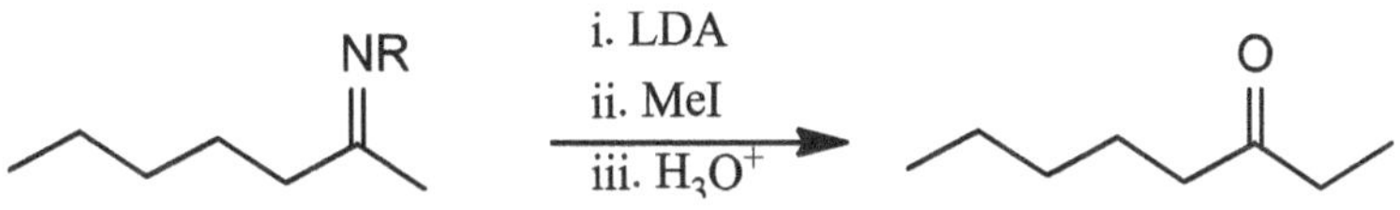

(Equation 29)

Rai et al synthesized series of ethyl-3,5-dicyano-4,6-diaryl-piperidine-2-one-5-carboxylates by the intermolecular condensation of imines and ethyl cycanoacetate in ethyl alcohol and sodium metal. A probable mechanism for the formation of products is as follows (Scheme 50) [2006BCC93].

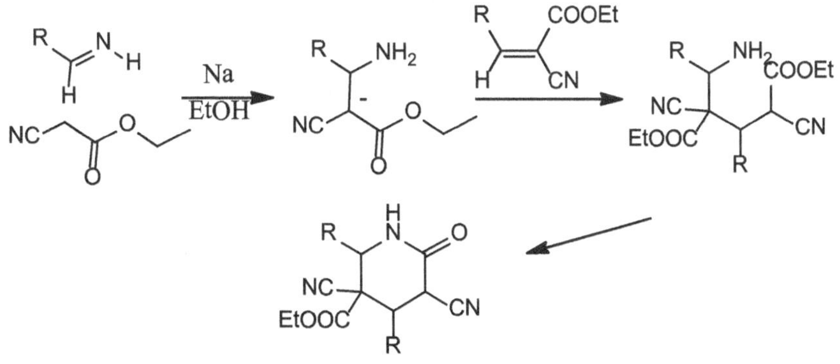

Scheme 50

2.5.1.1   Leuckart reaction:

The Leuckart reaction is the chemical reaction that converts aldehydes or ketones to amines by reductive amination in the presence of heat [1948JACS1187, 1951JOC661, 2007JCSCC3714]. It is carried out by heating the carbonyl compound with ammonium formate or formamide as the reducing agent. It requires high temperatures, usually between 120 and 130°C, although under the presence of formamide, the temperature can be greater than 165°C (Equation 30). This is a good means for preparing primary amines.

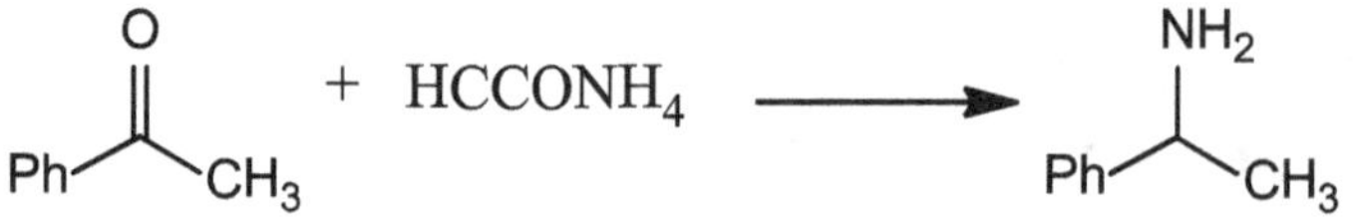

(Equation 30)

Proposed mechanism of the reaction is as follows. First ammonium formate dissociates into formic acid and ammonia. Ammonia then performs a nucleophilic attack on the carbonyl carbon to form a adduct, which then rearranges to α,α-hydroxyamino compound. The hydroxyl is protonated using hydrogen from formic acid, which allows for water molecule to leave. This forms a carbocation, which is resonance stabilized. The compound attacks hydrogen from the deprotonated formic acid from previous step, forming a carbon dioxide and an amine (Scheme 51).

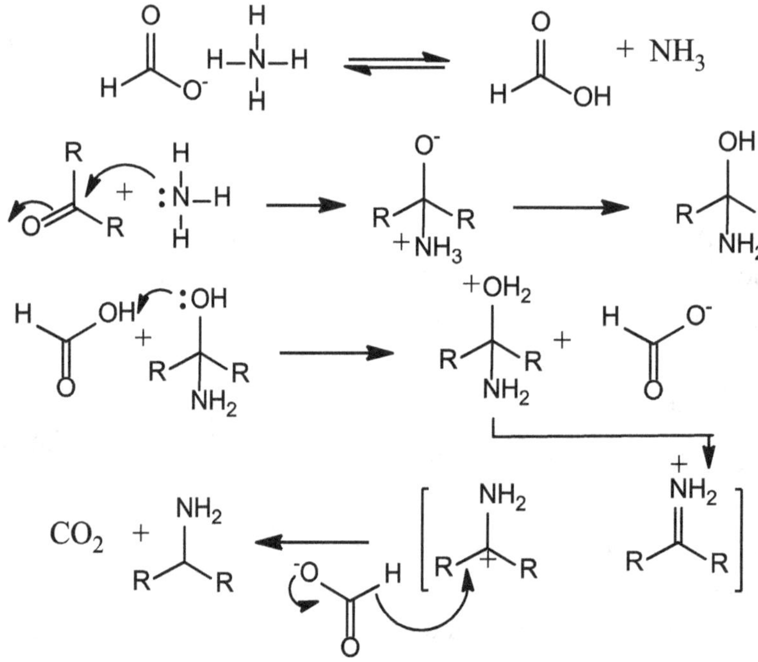

Scheme 51

A notable example of the Leuckart reaction is its use in the synthesis of tetrahydro-1,4 benzodiazepin-5-one, a molecule that is part of the benzodiazepine family. Many compounds in this family of molecules are central nervous system suppressants and are associated with therapeutic uses and a variety of medications, such as antibiotics, antiulcer, and anti-HIV agents. Researchers were able to synthesize tetrahydro-1,4-benzodiazepin-5-ones with excellent yields and purities by utilizing the Leuckart Reaction. Researchers performed the reaction via solid-phase synthesis and used formic acid as the reducing agent.

2.5.1.2   *The* Petasis reaction

The **Petasis reaction** is the chemical reaction of an amine, aldehyde, and vinyl- or aryl-boronic acid to form substituted amines [1993TL583, 1997JACS445, 1998JACS11798].

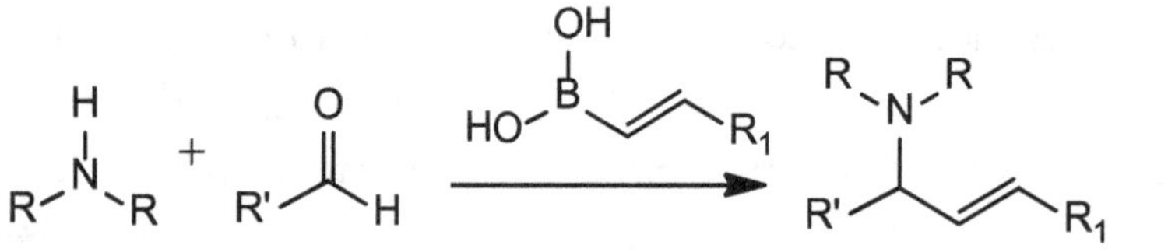

(Equation 31)

Proposed mechanism for this reaction involves the formation hemiaminal via the condensation of carbonyl compounds with secondary amines. Hemiaminal then react with boronic acid in reversible fashion via intermediate forming again electrophilic iminium ion accompanied by nucleophilic boronate (Scheme 52). The irreversible C-C bond migration between **13** and **14** then follows, furnishing desired product with loss of boric acid. All intermediates will ultimately lead to the final product, as the reaction between **13** and **14** is irreversible, pulling the equilibrium of the entire system towards the final product.

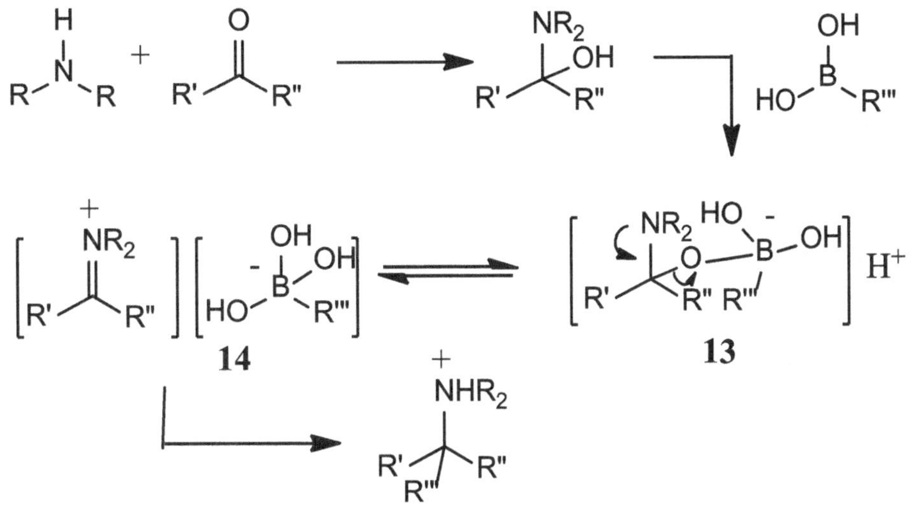

Scheme 52

The Petasis reaction proceeds under mild conditions, without the use of strong acids, bases, or metals. β,γ-unsaturated, N-substituted amino acids are conveniently prepared through the condensation of organoboronic acids, boronates, or boronic esters with amines and glyoxylic acids (Equation 32).

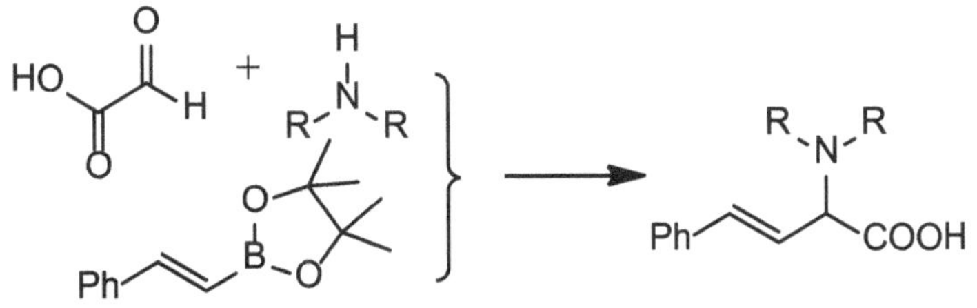

(Equation 32)

Imines in which the nitrogen atom carries electronegative groups are usually stable, example includes hydrazones and semicarbazones. These compounds are more stable than imines because the electronegative substituent can participate in delocalization of the imine double bond. Delocalization decreases the small positive changes on the carbon atom of the imine double bond and raises the energy of the LUMO, making it less susceptible to nucleophilic attack. Oximes, hydrazones and semicarbazones require acid or base catalysis to be hydrolyzed.

2.5.2 Oximes

Oximes are formed with hydroxylamine. Oximes are usually well defined crystalline solids, and may be used to identify carbonyl compounds. The absorption region of oxime (=N stretch) is 1690-30cm^{-1} and that of O-H stretch is 3650-30cm^{-1} [Scheme 53].

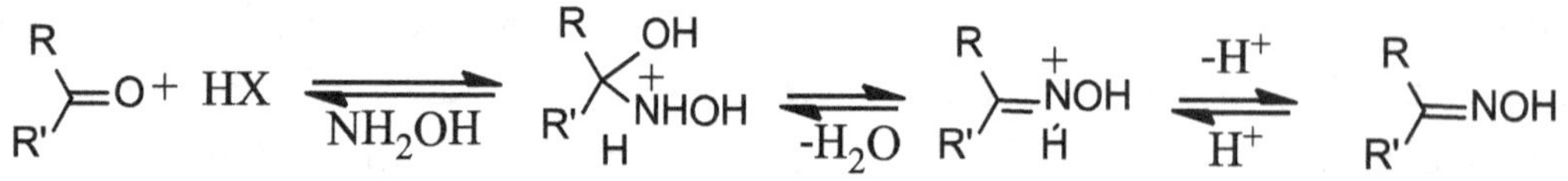

Scheme 53

Acid-catalyzed transformation of ketoximes to N-substituted amides is known as Beckmann rearrangement.

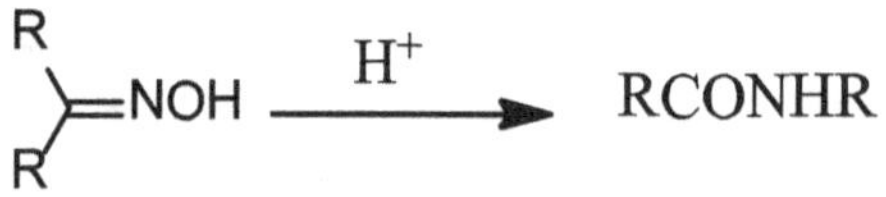

(Equation 33)

Usually this reaction is carried out either with PCl_5, H_2SO_4, P_2O_5, PPA, $SOCl_2$ etc. Under the conditions of the experiment, the aldoximes suffer dehydration and thus forms nitrile. Mechanism of Beckmann rearrangement belongs to that of a reaction which involves the rearrangement of an imino species, whose nitrogen bears a good leaving group that is stable anion or neutral molecule and is represented as follows. (Scheme 54)

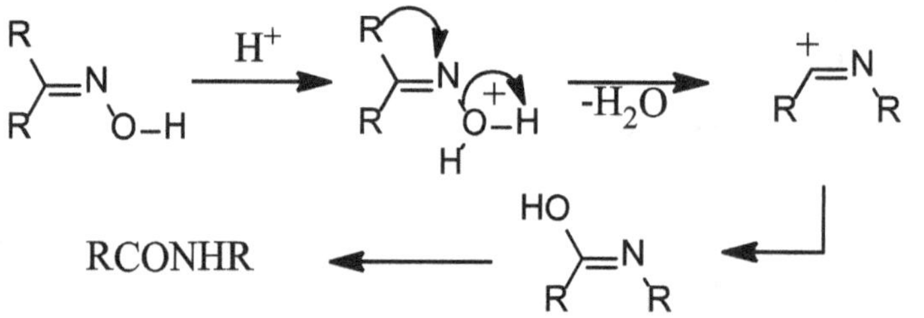

Scheme 54

It is now established that this rearrangement is highly stereospecific in that the migrating groups always approaches the nitrogen atom on the side opposite to the oxygen atom.

Aldoxime form cyanide when boiled with acetic anhydride whereas ketoximes form acetyl derivatives of the oxime.

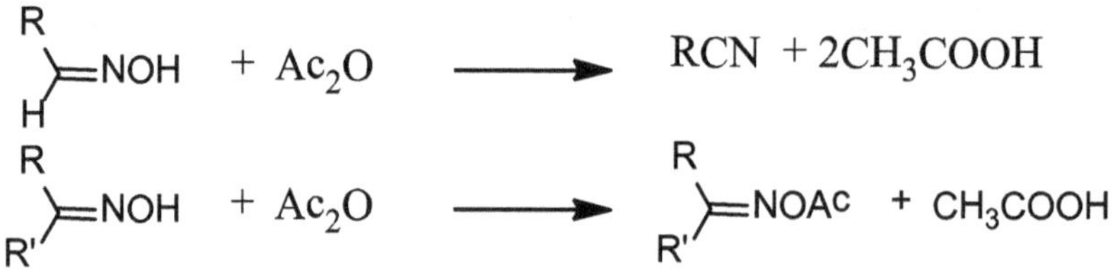

2.5.2.1   Neber reaction (*Neber rearrangement*):

When ketooxime tosylats bearing acicic hydrogen atom adjacent to CN function are treated with bases such as NaOEt or pyridine, rearrangement takes palce yielding α-aminoketones. This is called Neber rearrangement and involves the migration of nitrogen. Here R and R' can be alkyl or aryl (Equation 34) [1964CR81].

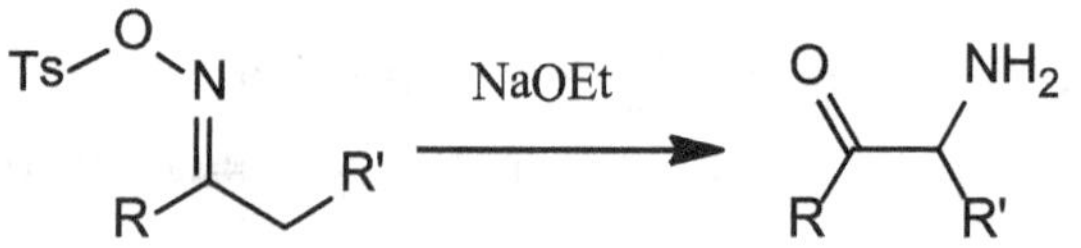

(Equation 34)

The first attempt regarding the elucidation of the mechanism of the reaction was made by Neber himself who proposed the sequence of reaction depicted below.

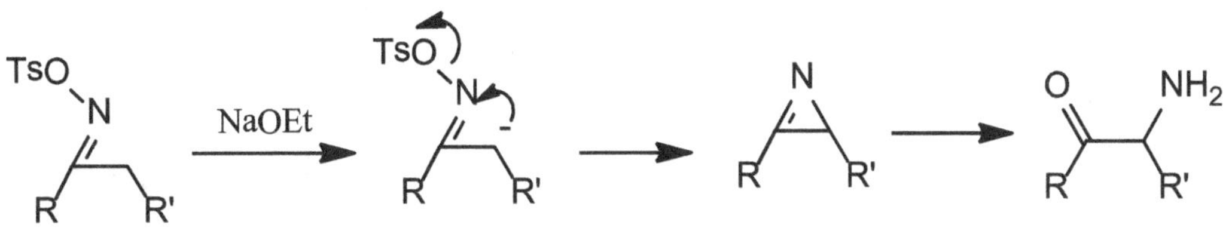

Scheme

Apparently a carbanion is generated in the first step by pulling the α-proton which makes a direct 1,3-intramolecular nucelophilic attack to displace the tosyl group and an azirine is formed as an intermediate. Ring opening by acid hydrolysis yields the α-aminoketone. The occuraence of the azirine ring has been proved by its isolation and conversion to products under experimental conditions. Cram and Hatch also supported this mechanism. This reaction is non-stereospecific unlike the Beckmann rearrangement. In otherwords, the conformation of the ketone derivatives (syn or anti) has little if any, influence on the direction of reaction.

House and Borkowitz subsequently proposed an alternative to the above mechanism by suggesting that azirine is formed via the intermediacy a nitrene and not directly as postulated above. Isolation of ethylenemide supports the formation of azirine system.

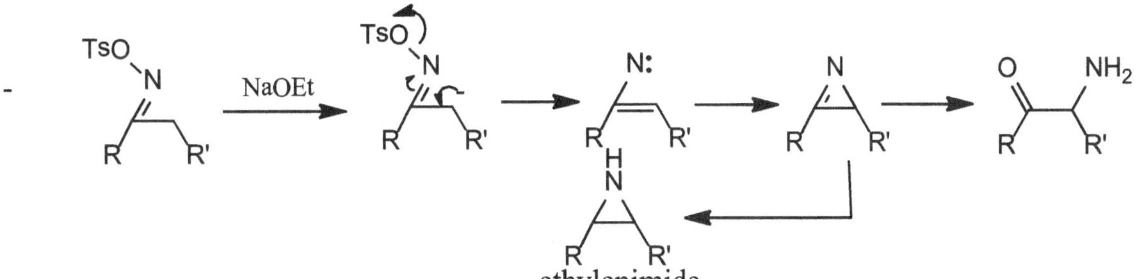

ethylenimide

Scheme

In unsymmetrical ketones possessing acidic hydrogen atoms on both the α-carbon atoms but slightly differing considerably in acidities, it has been established that the amino group in the product is attached exclusively to the carbon bearing the more acidic hydrogen, irrespective of the configuration of oxime tosylate. Depending upon the reaction conditions and method of workup, with NaOEt, the tosylate can give aminoketal, aminoketone hydrochloride or a pyrazine (the known roduct of ready cyclization and dehydration of α-aminoketones).

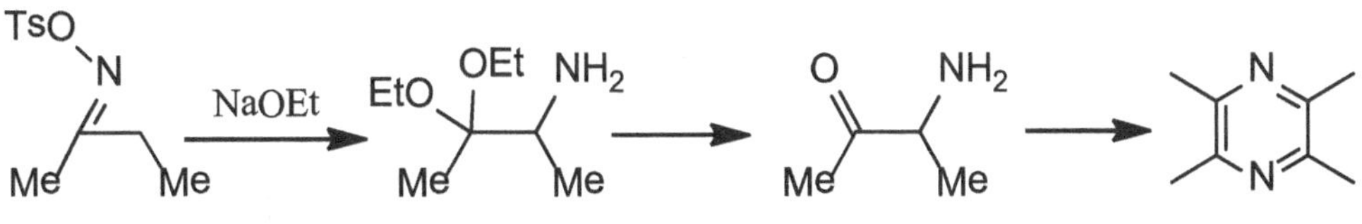

Scheme

2.5.2.2   [3+2]-1,3-dipolar cycloaddition reactions of nitrile oxides

All known methods for the synthesis of nitrile oxides start with organic system already containing -C-N-O sequence of the nitrile oxide structure. Many methods are reported to generate nitrile oxide [2008Mic69]. A few oxidative dehydrogenation methods of aldoximes using oxidants such as lead tetraacetate, alkali hypohalite, N-bromosuccinimide in dimethyl formamide followed by base treatment, 1-chlorobenzotriazole, chloramine-T, mercuric acetate are reported. In situ generation of nitrile oxide from aldoxime by potassium fericycanide require aqueous medium while that of ceric ammonium nitrate can be used only for aromatic aldoximes. Radhakrishna et al reported the use of hypervalent iodine compounds as an oxidizing agent for the in situ conversion of aldoximes to nitrile oxides. Since the workup requires alkaline condition, this method is limited to alkaline resistant compounds. Moreya et al reported the insitu generation of nitrile oxides by the reaction of aldoximes with tertiary butyl hypochlorite and bis (tributyltin) oxide. The reaction proceeded efficiently under mild condition in which O-stannylated aldoximes are thought to be the intermediate (Scheme 55)

Rai and Hassner's method not only allows in situ generation but also allows the isolation of nitrile oxides from aldoximes using chloramine-T as dehydrogenating reagent. This reaction is usually carried out by heating a mixture of aldoxime and an alkene in ethanol in the presence of chloramine-T. By employing this method, we have isolated and characterized the nitrile oxide, of which some are liquids and some are solids. The unstable compound identified by NMR spectrometry slowly dimerizes on standing it alone or in presence of added vinyl sulphone, undergo cycloaddition to yield isoxazoline in good yield.

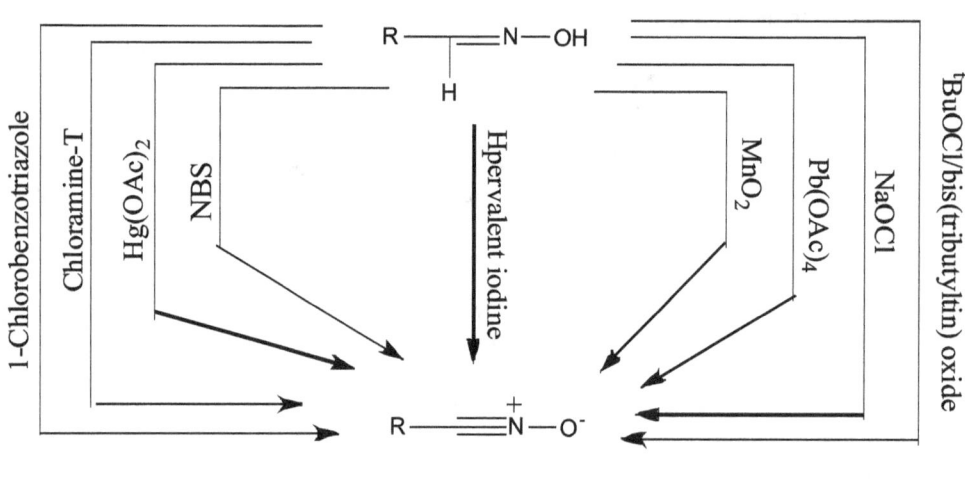

Scheme 55

The formation of nitrile oxide from the corresponding aldoxime by chloramine-T may be rationalized by the probable mechanism proposed below (Scheme 56) [1989Syn57].

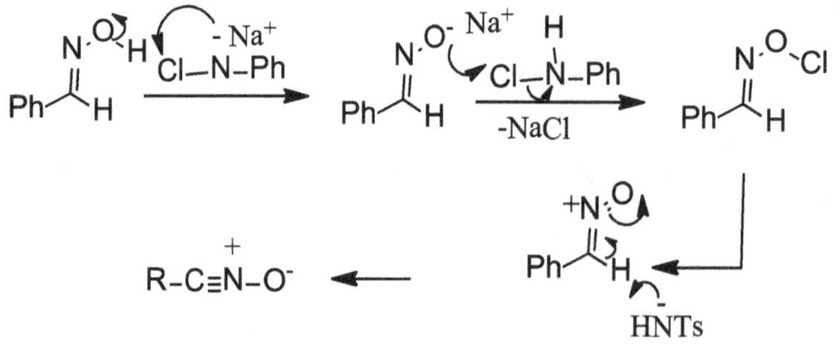

Scheme 56

Recently Rai et al [2019JCS 131] developed potassium iodate (KIO3) as new reagent for the insitu generation of nitrile oxide starting from aldoximes, which is an important intermediate for the synthesis of the valuable heterocycle like isoxazoline. The probable mechanism for the generation is shown in the scheme 56a.

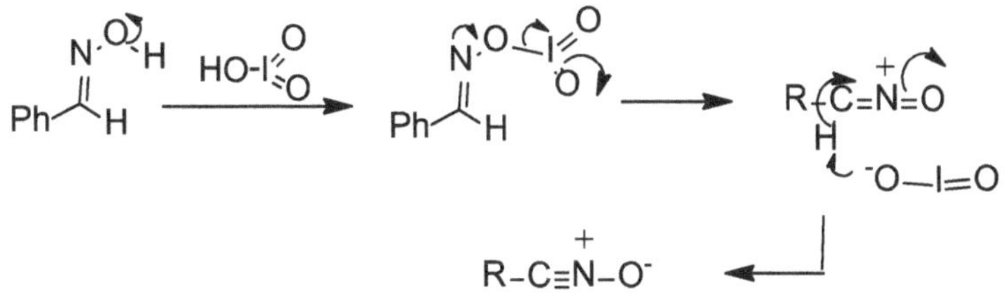

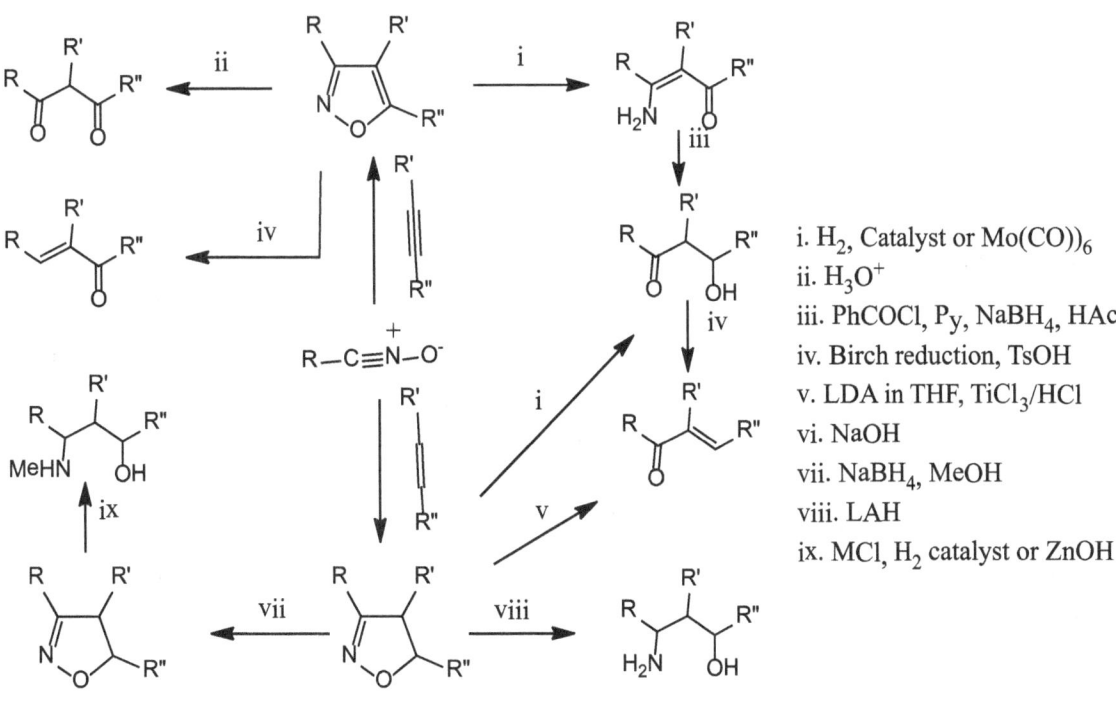

Scheme 56a

2.5.2.2.1 Application of nitrile oxide cycloaddition reactions:

1,3-Dipolar cycloaddition of nitrile oxide to C=C bond of dipolarophile is of considerable importance in organic synthesis, since this reaction yields 2-isoxazolines. Isoxazoles and isoxazolines were serves as an important building block in the construction of new molecular systems for several reasons. First of all, they can be very efficiently prepared from readily available precursors; secondly, they can be conveniently modified, thus allowing transformation of molecule with simple structure to functionally complex derivatives; thirdly, a suitable pattern of substituents makes the isoxazoline ring survive under a variety of chemical reaction conditions, thus allowing manipulation in other parts of the molecule; and finally the ability of the nitrogen-oxygen bond to catalytic or chemical reduction under mild conditions unravels a vast array of different functionalities (Scheme 57). Baraldi et al, synthesized β-hydroxy ketones from isoxazolines utilizing molybdenum hexacarbonyl as catalyst for the reductive cleavage of O-N bond.

i. H_2, Catalyst or $Mo(CO))_6$
ii. H_3O^+
iii. $PhCOCl$, P_y, $NaBH_4$, HAc
iv. Birch reduction, TsOH
v. LDA in THF, $TiCl_3/HCl$
vi. NaOH
vii. $NaBH_4$, MeOH
viii. LAH
ix. MCl, H_2 catalyst or ZnOH

Scheme 57

For instsance, Hassner et al [1984SC1669] shown that α-bromoaldoxime synthon {dervived from the α-bromination of aliphatic aldoximes by NBS and benzoyl peroxide in carbon tetrachloride} can be converted into alkenyl amino or alkenyoxy oximes by reaction with an unsaturated amine or alcohols respectively (Scheme 58). Oxidation of oximes (X=O) to nitrile oxides provides a stereoselective route to functionalized tetrahydrofuran derivative via an INOC reaction. The preferred stereoisomer in the formation of five membered ring ether is trans whereas in the six membered ring ether the cis isomer predominates.

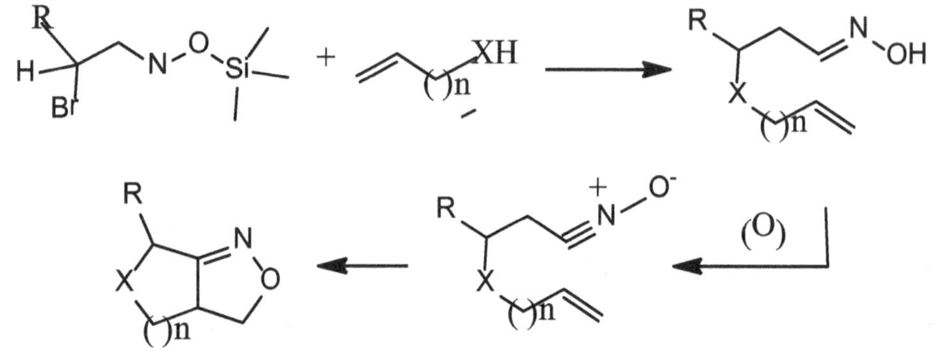

Scheme 58

Aldehyde and ketones combine with N-substituted hydroxylamines to form nitrones, which readily undergo 1,3-cycloaddition with alkenes to form isoxazolidines, while when it react with HCN, it forms cyanoimine (Scheme 59).

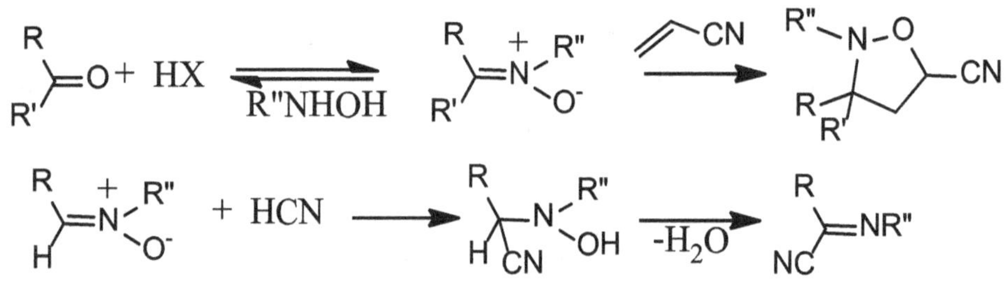

Scheme 59

2.5.2.3   [4+2] Cycloaddition of nitrosoalkenes

The α-nitrosoalkenes are very useful synthetic intermediates because of double bond in conjugation with nitroso group. These reactive species along with nitroso carbonyls constitute the two major sources for the construction of the 1,2-oxazine structure. Nitrosoalkenes are unstable and highly reactive. Normally they are observed only in solution, their presence sometimes being detectable by blue coloration (they have a λ_{max} close to 700 nm). Nitrosoalkenes have been isolated in only a few cases.

The usual method of generating nitrosoalkenes is by the elimination of hydrogen halide from α-monohaloketooximes in presence of base. The generated nitrosoalkenes are trapped by alkenes to produce 1,2-oxazine derivatives (Scheme 60).

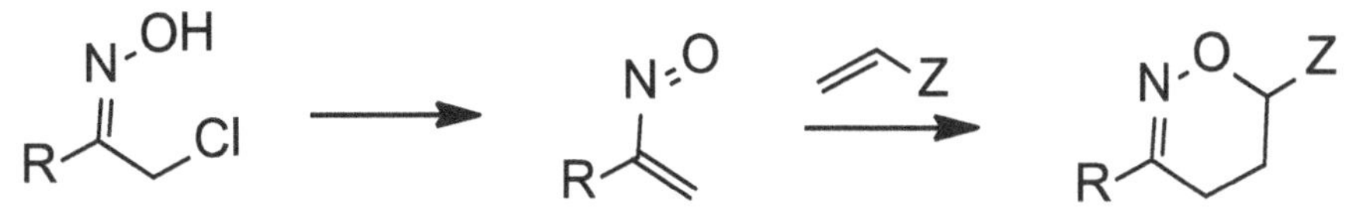

Scheme 60

Recently Rai et al [2005JHC877] utilized chloramine-T for the generation of nitrosoolefin starting from ketooxime. The resultant nitrosoolefins cycloadd with dienophile leads to the formation of oxazine derivative almost in quantitative yield (Scheme 61).

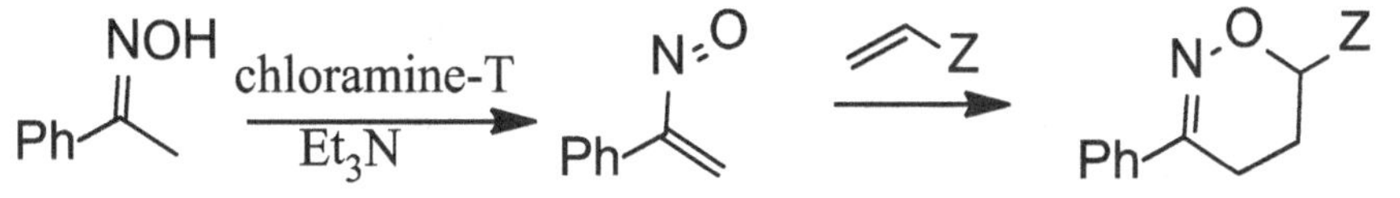

Scheme 61

2.5.2.3.1 Mechanism for the generation of α-nitrosoolefins

When ketooxime containing α-methylene group was treated with chloramine-T produced a blue coloration suggestive of formation of α-chloronitroso compound, which on treatment with triethylamine generates α-nitrosoolefins. The generated α-nitrosoolefin was further treated with alkene to produce 1,2-oxazine derivatives. This was further confirmed by comparing with the 1,2-oxazine derivatives synthesized using literature procedure. The formation of α-nitrosoolefin may be rationalized by the probable mechanism proposed below (Scheme 62).

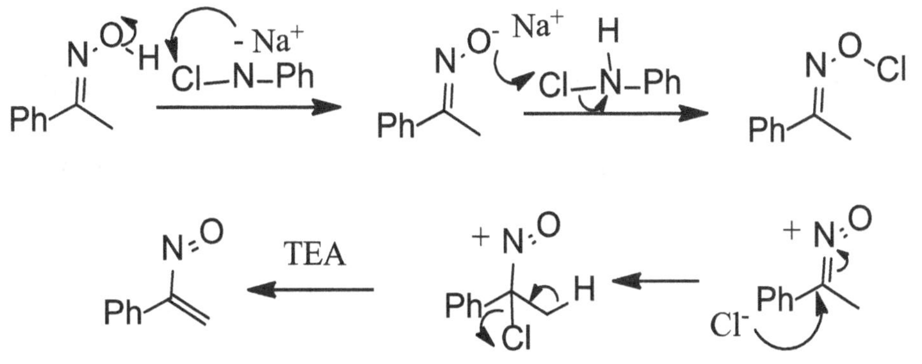

Scheme 62

2.5.3 Hydrazones

Reaction with hydrazine, carbonyl compounds form hydrazones and azines while with phenyl hydrazine, phenyl hydrazone is formed (Scheme 63).

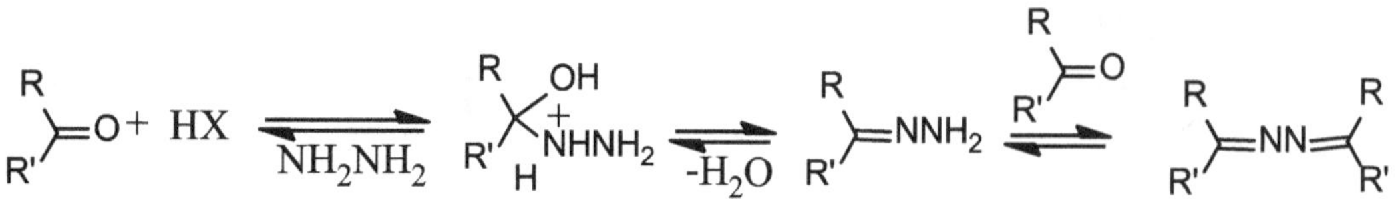

Scheme 63

The mechanism for the formation of semicarbazones and oximes has been studied in great detail. These reactions are catalyzed by acids and it has been found that the rates of formation of the products reach a maximum at some particular pH. According to Jencks et al (1960), the pH optima observed are due to transition in the rate determining step with changing concentration of acid.

Jencks showed by means of UV and IR studies that at pH 7, the formation of "**5**" is fast and that as the acidity increases the overall rate of reaction increases, since the dehydration step which is the rate determining step also increases (through protonation of –OH) (Scheme 44) where G= $NHCONH_2$ or OH.. However as the acidity increases, the addition step becomes progressively slower because G-NH_2 is decreasing in concentration due to its conversion into the conjugate acid, G-NH_3^+. The latter no longer has a lone pair of electrons on the N-atom and so is not a nucleophile. Hence, at sufficiently high acidity, the addition step is slower down to such an extent that this now becomes the rate determining step in the reaction.

Oximes and hydrazones regenerate the carbonyl compounds when refluxed with dil. HCl acid. Regeneration from phenyl hydrazones by this method is usually difficult. One fairly successful method is by exchange of the phenylhydrazino group with another oxo compound, e.g., pyruvic acid. The most satisfactory method however appears to be the use of acetyl acetone in faintly acid solution, the product is 3,5-dimethyl-1-phenyl pyrazole (Equation 35).

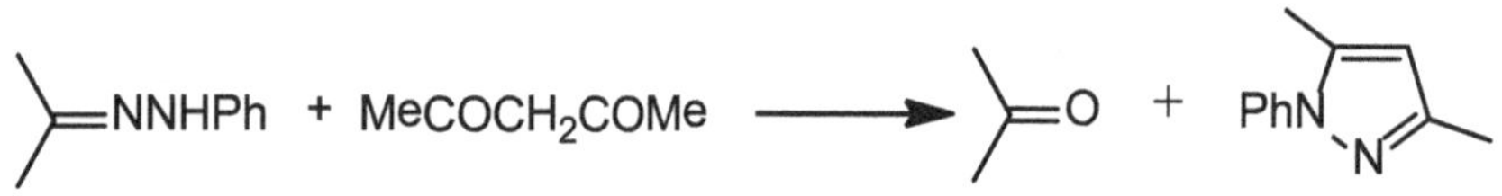

(Equation 35)

The **Shapiro reaction** is an organic reaction in which a ketone or aldehyde is converted to an alkene through an intermediate hydrazone in the presence of 2 equivalents of strong base (Scheme 64). The reaction was discovered by Robert H. Shapiro in 1967 [1967JACS5734, 1975TL1811, 1976OR405, 1983SCR55].

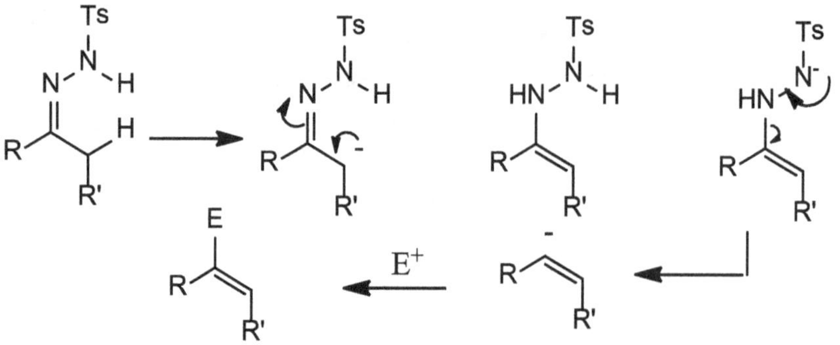

Scheme 64

The probable mechanism for this involves the first formation of tosyl hydrazone by the reaction of ketone or aldehydes possessing α-hydrogen with p-toluenesulfonylhydrazide. Two equivalents of a strong base, such as n-butyllithium, then abstract first the proton from the α to the hydrazone carbon and then the less acidic proton from tosyl hydrazide nitrogen results in the expulsion of the diazonium anion, leaving a vinyl lithium at the position where the nitrogen had been attached. This organolithium carbon is both nucleophilic and basic. It can be reacted with various electrophiles or simply neutralized with water or an acid.

The tosylhydrazone prepared from epoxy ketone is unstable, which underwent rearrangement to generate zwitterion and this ion is ready for fragmentation. The "push" comes from the newly created hydroxyl group and the "pull" from the irresistible concerted loss of a good leaving group (Ts⁻) and an even better one (N_2). Leads to the formation of keto alkyne, which is the precursor for the synthesis of exo-brevicomin (Scheme 65).

exo-brevicomin

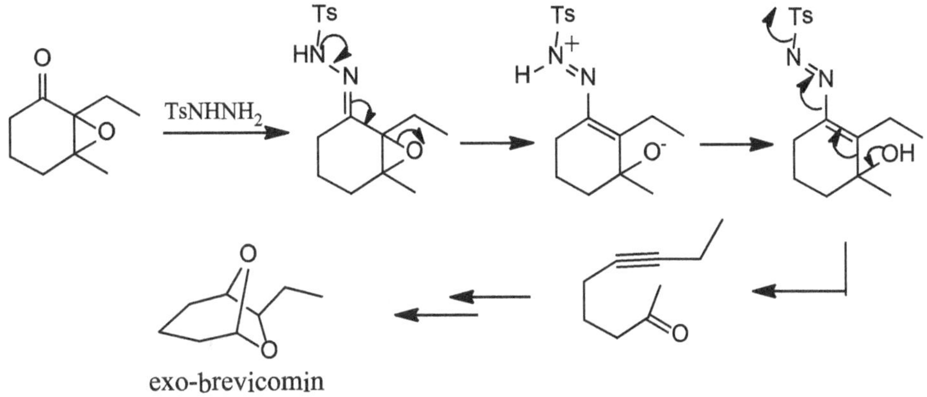

Scheme 65

2.5.3.1 [3+2]-1,3-dipolar cycloaddition reactions of nitrile imines

Nitrile imines formed by the oxidative dehydrogenation of phenylhydrazone derived from aldehydes using oxidants such as lead tetraacetate, alkali hypohalite, N-bromosuccinimide in dimethyl formamide followed by base treatment, 1-chlorobenzotriazole, chloramine-T or mercuric acetate undergo 3+2 cycloaddition with alkenes produces pyrazolines while with alkyne yiled pyrazole derivatives (Scheme 66).

Scheme 66

Alkenes and alkynes serve as an excellent dipolarophiles. Cycloaddition of nitrile imines to olefins yield 2-pyrazolines, whereas addition of nitrile imines to a C-C triple bond yields aromatic pyrazole system directly (Scheme 67).

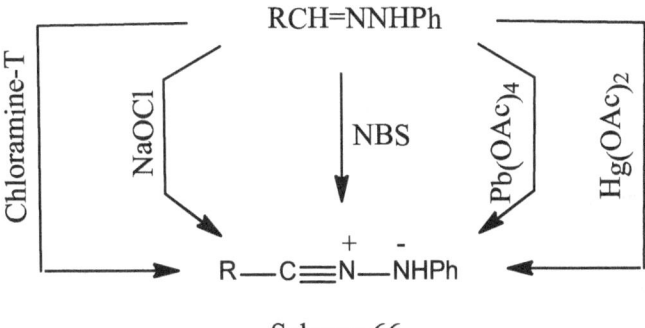

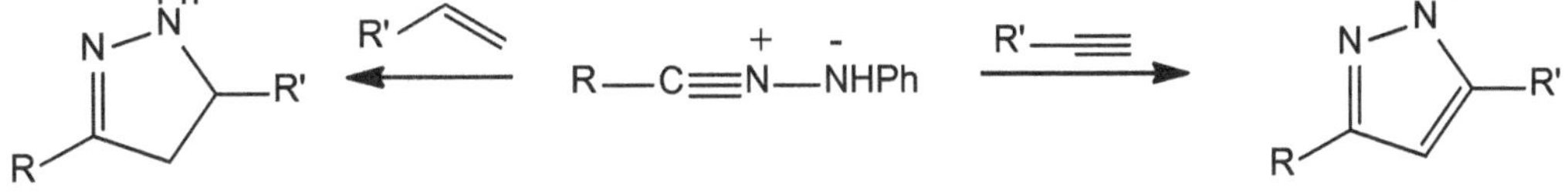

Scheme 67

Rai et al synthesized bishydrazones (aldazines) by thermolysis of aldehyde semicarbazones under reduced pressure in a sealed tube at 120-30°C (Scheme 68) [99IJC1126].

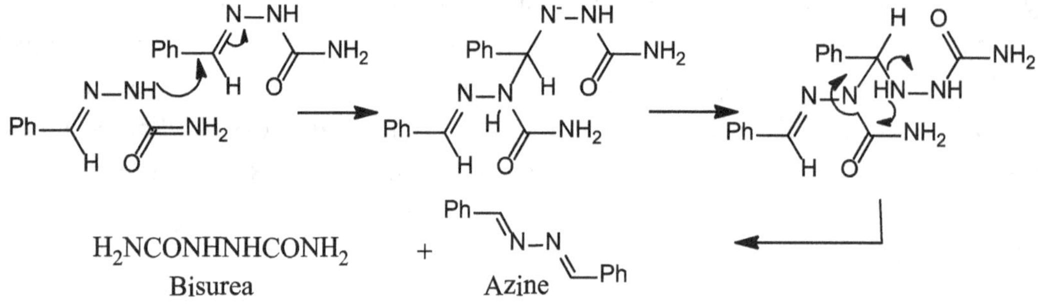

H$_2$NCONHNHCONH$_2$ +   Azine
Bisurea

Scheme 68

2.5.3.2   [4+2] Cycloaddition of azoalkenes

Azoalkenes acts as 4π-electron component in [4+2]- cycloaddition reactions. [4+2]- Cycloaddition reactions of azoalkenes with olefinic compounds are of synthetic interest since tetrahydropyridazine derivatives formed are important in the synthesis of an antihypertensive agent, vasodilators, glycosidase inhibitors etc. Azoalkenes are unstable and highly reactive. Normally they are observed only in solution,

The usual method of generating azoalkenes is by the elimination of hydrogen halide from the hydrazones of α-monohaloketones in presence of base. The generated azoalkenes are trapped by alkenes to produce pyrazine derivatives (Scheme 69).

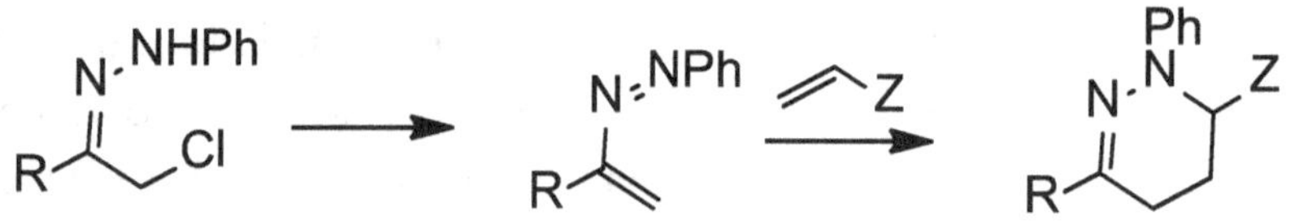

Scheme 69

Recently Rai et al utilized chloramine-T for the generation of azoolefins starting from ketohydrozones. The resultant azoolefins cycloadd with dienophile leads to the formation of tetrahydropyridazine derivative almost in quantitative yield (Scheme 70) [2005TL5969].

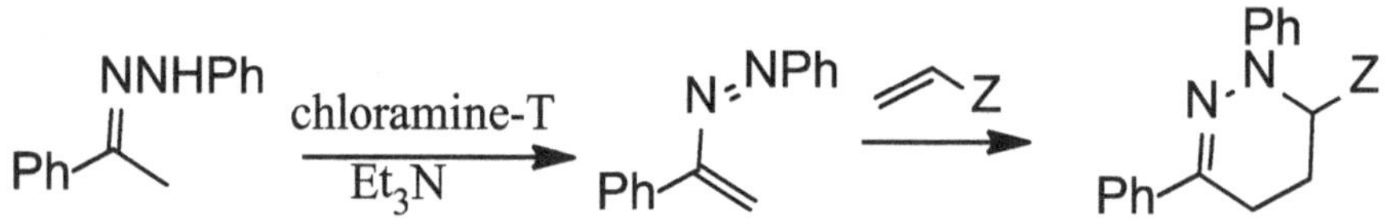

Scheme 70

2.5.4   Semicarbazones

A semicarbazone is a derivative of imines formed by a condensation reaction between a ketone or aldehyde and semicarbazide. They are classified as imine derivatives because they are formed from the reaction of an aldehyde or ketone with the terminal -NH$_2$ group of semicarbazide, which behaves very similarly to primary amines.

Some semicarbazones, such as nitrofurazone, and thiosemicarbazones are known to have anti-viral and anti-cancer activity, usually mediated through binding to copper or iron in cells. Many semicarbazones are crystalline solids, useful for the identification of the parent aldehydes/ ketones by melting point analysis [1999Misc426].

Oxidation of aldehyde semicarbazones by mild oxidant like chloramine-T afforded 5-aryl-2-aminooxadiazole (Scheme 71) [96JSST71]. Series of N-alkylated-5-aryl-2-amino-1,3,4-oxadiazoles were synthesized via alkylation of 5-aryl-2-amino-1,3,4-oxadiazoles under phase transfer condition and screened for their antimitotic acitivity by onion root tip method [2000IF389].

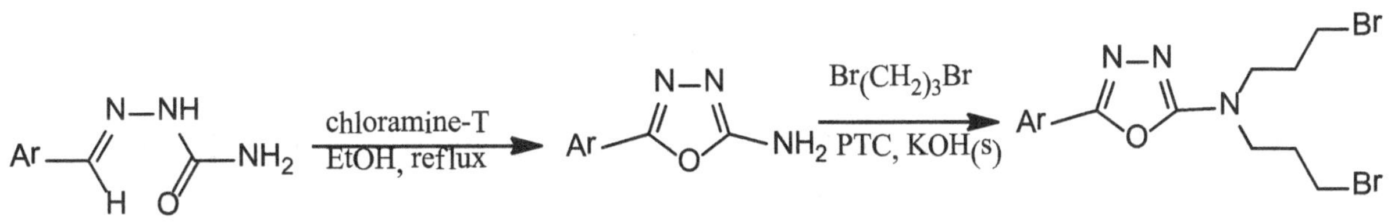

Scheme 71

2.5.5 Schmidt reaction

The Schmidt reaction is an organic reaction in which an hydrazoic acid reacts with a carbonyl group to give an amine or amide in the presence of conc. H_2SO_4 acid, with expulsion of nitrogen [2006CSR146]. It is named after Karl Friedrich Schmidt (1887–1971), who first reported it in 1924 by successfully converting benzophenone and hydrazoic acid to benzanilide. Aldehydes give a mixture of cyanide and formyl derivatives while ketones give amide [2011JOC0001, 1946OR307].

$$RCHO + HN_3 \xrightarrow{H^+} RCN + RNHCHO + N_2 \qquad \text{(Equation 36)}$$

$$R_2CO + HN_3 \xrightarrow{H^+} RCONHR + N_2 \qquad \text{(Equation 37)}$$

The mechanism of the reaction is uncertain. It has been shown to be intramolecular and Schmidt has proposed the following mechanism, which is an example of the 1,2-alkyl/aryl shift (from carbon to nitrogen) to form ketone (Scheme 72). In ketones if the two groups are not identical, then two geometrical isomers of "y" are possible. It is also reasonable to suppose that anti group (to the diazonium nitrogen) is the group that migrates. In this it is possible to explain how steric factors influence the isomer ratio of amides formed

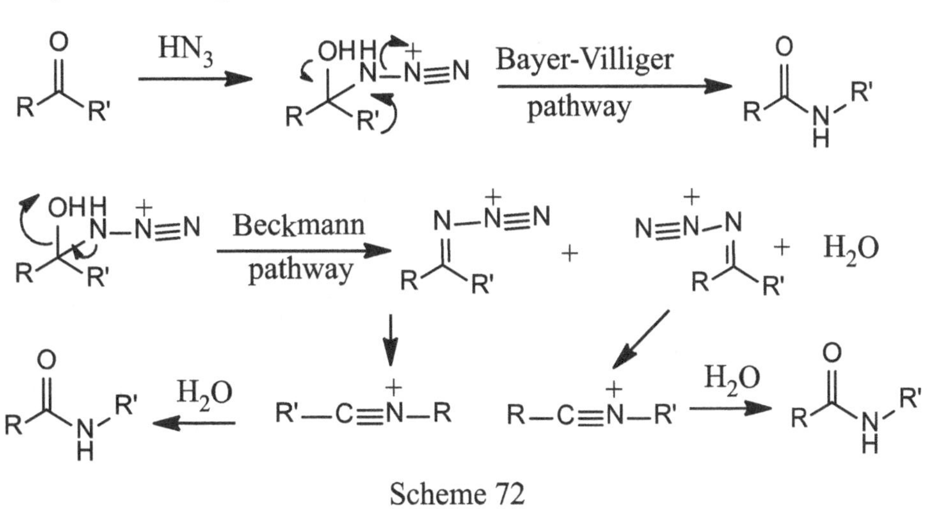

Scheme 72

For aldehyde,

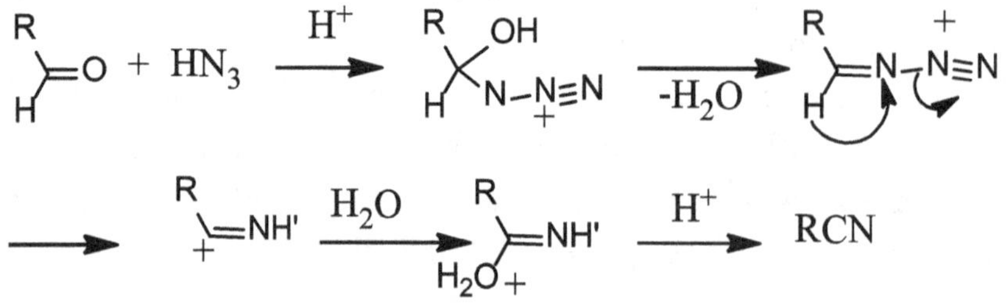

Scheme 73

Steric hindrance to nucleophilic addition lowers yields somewhat, but site selectivity is improved when the substituents of the carbonyl are of different sizes (Equation 38). The less substituted carbon rarely migrates.

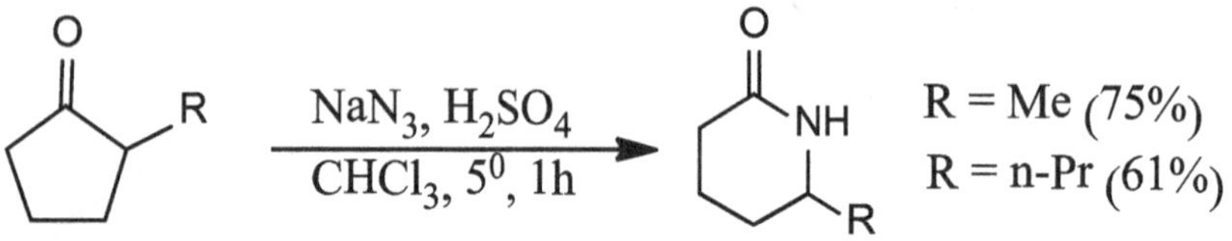

$R = Me$ (75%)
$R = n\text{-}Pr$ (61%)   (Equation 38)

Aromatic groups often migrate in preference to alkyl groups in reactions of ketones, particularly if the ring is electron rich.

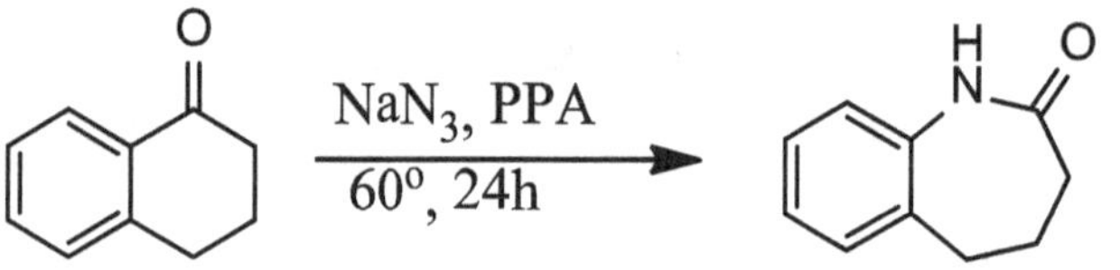

(Equation 39)

Intermolecular Schmidt reactions of alkyl azides are plagued by poor site selectivity and strong sensitivity to steric effects. However, intramolecular reactions of alkyl azides are among the most synthetically useful applications of the Schmidt reaction. The range of ketones that engage in the intramolecular Schmidt reactions is exceptionally broad. For these reactions, the azide and carbonyl carbon must be separated by four or five atoms to facilitate cyclization, with four preferred.

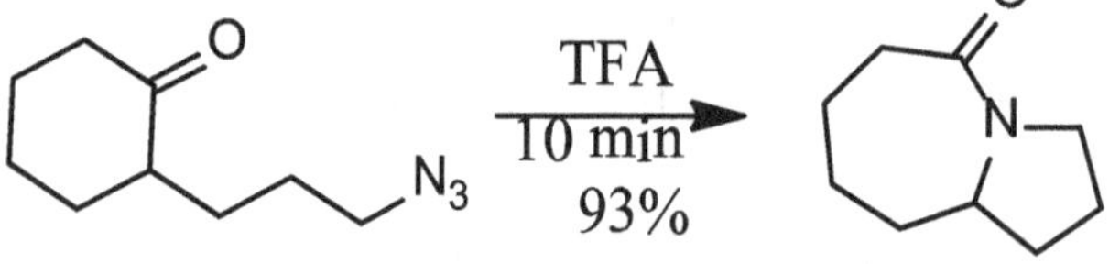

(Equation 40)

2.6   ADDITION OF PHOSPHOROUS NUCLEOPHILE

Addition of a phosphine to an unconjugated carbonyl compounds affords no opportunity for delocalization of the negative charge and these additions, which theoretically can take place on oxygen or carbon are reversible with the equilibrium lying well to the left (Scheme 74) [74P109]. The reaction pathway followed appears to depend on the nature of the carbonyl compound and may be governed to some extent by the possibility of electron delocalization in the transition state of the reaction or in intermediates 15 and 16.

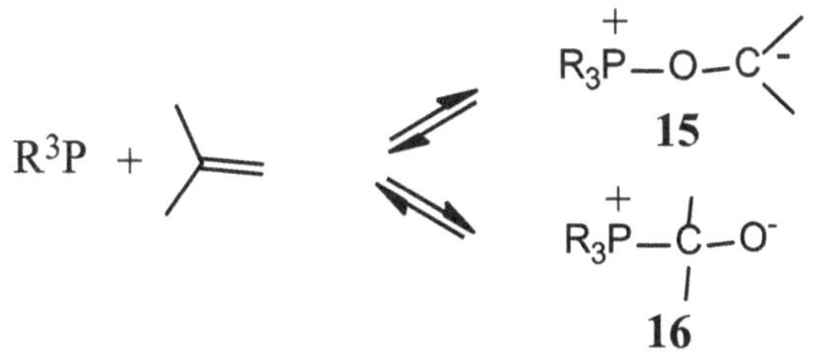

Scheme 74

Although most reactions proceed by nucleophilic attack at carbon, exceptions can be formed where addition at oxygen gives anions which are stabilized by resonance. An excellent example is the formation of betaine (17) from the reaction of triphenyl phosphine with the cyclopentadiene.

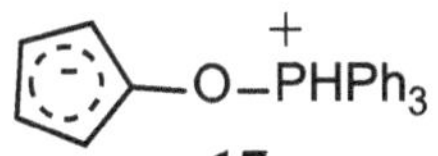

17

This interesting reaction represents a novel approach to the formation of carbon-carbon bonds, and is one of great synthetic potential since the products may undergo further reaction leading to loss of phosphorous; for example the adduct decompose in boiling ethanol to afford stilbine oxide (Scheme 75) [66JACS1830].

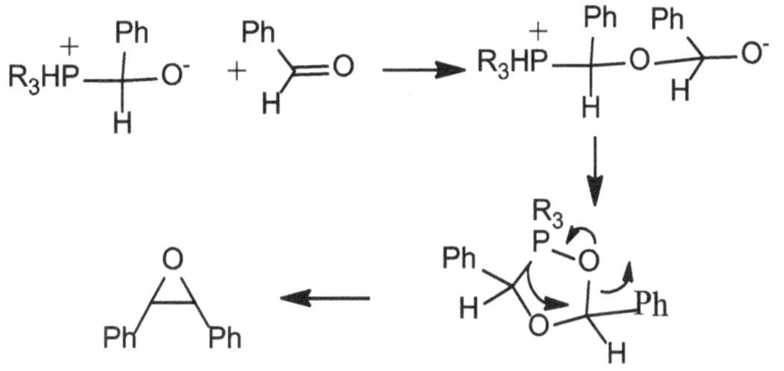

Scheme 75

2.7   ADDITION OF CARBON NUCLEOPHILE

2.7.1   Addition of hydrogen cyanide

Hydrogen cyanide adds to many aldehydes and ketones to give α-cyanoalcohol, usually called cyanohydrins (Scheme 76). Cyanohydrins are important compounds in organic synthesis since they are readily hydrolyzed to α-hydroxy acids.

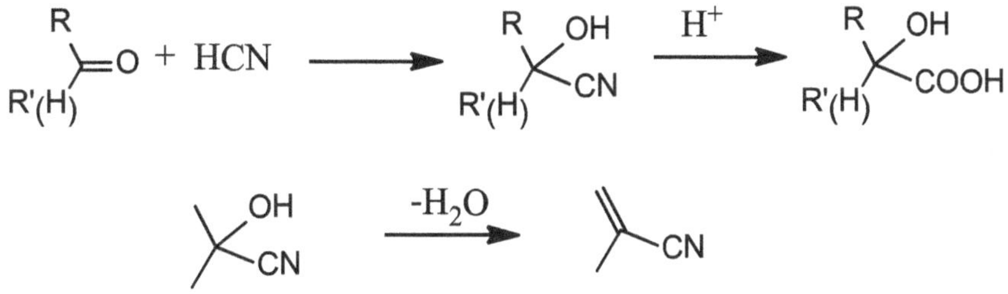

Scheme 76

An important feature of cyanohydrin formation is that it requires a basic catalyst. In the absence of base, the reaction does not proceed or is at best very low. In principle the basic catalyst might activate either the carbonyl group or HCN. With hydroxide ion as the base, one reaction to be expected is a reversible addition of the hydroxide to the carbonyl group. However, such addition is not likely to facilitate formation of cyanohydrin because it represents a competitive saturation of the carbonyl double bond.

Hydrogen cyanide itself has no unshared electron pair on carbon and is unable to form a carbon-carbon bond to a carbonyl carbon. However, an activating function of hydroxide ion is clearly possible through conversion of hydrogen cyanide to cyanide ion which can function as a nucleophile toward carbon. A complete reaction sequence for cyanohydrin formation is thus as follows. The last step regenerates the base catalyst. All steps of the overall reaction are reversible (this is because cyanide is a good leaving group) but with aldehydes and most non-hindered ketones, formation of cyanohydrin is reasonably favorable (Scheme 77).

$$HCN + {}^-OH \longrightarrow {}^-CN + H_2O$$

Scheme 77

This equilibrium is more favourable for aldehyde cyanohydrins than ketone cyanohydrins, and the reason is the size of the groups attached to the carbonyl carbon atom. As the carbonyl carbon atom changes from sp2 to sp3, its bond angles changes the carbonyl carbon atom change from about 120° to about 109° – in other words, the substituents it carries move closer together. This reduction bond angle is not a problem for aldehydes, because one of the substituents is just hydrogen atom, but for ketones, especially ones that carry larger alkyl groups, this effect can disfavor the addition reaction. Effects that result from the size of substituents and the repulsion between them are called steric effects, and we call the repulsive force experienced by large substituents steric hindrance. Steric hindrance is a consequence of repulsion between the elctrons in all the field orbitals of the alkyl substituents.

As usual, the extent of cyanohydrin formation depends on the steric and stereoelectronic factors. The cyanohydrin formation constants can well explain the fact.

Table: cyanohydrin formation constants

Carbonyl compounds	$K_{formation}$
Acetaldehyde	High
p-nitrobenzaldehyde	1430
p-methoxybenzaldehyde	210
p-dimethylaminobenzaldehyde	2.50
Cyclohexanone	10,000
Ethyl methyl ketone	38
Acetophenone	0.77
Benzophenone	Low

-I and effects of nitro group in p-nitrobenzaldehyde increases the electrophilicity of the aldehydic C and thus increases the value of formation constant. +R groups, p-dimethyl amino and p-methoxy, decrease the electrophilicity of the aldehydic C and thereby decrease the value of formation constant. In the cases of ethyl methyl ketone, acetophenone and benzophenone, the steric congestion around the carbonyl; C decreases the reactivity of the carbonyl compounds and will get a small formation constant for the former and for the latter two, $K_{formation}$ is immeasurably small.

The reversibility of cyanohydrin formation is of more than theoretical interest. In parts of Africa, the staple food is cassiva. This food contains substantial quantities of the glycoside of acetone cyanohydrin (Scheme 78). The glucoside is not poisonous in itself but enzymes in the human gut break it down and release HCN. Eventually 50mg HCN per 100g of cassiva can be released and this is enough to kill a human being after a meal of unfermented cassava. If the cassava is crushed with water and allowed to stand (ferment), enzymes in the cassava will do the same job and then the HCN can be washed out before the cassava is cooked and eaten.

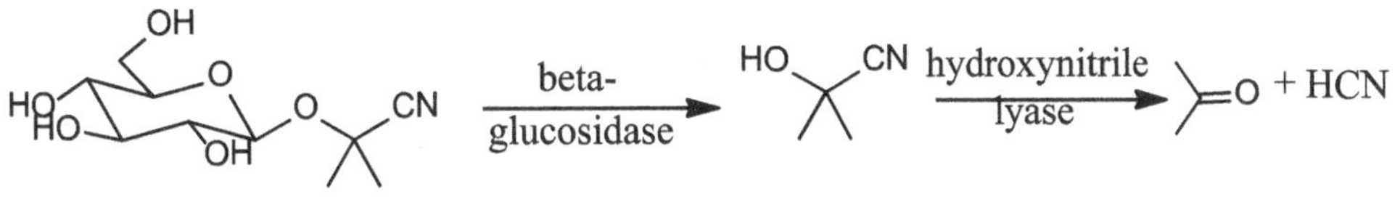

Scheme 78

2.7.1.1   Benzoin condensation:

The self-condensation of an aromatic aldehyde specifically catalyzed by cyanide ion in is usually referred to as benzoin condensation [1991Misc541]. This reaction is usually carried out by refluxing the aldehyde with aqueous ethanolic potassium cyanide. The product formed is α-hydroxy ketone, a dimer known as benzoin. An early version of the reaction was developed in 1832 by Justus von Liebig and Friedrich Wohler during their research on bitter almond oil. The catalytic version of the reaction was developed by Nikolay Zinin in the late 1830s, and the reaction mechanism for this organic reaction was proposed in 1903 by A. J. Lapworth.

Here two molecules of the aldehyde react differently. One of them donates the aldehydic H atom to the other which accepts it and consequently the former is known as a donor and the latter called an acceptor.

Benzoin condensation follows the rate equation: rate = k[(ArCHO)²][⁻CN]

It is supposed that the rate determining step involves three species i.e., the reaction is termolecular. The general path of the accepted mechanism may be represented in four steps. In the first step in this reaction, the cyanide anion reacts with the aldehyde in a nucleophilic addition to the carbonyl carbon atom of the donor molecule. In the second step, rearrangement of the intermediate results in the formation of cyanohydrin carbanion. Next step involves the nucleophilic addition of the resonance stabilized cyanohydrin carbanion to the second carbonyl group of the acceptor molecule. Final step involves the proton transfer and elimination of the cyanide ion affords benzoin as the product (Scheme 79). This is a reversible reaction.

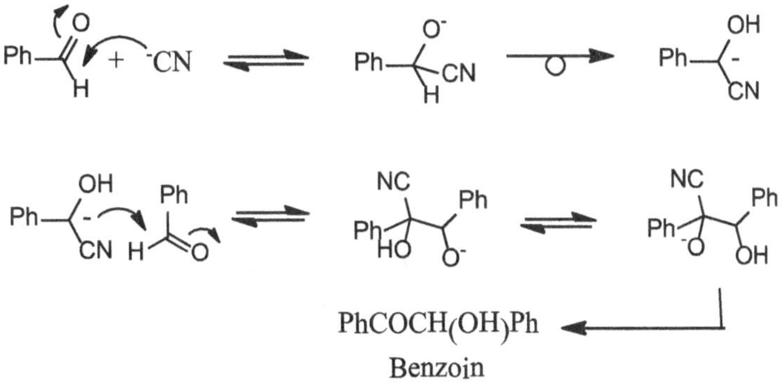

PhCOCH(OH)Ph
Benzoin

Scheme 79

The cyanide ion serves three different purposes in the course of this reaction. It acts as a nucleophile, facilitates proton abstraction, and is also the leaving group in the final step. The benzoin condensation is in effect a dimerization and not a condensation because a small molecule like water is not released in this reaction. For this reason the reaction is also called a benzoin addition. 4-Dimethylaminobenzaldehyde is an efficient proton donor while benzaldehyde is both a proton acceptor and donor. In this way it is possible to synthesize mixed benzoins, i.e. products with different groups on each half of the product.

The reaction is unsuccessful with aliphatic aldehydes having α-H atom. Perhaps the enolization of aldehyde is responsible for this. The presence of strongly electron donating or withdrawing groups at the para positions makes the reactions unsuccessful. For instance, when the strongly electron donating $-NMe_2$ group is substituted para to the $-CHO$ group in benzaldehyde, the reaction fails. Due to conjugation, the carbonyl group acquires additional electron density, that is, it becomes less electrophilic.

The benzoin condensation is also inhibited by electron attracting groups; p-bromobenzaldehyde forms a benzoin slowly and incompletely, and p-nitrobenzaldehyde does not undergo the benzoin condensation. Both of these aldehydes readily form cyanohydrins, but in the anions, electron density has been pulled away from the attacking carbon atom, making it is a far less effective attacking site. The p-nitro group is particularly effective in this respect for here the nitro group and the attacking carbon are in conjugation.

On the other hand, o-nitrobenzaldehyde undergoes benzoin condensation presumably because the conjugation between the $-NO_2$ group and the cyanohydrin anion group, which prohibits the condensation of the p-isomer, is rendered ineffective in the anion of the ortho isomer because coplanarity between the two groups is no longer possible.

Instead of self-condensation, two different aldehydes, in which one can acts as donor only, undergo benzoin condensation and the reaction is called the crossed benzoin condensation (Equation 41).

2.7.2   Addition of organometallic reagents to aldehydes and ketones

Compounds have a carbon-metal bond, lithium and magnesium are very electropositive metals and the Li-C or Mg-C bonds in organolithium or organomagnesium reagents are highly polarized towards carbon. They are therefore very powerful nucleophiles and attack the carbonyl group to give alcohols forming a new C-C bond. For instance, methyl lithium which is commercially available as a solution in Et_2O reacts with aldehydes to form alcohols (Scheme 80). The orbital diagram of the addition step shows how the polarization of the C-Li bond means that it is the carbon atom of the nucleophile that attacks the carbon atom of the electrophile and we get a new C-C bond.

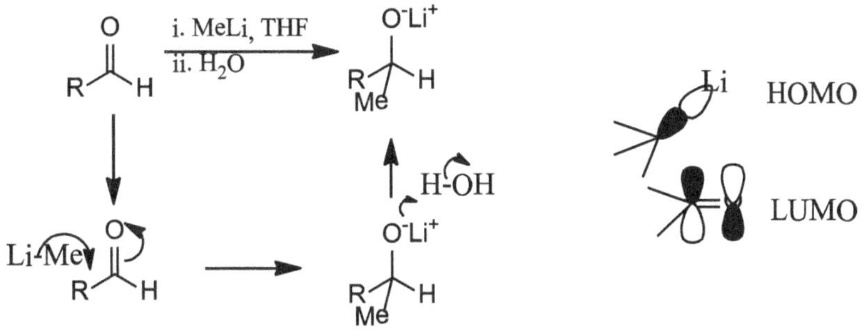

Scheme 80

Because they are so reactive, organolithiums are usually used at low temperature, often -78°C (the sublimation temperature of solid CO_2), in aprotic solvents such as Et_2O, THF. Protic solvents such as water or alcohols have acidic protons but nonprotic solvents such as ether do not. Organolithiums also react with oxygen, so they have to be handled under a dry, inert atmosphere of nitrogen or argon.

The reaction of lithium reagents as well as sodium acetylide reagents with aldehydes and ketones are fundamentally similar to the Grignard reaction.

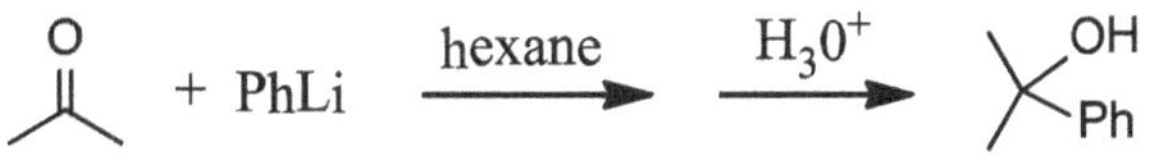

(Equation 42)

2.7.2.1   Addition of acetylides:

Acetylide ions are strong nucleophiles and readily attack the electron-deficient carbonyl carbon atom to give metal ynolate salts which on hydrolysis form ynols.

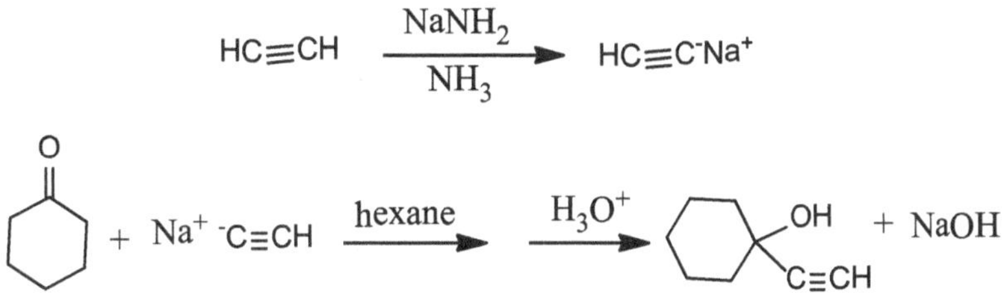

Scheme 81

Ynols may be converted to other compounds of synthetic value. For instance, the addition product of acetone and sodium acetylide may be converted to isoprene (Scheme 82), which polymerizes to a substance resembling natural rubber.

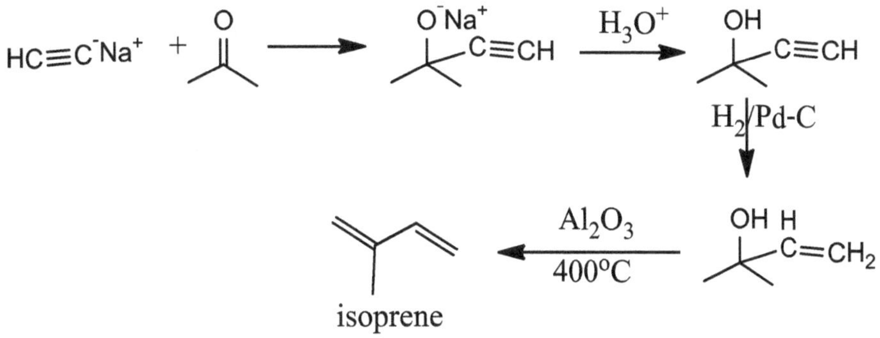

isoprene

Scheme 82

The **Favorskii reaction** is the nucleophilic attack of an acetylide on a carbonyl group. The resulted molecule undergo rearrangement to yield α,β-unsaturated carbonyl compounds (Scheme 82a) [2007Misc1359].

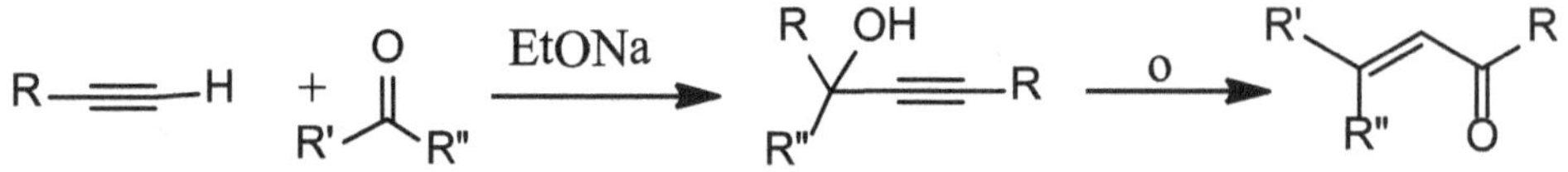

Scheme 82a

2.7.2.2   Reaction with Grignard reagent

The **Grignard reaction** is an organometallic chemical reaction in which alkyl, vinyl, or aryl-magnesium halides (**Grignard reagents**) add to a carbonyl group in an aldehyde or ketone. This reaction is an important tool for the formation of carbon–carbon bonds. The reaction of an organic halide with magnesium is not a Grignard reaction, but provides a Grignard reagent.

A Grignard reagent is prepared by shaking an ethereal solution of halogen derivative of a hydrocarbon with pure and dry pieces of magnesium in ether in the presence of piece of iodine which initiates the reaction. The apparatus, the reactants and the solvent should be free from moisture, alcohol and anything which reacts with the Grignard reagent to be formed. For this reason all these are washed, purified and dried perfectly. However, at the start, the ether becomes cloudy and begins to boil gently and then the ether refluxes. When the reaction is complete, a clear solution of the Grignard reagent in ether is formed and this is then treated with any chosen reagent in situ to carry out the synthesis of a required compound. After the reaction is over, in many a case, the product is treated with acidulated water or aqueous ammonium chloride solution.

Addition of Grignard reagent to a ketones or aldehydes followed by hydrolysis gives tertiary or secondary alcohol respectively (Scheme 83). The reaction with formaldehyde leads to a primary alcohol.

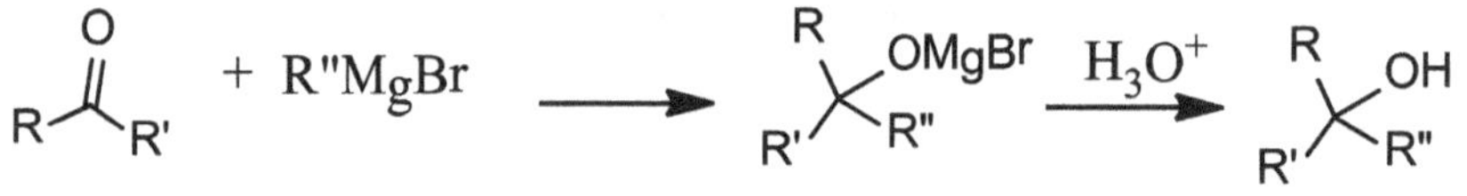

Scheme 83

Reaction mechanism

The reaction of a Grignard reagent with an aldehydes or ketones is another example of carbonyl addition. The magnesium of the Grignard reagent, a Lewis acid, coordinates with the carbonyl oxygen and proceeds through a six-membered ring transition state (Scheme 84). This coordination increases the positive charge on the carbonyl carbon, which is attacked by the nucleophilic organic group of the Grignard reagent.

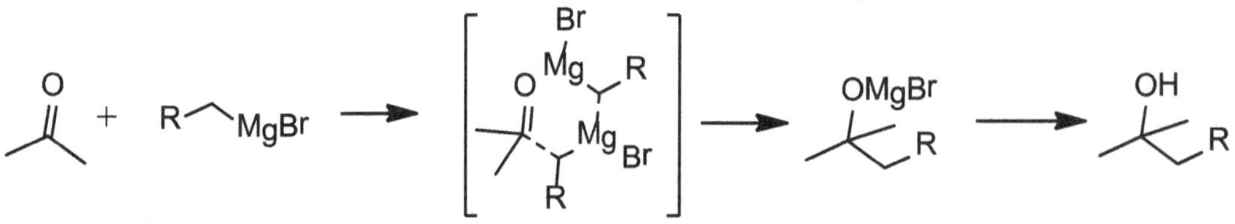

Scheme 84

Addition of the Grignard reagent to the carbonyl group of an aldehyde or ketone gives a bromomagnesium alkoxide. Like other alkoxides, this compound is very basic and protonates on oxygen to yield the alcohol when water is added to the reaction mixture.

It is important to notice the role of the metal ion in lithium aluminium hydride, Grignard and related reactions. The metal ion does not "stand idle by" while its nucleophilic partner attacks the carbonyl group. Rather, the metal plays an active role by coordinating with the carbonyl oxygen, thus giving the carbonyl carbon some carbonium ion character and increasing its reactivity toward nucleophiles. In this respect we can think of the metal ion as a "fast proton". That is, it plays the same role in these reactions as the proton plays in acid-catalyzed carbonyl additions in aqueous solution.

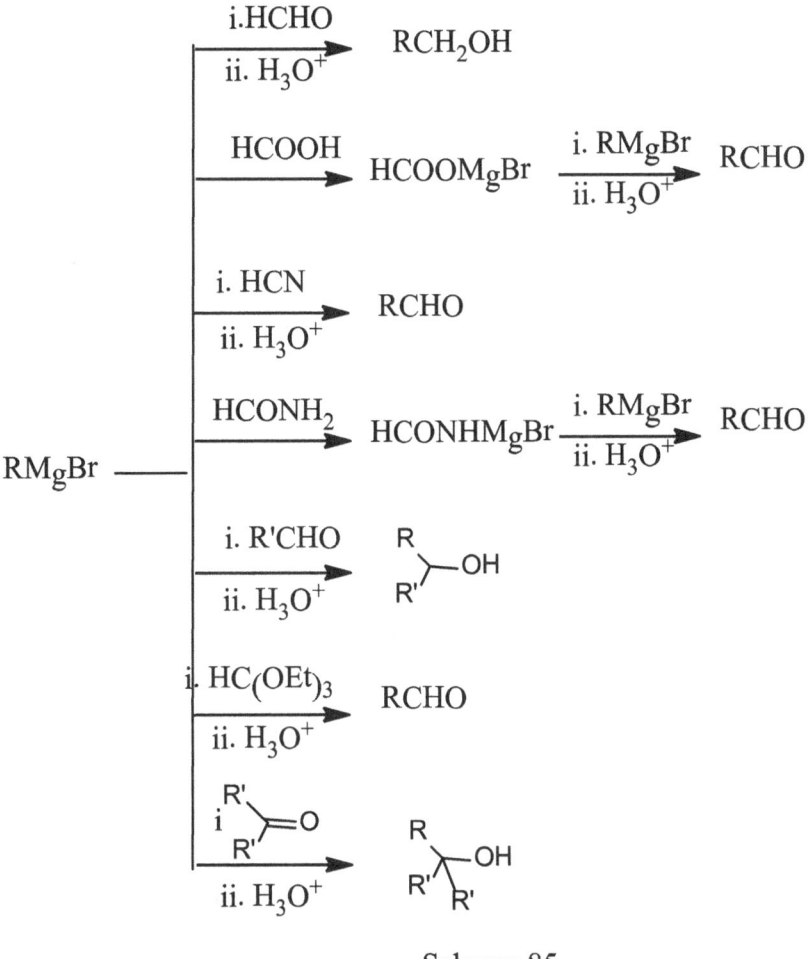

Scheme 85

The **Cram's rule** of asymmetric induction developed by Donald J. Cram in 1952 [1952JACS5828] is an early concept relating to the prediction of stereochemistry in certain acyclic systems. In full the rule is: In certain non-catalytic reactions that diastereomer will predominate, which could be formed by the approach of the entering group from the least hindered side when the rotational conformation of the C-C bond is such that the double bond is flanked by the two least bulky groups attached to the adjacent asymmetric center. The rule indicates that the presence of an asymmetric center in a molecule induces the formation of an asymmetric center adjacent to it based on steric hindrance.

For instance, reaction of 2-phenylpropionaldehyde with phenyl magnesium gave 1,2-diphenyl-1-propanol (**18**) as a mixture of diastereomers, in which threo isomer predominates.

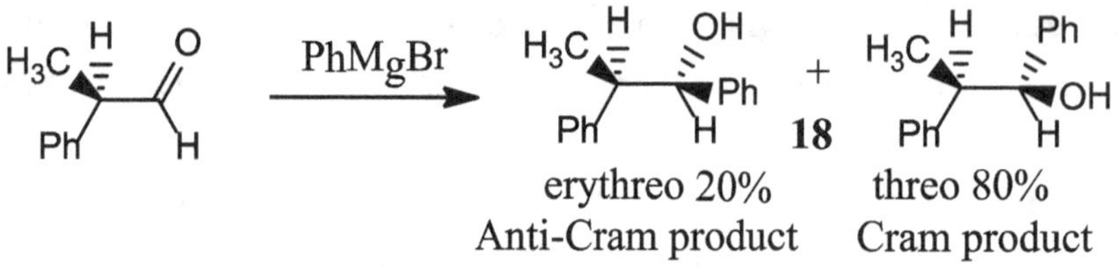

$$\text{(Equation 43)}$$

erythreo 20%   threo 80%
Anti-Cram product   Cram product

The preference for the formation of the threo isomer can be explained by the rule stated above by having the active nucleophile in this reaction attacking the carbonyl group from the least hindered side (see Newman projection) when the carbonyl is positioned in a staggered formation with the methyl group and the hydrogen atom, which are the two smallest substituents creating a minimum of steric hindrance, in a gauche orientation and phenyl as the most bulky group in the anti conformation (Scheme 86).

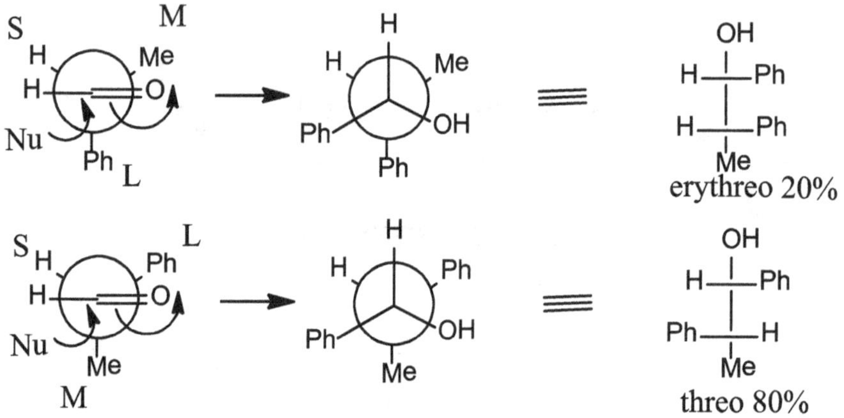

Scheme 86

When an α-substituent to carbonyl compound is capable of coordination with metal ion, the "cyclic chelate model" is invoked. The favoured conformation results from formation of a chelate between the cationic reagent, the carbonyl oxygen and the coordinating substituent. Addition then occurs from the least hindered face [1959JACS2748, 1963JACS1245].

The addition of Grignard reagents to chiral α-alkoxy α-halo ester reacts with acyclic ketones is one of the most thoroughly studied examples of a chelation-controlled. Under these conditions, these reactions proceed with very high stereoselectivities, which can be explained by the cyclic chelate model illustrated in equation 44 [1980TL1031].

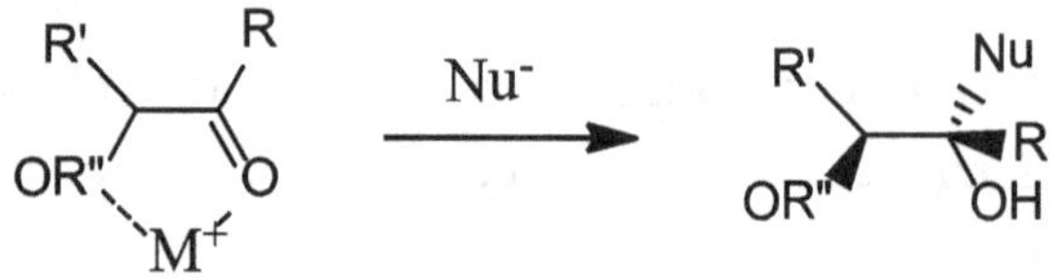

$$\text{(Equation 44)}$$

The nature of the solvent and the organometallic reagent has a profound effect on the degree of selectivity observed (Equation 45).

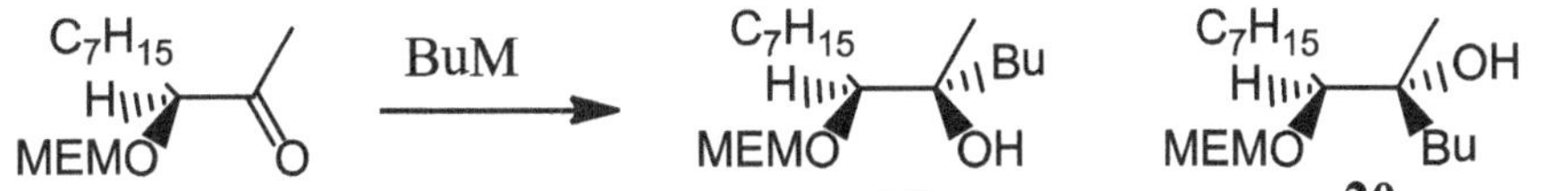

19 **20**

(Equation 45)

	M = Li	M = MgBr
Pentane	63:33	90:10
CH_2Cl_2	75:25	93:7
Et_2O	50:50	90:10
THF	41:59	99:1

2.7.2.3 Reformatsky reaction:

An α-haloester reacts with an aldehyde or ketone in the presence of zinc powder in benzene as solvent followed by hydrolysis yield β-hydroxy ester as the product (Equation 46). This is called the Reformatsky reaction, discovered in 1887 by Sergius Reformatsky.

$$\text{PhCHO} + \text{BrCH}_2\text{COOEt} \xrightarrow[\text{benzene}]{\text{Zn}} \xrightarrow{\text{H}_3\text{O}^+}$$

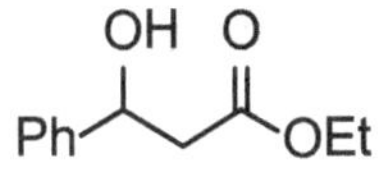

(Equation 46)

The reaction is very similar to the Grignard reaction in that an organometallic compound is formed, in this case, an organozinc compound. This can be regarded as the zinc salt of the ester enolate ion. This enolate ion adds to the carbonyl group of the aldehyde or ketone in much the same as a Grignard reagent to give, after addition of water, the β-hydroxyester (Scheme 87).

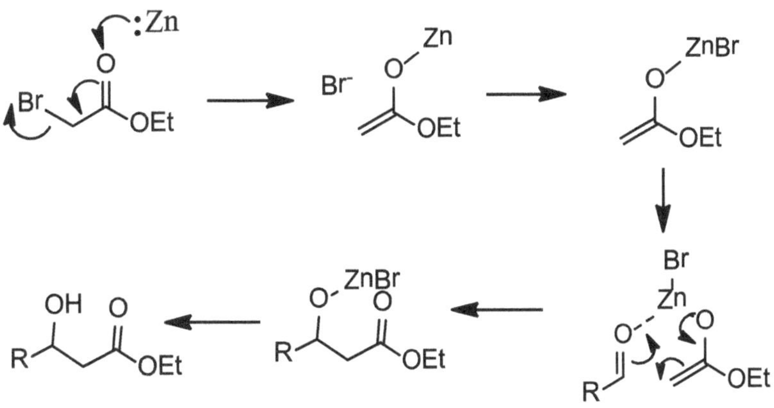

Scheme 87

There is no danger of self condensation with zinc enolates as they do not react with esters, for they do react cleanly with aldehydes and ketones to give aldol on work up.

Parrirsh et al [2003OL3615] promoted the synthesis of β-hydroxyester via Titanocene(III)-Promoted Reformatsky additions reaction of aldehyde with α-halo esters in the presence of Cp_2TiCl_2 and Mn in THF as solvent.

2.7.3 Wittig reaction

The Wittig reaction is a chemical reaction of an aldehyde or ketone with a triphenyl phosphonium ylide (often called a Wittig reagent) to give an alkene and triphenylphosphine oxide (Equation 47). The Wittig reaction was discovered in 1954 by Georg Wittig, for which he was awarded the Nobel Prize in Chemistry in 1979. It is widely used in organic synthesis for the preparation of alkenes.

(Equation 47)

Because phosphorous, like sulphur can accomdate more than eight valence electrons, a phosphorous ylide has a neutral resonance structure.

(ten electrons around phosphorous)

This structure shows that the carbon has a negative charge and a lone electron pair. It should therefore be a base and a nucleophile. The Wittig reaction involves the nucleophilic attack of the anionic ylide carbon on the carbonyl group of the aldehyde or ketone (Scheme 88). Formation of an oxophosphetane completes a carbonyl addition. Under the usual reaction conditions, oxophosphetane spontaneously breaks down to the alkene and the by-product triphenylphosphine oxide.

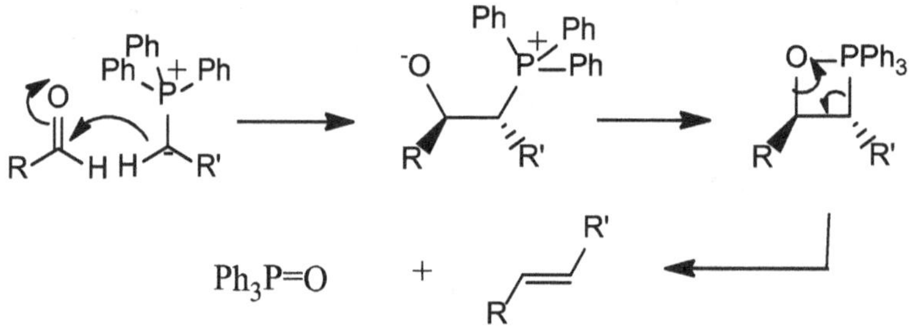

Scheme 88

With simple Wittig reagents, the first step occurs easily with both aldehydes and ketones, and the decomposition of the betaine is the rate-determining step. However, with stabilized ylides (where R_1 stabilizes the negative charge) the first step is the slowest step, so the overall rate of alkene formation decreases and a bigger proportion of the alkene product is the E-isomer. This also explains why stabilized reagents fail to react well with sterically hindered ketones.

The ylide used as starting materials in the Wittig reaction are formed from phosphonium salts, which in turn are prepared by the S_N2 reaction of an alkyl halide with triphenyl phosphine. The phosphonium salt is usually isolated. Reaction of the phosphonium salt with a strong base such as a lithium reagent gives the ylide, which is then immediately in the Wittig reaction (Scheme 89).

$$Ph_3P \ + \ CH_3Br \ \longrightarrow \ \overset{+}{Ph_3P}\text{-}CH_3 \ \overset{Br^-}{} \ \xrightarrow{PhLi} \ \overset{+}{Ph_3P}\text{-}\overset{-}{CH_2} \ + \ PhH$$

Scheme 89

In the Wittig reaction, the α-carbon of the alkyl group used to make the phosphonium salt ends up as one of the alkene carbon atoms. The other alkene carbon atoms are derived from the carbonyl carbon of the aldehyde or ketone. Homologation of aldehydes can be done employing Wittig reaction. The synthetic scheme for this reaction is as follows (Scheme 90).

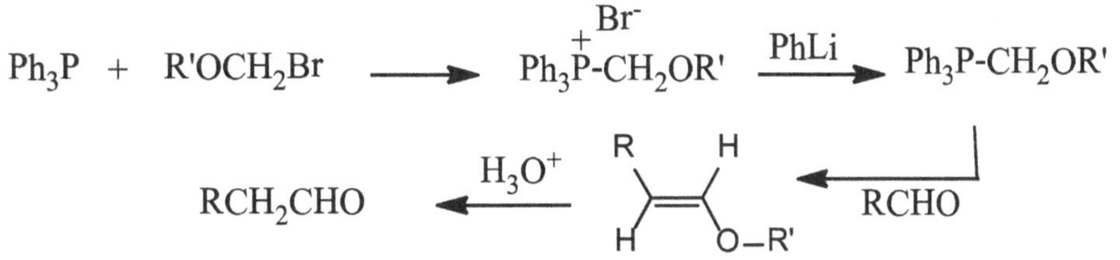

$$Ph_3P \ + \ R'OCH_2Br \ \longrightarrow \ \overset{+}{Ph_3P}\text{-}CH_2OR' \ \overset{Br^-}{} \ \xrightarrow{PhLi} \ Ph_3P\text{-}CH_2OR'$$

$$RCH_2CHO \ \xleftarrow{H_3O^+} \ \underset{H \quad O-R'}{\overset{R \quad H}{\diagup\diagdown}} \ \xleftarrow{RCHO}$$

Scheme 90

The Wittig reaction is important because it gives alkenes in which there is no ambiguity about the position of the alkene double bond; it is a completely regioselective reaction. The Wittig reaction is particularly useful for the synthesis of relatively unstable alkene isomers that are hard to obtain by other means.

2.7.4 Horner–Wadsworth–Emmons reaction

The **Horner–Wadsworth–Emmons reaction** is a chemical reaction used in organic chemistry of stabilized phosphonate carbanions with aldehydes (or ketones) to produce predominantly E-alkenes (Equation 48). In contrast to phosphonium ylides used in the Wittig reaction, phosphonate-stabilized carbanions are more nucleophilic but less basic. Likewise, phosphonate-stabilized carbanions can be alkylated. Unlike phosphonium ylides, the dialkylphosphate salt byproduct is easily removed by aqueous extraction.

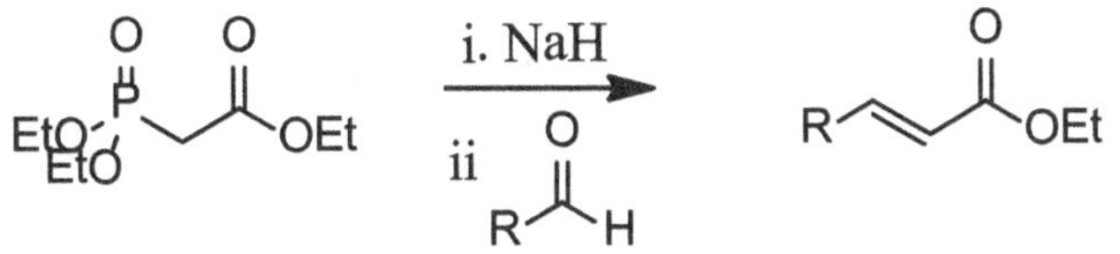

$$\underset{EtO}{\overset{O \quad O}{EtO}}\overset{\|}{\underset{}{P}}\diagup\diagup\diagdown OEt \quad \xrightarrow[\underset{R \diagdown H}{\overset{O}{ii}}]{i. \ NaH} \quad R\diagup\diagdown\diagup\overset{O}{\diagdown}OEt \qquad \text{(Equation 48)}$$

The Horner–Wadsworth–Emmons reaction begins with the deprotonation of the phosphonate to give the phosphonate carbanion **21**. Nucleophilic addition of the carbanion onto the aldehyde **22** (or ketone) producing **23a** or **23b** is the rate-limiting step. If $R^2 = H$, then intermediates **23a** and **24a** and intermediates **23b** and **24b** can interconvert with each other (Scheme 91). The final elimination of **23a** and **23b** yield (E)-alkene **25** and (Z)-alkene **26**. The ratio of alkene isomers **25** and **26** is dependent upon the stereochemical outcome of the initial carbanion addition and upon the ability of the intermediates to equilibrate.

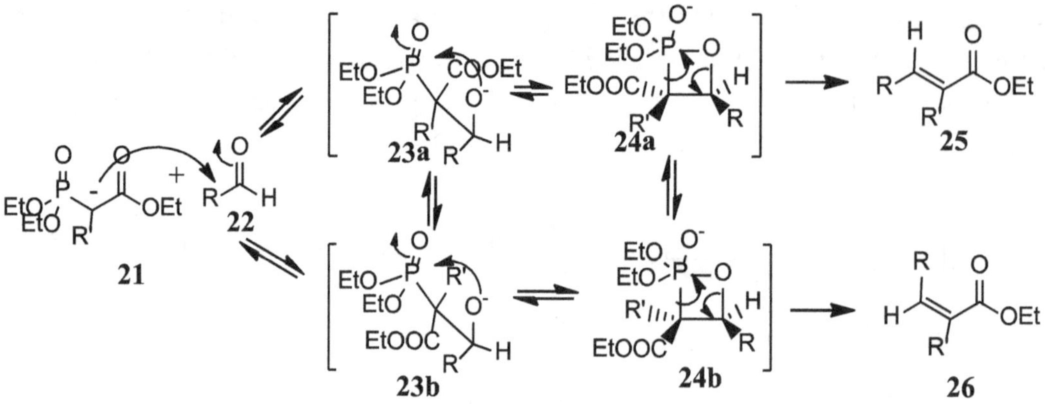

Scheme 91

The electron-withdrawing group (EWG) alpha to the phosphonate is necessary for the final elimination to occur. In the absence of an electron-withdrawing group, the final product is the α-hydroxy-phosphonate **23a** and **23b**. However, these α-hydroxyphosphonates can be transformed to alkenes by reaction with diisopropylcarbodiimide. The Horner–Wadsworth–Emmons reaction favours the formation of (E)-alkenes.

2.7.5　Johnson–Corey–Chaykovsky reaction

The **Johnson–Corey–Chaykovsky reaction** is a chemical reaction used in organic chemistry for the synthesis of epoxides, aziridines, and cyclopropanes [2003CC2644, 2004ACR611]. It was discovered in 1961 by A. William Johnson and developed significantly by E. J. Corey and Michael Chaykovsky. The reaction involves addition of a sulfur ylide to a ketone, aldehyde, imine, or enone to produce the corresponding 3-membered ring (Scheme 92). The reaction is diastereoselective favoring trans substitution in the product regardless of the initial stereochemistry. The synthesis of epoxides via this method serves as an important retrosynthetic alternative to the traditional epoxidation reactions of olefins.

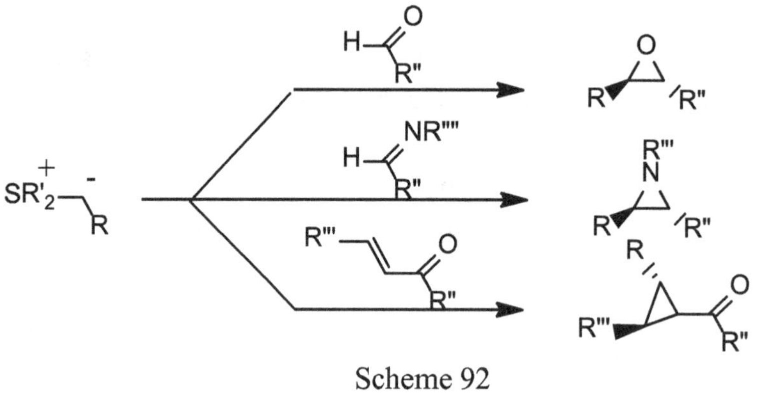

Scheme 92

The reaction is most often employed for epoxidation via methylene transfer, and to this end has been used in several notable total syntheses (See Synthesis of epoxides below). Additionally detailed below are the history, mechanism, scope, and enantioselective variants of the reaction. Several reviews have been published.

The subsequent development of (dimethyloxosulfaniumyl)methanide, $(CH_3)_2SOCH_2$ and (dimethyl-sulfaniumyl)methanide, $(CH_3)_2SCH_2$ (known as Corey–Chaykovsky reagents) by Corey and Chaykovsky as efficient methylene-transfer reagents established the reaction as a part of the organic canon (Equation 49).

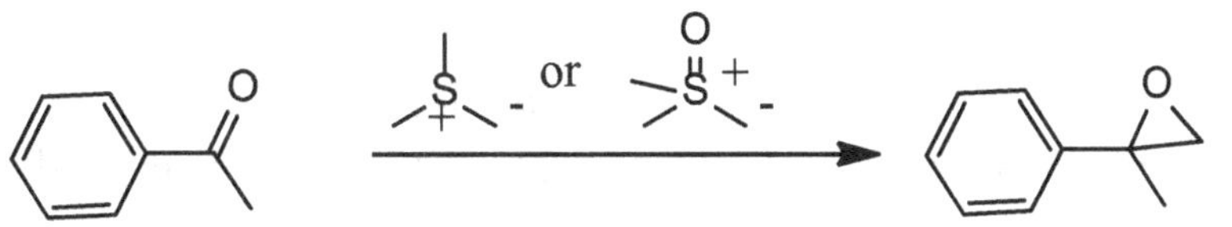

(Equation 49)

The reaction mechanism for the reaction consists of nucleophilic addition of the ylide to the carbonyl or imine group. A negative charge is transferred to the heteroatom and because the sulfonium cation is a good leaving group it gets expelled forming the ring (Scheme 93). In the related Wittig reaction, the formation of the much stronger phosphorus-oxygen double bond prevents oxirane formation and instead, olefination takes place through a 4-membered cyclic intermediate.

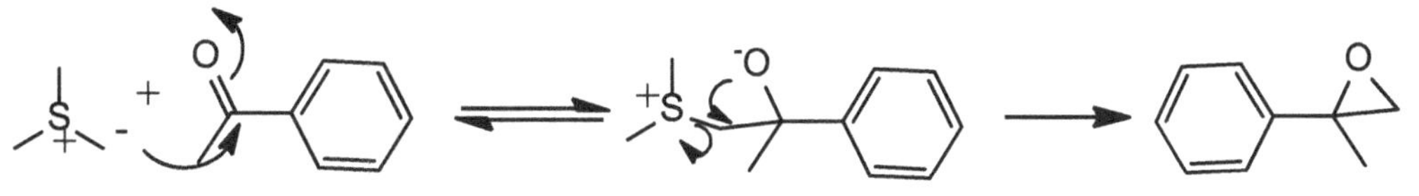

Scheme 93

The trans diastereoselectivity observed results from the reversibility of the initial addition, allowing equilibration to the favored anti betaine over the syn betaine. Initial addition of the ylide results in a betaine with adjacent charges; density functional theory calculations have shown that the rate-limiting step is rotation of the central bond into the conformer necessary for backside attack on the sulfonium (Scheme 94).

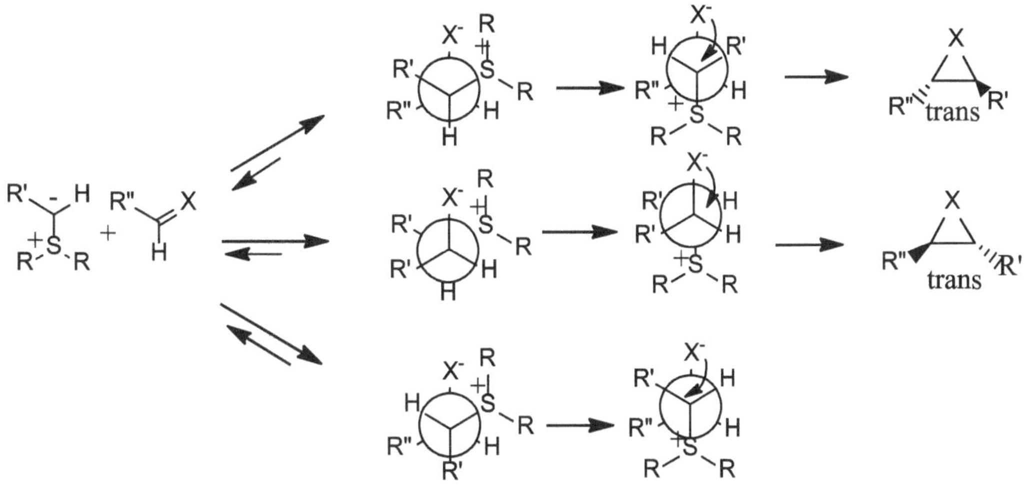

Scheme 94

The degree of reversibility in the initial step (and therefore the diastereoselectivity) depends on four factors, with greater reversibility corresponding to higher selectivity:

i. Stability of the substrate with higher stability leading to greater reversibility by favoring the starting material over the betaine.

ii. Stability of the ylide with higher stability similarly leading to greater reversibility.

iii. Steric hindrance in the betaine with greater hindrance leading to greater reversibility by disfavoring formation of the intermediate and slowing the rate-limiting rotation of the central bond.

iv. Solvation of charges in the betaine by counter ions such as lithium with greater solvation allowing more facile rotation in the betaine intermediate, lowering the amount of reversibility.

Reaction of S,S-dialkyl-N-(p-toluenesulfonyl) sulfoximine with sodium hydride or n-BuLi affords sulfoximidyl stabilized carbanion, which acts as nucleophilic alkylidine transfer reagent [73JACS4287]. The reagent reacted with substrate containing an electrophilic double bond or a ketonic group to yield a cyclopropane or an oxirane respectively. The reaction is best explained by the nulceophilic attack of sulfoximidyl carbanion at the carbonyl group (Scheme 95).

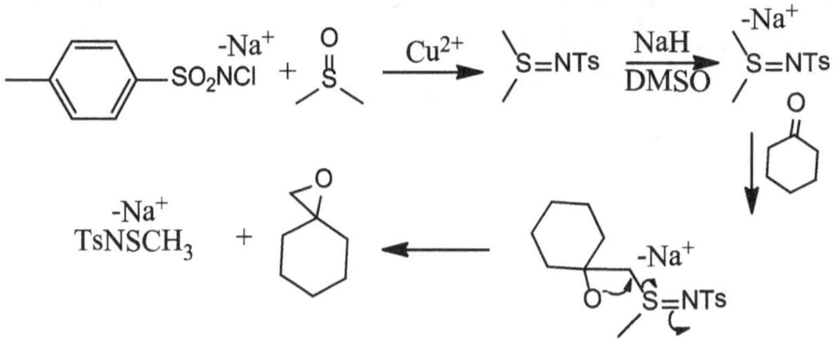

Scheme 95

With the aim to obtain oxirane derivatives from podophyllotoxone, Rai et al [1995IJHC63} treated podophyllotoxone in DMSO with S,S-dialkyl-N-(p-toluenesulfonyl) sulfoximine with sodium hydride at room temperature. Regarding the sterochemical approach of the sulofimidyl carbanion, it would appear, the anion approaches the ketonic carbonyl group of the substrate from the less hindered front side, anti to the axial bulky trimethoxyphenyl ring. Hence in all probability, the methylene group could be situated anti to trimethoxyphenyl ring at the spiro centre C_4, in the oxirane (Equation 50).

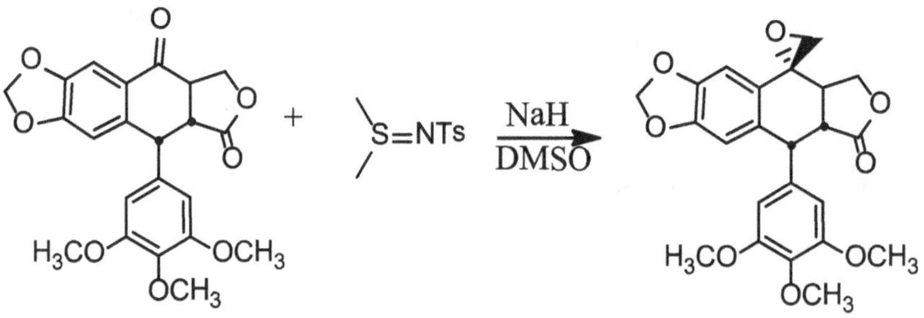

(Equation 50)

2.7.6 Julia olefination

The Julia olefination is the chemical reaction used involves the reaction of phenyl sulphone with aldehyde or ketones in presence of strong base followed by alkylation with alkyl halide. The resulted alkoxy ether on reductive elimination using sodium amalgam gave the required alkenes in almost quantitative yields. The reaction is named after the French chemist Marc Julia (Scheme 96) [1973TL4833, 1978JCSPT829]. This transformation highly favors formation of the E-alkene.

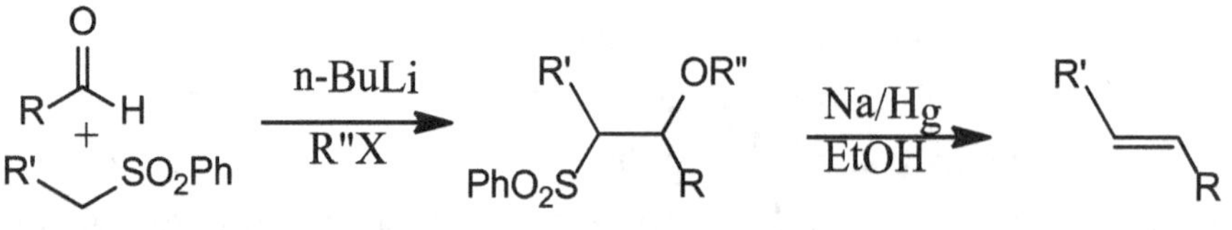

Scheme 96

2.7.7 Peterson olefination

The Peterson olefination concerns the construction of double bonds from trialkyl silyl-substituted organometallics and carbonyls [1967JOC780]. The reaction involves the formation of a hydroxysilane, which then undergoes elimination to give the alkene. Elimination can take place under either acidic or basic conditions (Scheme 97).

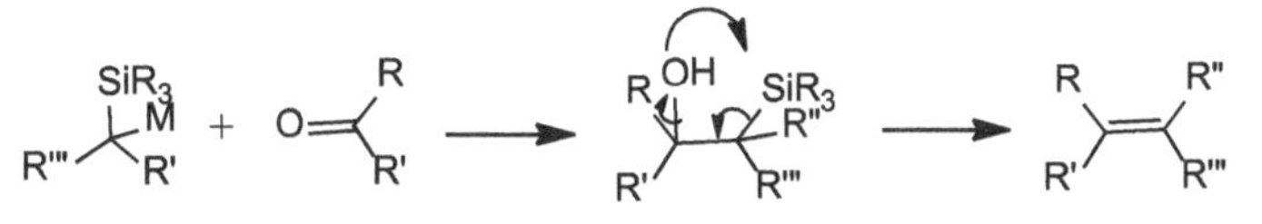

Scheme 97

Due to the mechanism of action, the Peterson olefination is a very versatile method of forming carbon-carbon double bonds. In particular, the reaction can be carried out under basic or acidic conditions giving rise to either the trans or cis isomer respectively. The mechanism begins with the addition of a silyl-substituted carbanion to a carbonyl compound leads to a diastereomeric mixture of α-hydroxyalkylsilanes, often isolable and sometimes separable. The stereo-selectivity of the reaction can be controlled by the steric demands of the silyl group; the use of more sterically demanding silyl groups results in the erythro isomer as the major product. A subsequent basic elimination can proceed via deprotonation of the hydroxyl followed by a silyl 1,3-shift which then collapses to give the alkene product.

2.7.8 Tebbe's reagent

The Tebbe's reagent is the organometallic compound with the formula $(C_5H_5)_2TiCH_2ClAl(CH_3)_2$. It is used in the methylenation of carbonyl compounds, that is, it converts organic compounds containing the $R_2C=O$ group into the related $R_2C=CH_2$ derivative [1978JACS3611]. It is a red solid that is pyrophoric in the air, and thus is typically handled with air-free techniques. Tebbe's reagent contains two tetrahedral metal centers linked by a pair of bridging ligands. The titanium has two cyclopentadienyl ([C5H5]–, or Cp) rings and aluminium has two methyl groups. Both titanium and aluminium atoms are linked together by a methylene bridge (-CH$_2$-) and a chloride atom in a nearly square-planar (Ti–CH$_2$–Al–Cl) geometry [2014OrgMet429]. Tebbe's reagent itself does not react with carbonyl compounds, but must first be treated with a mild Lewis base, such as pyridine, which generates the active Schrock carbene.

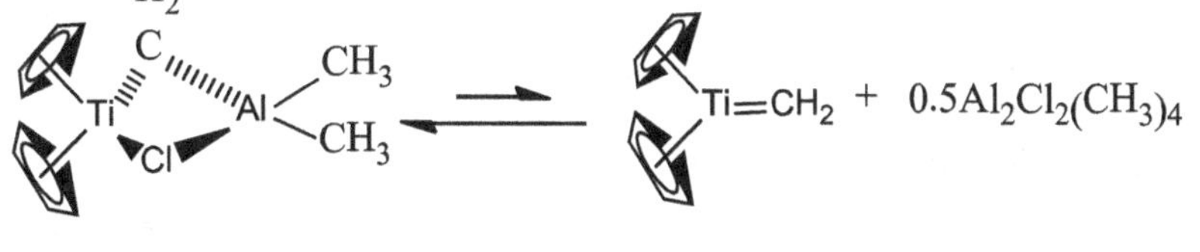

Also analogous to the Wittig reagent, the reactivity appears to be driven by the high oxophilicity of Ti(IV). The Schrock carbene reacts with carbonyl compounds to give a postulated oxatitanacyclobutane intermediate (Scheme 98). This cyclic intermediate breaks down immediately to the produce the desired alkene.

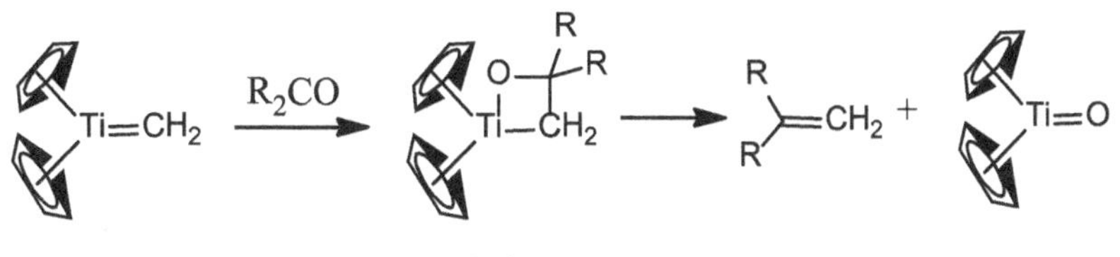

Scheme 98

The Tebbe's reagent converts esters and lactones to enol ethers and amides to enamines. If the compounds containing both keto and ester groups, the ketone selectively reacts in the presence of one equivalent of the Tebbe's reagent.

2.7.9 Sakurai allylation reaction

The Sakurai reaction can be defined as the allylation of a carbonyl compound, or equivalent thereof, performed with an allylsilane and promoted by a Lewis acid (Equation 51). The Sakurai reaction is most commonly performed in dichloromethane at -78°C with one equivalent of Lewis acid. Diethyl ether and tetrahydrofuran are rarely used as solvents, because they sequester the Lewis acid as ethereal complexes.

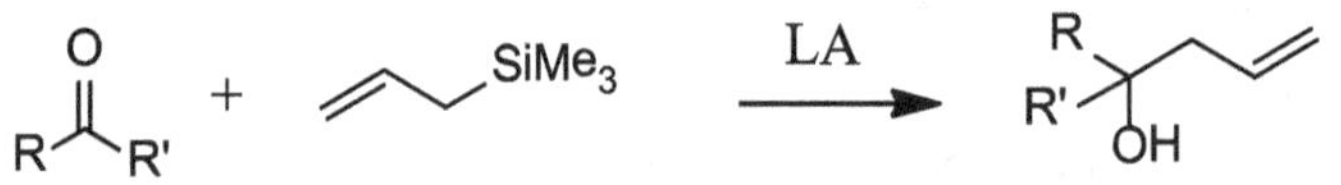

$$R\overset{O}{\underset{R'}{\big\|}} \quad + \quad \diagdown\!\!\diagup\!\!\diagdown SiMe_3 \quad \xrightarrow{\text{LA}} \quad R'\overset{R}{\underset{OH}{\diagdown}}\diagup\diagdown \qquad \text{(Equation 51)}$$

The recurrent theme in the above reactions is the activation of a carbonyl compound 4 (or equivalent) by a Lewis acid, followed by the addition of the allylsilane. The reaction proceeds through a β-silyl cation 27 that is stabilized by hyperconjugation, as represented by the resonance form 28. The hyperconjugation can also be depicted by structure 29, which is another way to indicate that the electrons forming the C-Si bond are partially delocalized into the vacant p orbital of the carbocation (Scheme 94). This effect accelerates the reaction by a factor of 30,000-200,000 (AAG* =4.2 kcal/mol).15·31 the subsequent elimination of the silyl group generates a new double bond.

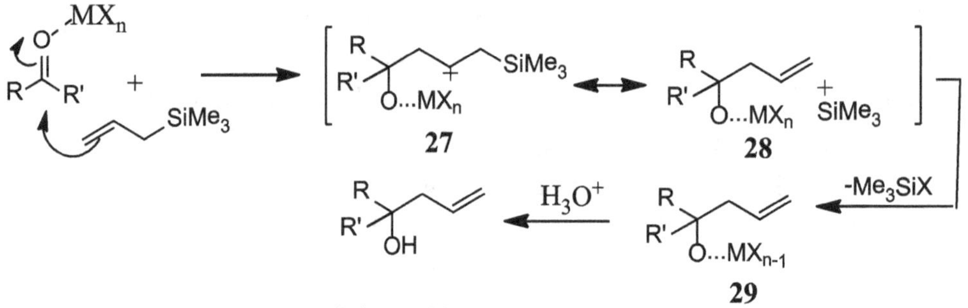

Scheme 99

2.7.10 The Darzen's glysidic ester condensation

The Darzen's glysidic ester condensation is an aldol type reaction between carbonyl compound and an α-halo ester containing α-hydrogen atoms in the presence of a strong base (NaOEt, NaNH$_2$ etc.,) followed by an internal nucleophilic substitution which leads to the formation of a glysidic ester (an α,β-epoxy ester). The first report of this transformation was published by Erlenmeyer and he described the condensation of benzaldehyde with ethyl chloroacetate in the presence of sodium metal. For this reaction one can propose two different mechanism namely ionic or carbene mechanism.

Ionic mechanism involves the following steps: first step involves the formation of a resonance stabilized α-carbanion from the α-halo ester by the abstraction of α-H as H$^+$ by a strong base (Scheme 100). In the second step, the carbanion attack nucleophilically to the carbonyl carbon possessing no α-hydrogen followed by internal S$_N$2 reactions with the loss of halide ion gave the expected α,β-epoxy ester.

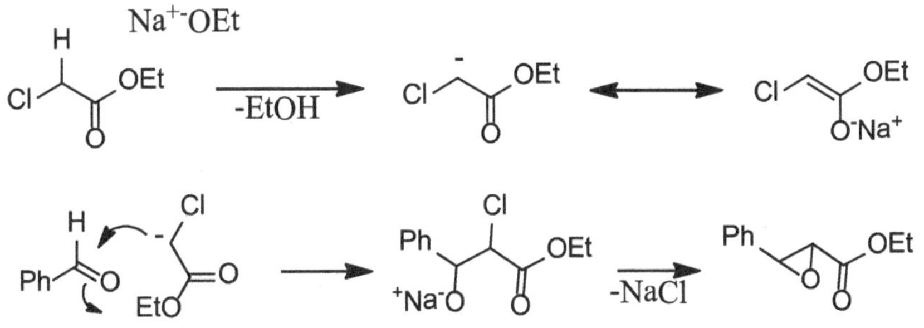

Scheme 100

Carbene mechanism involves the following steps: first step involves the formation of a resonance stabilized α-carbanion from the α-halo ester by the abstraction of a α-H as H[+] by a strong base (Scheme 101). Next involves the elimination of halide ion with the formation of carbene derivative which then undergo 2+1 cycloaddition with the carbonyl group yield the expected α,β-epoxy ester.

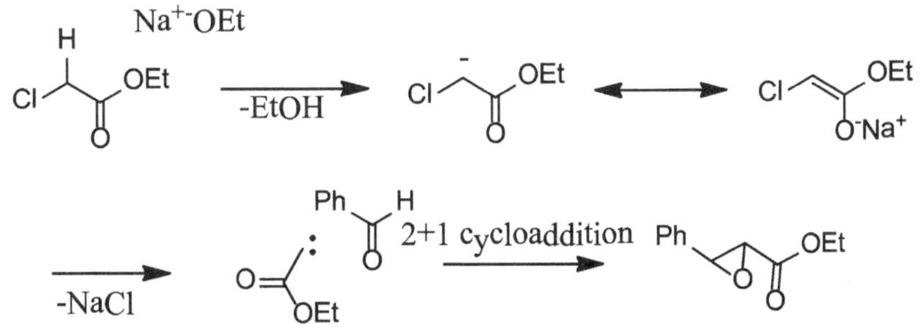

Scheme 101

Zimmerman and Ahramjian rejected the carbene mechanism for the Darzen's glysidic ester condensation based on the following experiment. These workers treated one molecule of ethyl-α-chlorophenyl acetate with one molecule of p-nitrobenaldehyde and one molecule of anisaldehyde in the presence of potassium tertiary butoxide. The electrophilic nature of carbene is well known and if phenyl carbethoxy carbene were actually the species attacking the aldehyde carbonyl carbon, it would prefer the relatively electron-rich carbonyl carbon of anisaldehyde. On the other hand, if the attacking species were the nucelophilic enolate ion, then the more electron deficient p-nitrobenzaldehyde carbonyl carbon would be attacked instead. A careful analysis of the reaction product showed the presence of only the ethyl-2-phenyl-p-nitrophenyl-2,3-epoxypropionate and none (**30**) in the reaction mixture (Scheme 102). On this basis a carbene mechanism is rejected and is in favour of enolate ion intermediate.

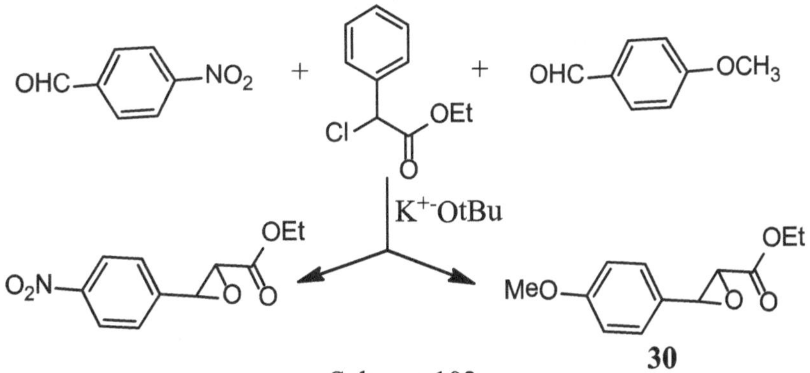

30

Scheme 102

2.7.11 Perkin reaction

The condensation of aromatic aldehydes with the anhydrides of carboxylic acids in the presence of a weak base to afford α,β-unsaturated carboxylic acids is known as the Perkin reaction. The general features of the transformation are the aldehyde component is most often aromatic but aliphatic aldehydes with no α-hydrogens as well as certain α,β-unsaturated aldehydes can also be used. Thus benzaldehyde condenses with acetic anhydride containing two α-hydrogen atoms in the presence of sodium acetate to give cinnamic acid.

The reaction is carried out by heating the aldehyde with an excess acid anhydride containing corresponding carboxylate salt at 180°C. Usually dehydration takes place under the conditions of the reaction and a mixed anhydride results. The unreacted aldehyde is removed by distillation in steam and the β-aryl acrylic acid is obtained by hydrolysis of the mixed anhydrides with dil. HCl acid.

Generally accepted mechanism of the Perkin reaction involves the following steps: first step involves the formation of a resonance stabilized α-carbanion from the anhydride by the abstraction of a α-H as H⁺ by a strong base (Scheme 103). In the second step, the carbanion attack nucleophilically to the carbonyl carbon possessing no α-hydrogen followed by internal transfer of the acetyl group from the carboxyl "O" atom to the alkoxy O atom via cyclic intermediate.

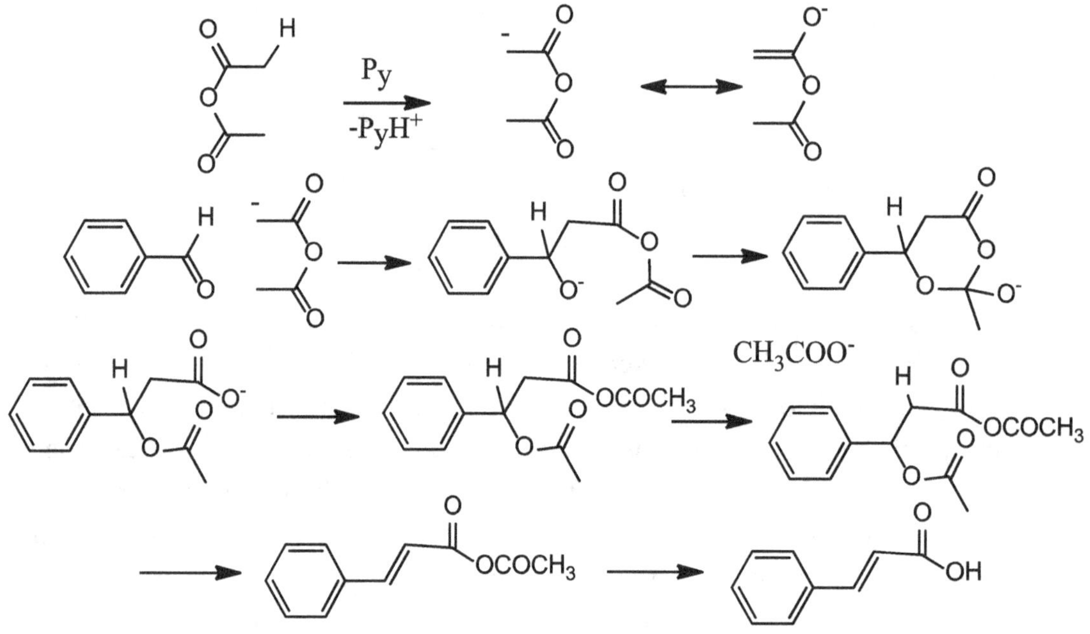

Scheme 103

Acetic anhydride condenses with benzaldehyde in the presence of pyridine as base to yield cinnamic acid after hydrolysis; whereas sodium acetate does not condense with benzaldehyde in the presence of pyridine. This supports the formation of carbanion from the anhydride and rules out the possibility of the formation of the same from sodium acetate.

2.7.12 Aldol condensation

Acetaldehyde in the presence of dilute NaOH or K_2CO_3 or HCl undergoes condensation to form a syrupy liquid known as aldol. On heating, aldol eliminates water to form unsaturated compound called crotanaldehyde (2-butenal) (Scheme 104).

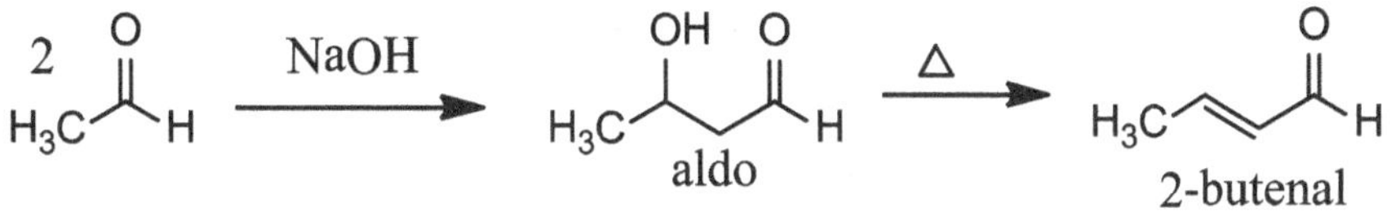

Scheme 104

In many cases it is the unsaturated compound is isolated and not the aldol. This is generally the case with acid catalysts, since these can readily being about dehydration of alcohols.

The aldol condensation can occur between two aldehydes (identical or different) having α-hydrogen, or between two ketones (identical or different) having α-hydrogen or between an aldehyde and a ketone. Whatever the nature of the carbonyl compound, it is only the α-hydrogen atoms which are involved in the aldol condensation.

Generally with different aldehydes all four possible condensation products are obtained, but by using different catalysts one product may be made predominant in the mixture (Scheme 105).

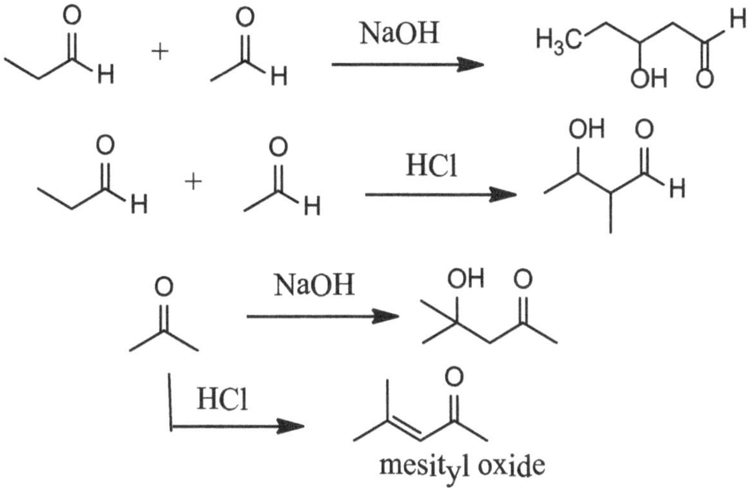

Scheme 105

Formation of enolate anion by removal of α-hydrogen by base is the first step in the reaction (Scheme 106). This anion then adds to the carbonyl bond in a manner analogous to addition of cyanide ion in cyanohydrin formation. It will be expected from consideration of the two resonance form of the enolate anion, that addition might takes place in either of two ways. The anion may add attack to form a carbon-carbon or carbon-oxygen bond, leading ultimately to the aldol or α-hydroxy ethyl vinyl ether.

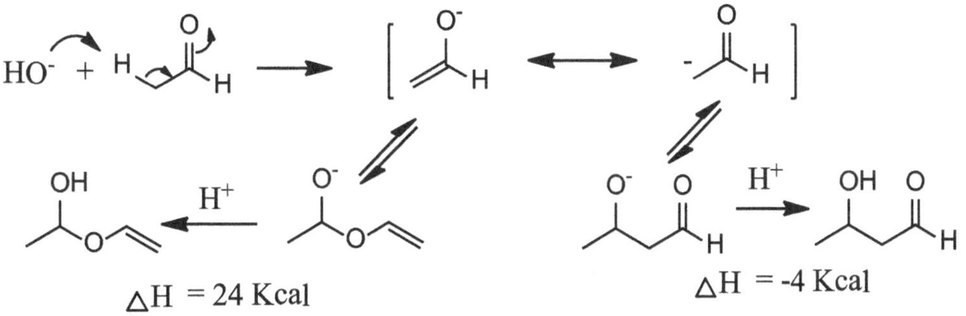

Scheme 106

Although the formation of vinyl ether is mechanistically reasonable, it is much less so on thermodynamic grounds.

The equilibrium constant is favourable for the aldol addition of acetaldehyde, as in fact it is for most aldehydes. For ketones however, the reaction is less favourable. With acetone, only a few % of addition product 'diacetone alcohol' is present at equilibrium. This is understandable on the basis of steric hindrance and the fact that the ketone- carbonyl bond is about 3 Kcals stronger than the aldehyde carbon.

The ingredients in the key step in aldol addition are fundamentally an electron pair donor and an electron-pair acceptor. In the formation of acetaldol and diacetone alcohol, both roles are played by one kind of molecule, but there is no reason why there should be a necessary condition for reaction. Many kinds of mixed conditions are possible. Consider the combination of formaldehyde and acetone: formaldehyde cannot form an enolate anion because it has no α-hydrogen but it is expected to be a particularly good electron pair acceptor because of freedom from steric hindrance and the fact that it has an usually week carbonyl bond (166 Kcal vs 179 Kcals for acetone). Acetone form an enolate anion easily but is relatively poor an acceptor. Consequently the addition of acetone to formaldehyde should and does occur readily.

The problem is not to get addition, but rather to keep in from going too far. Indeed, all six α-hydrogen can be easily replaced –CH_2OH group (Scheme 107).

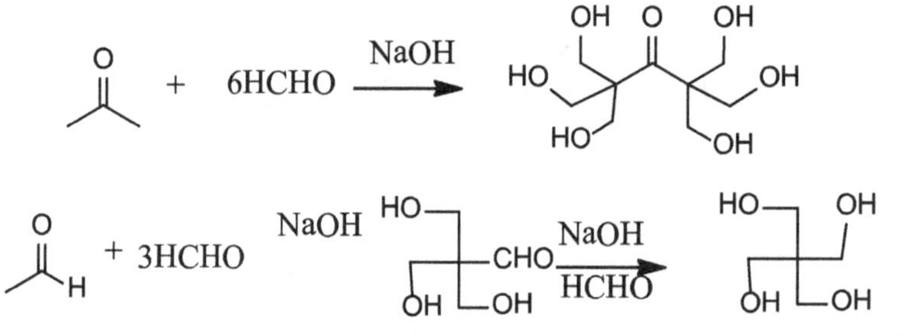

Scheme 107

Pentaerythritol is widely used in the preparation of surface coating and in the formation of tetra nitrate ester, pentaerythritol tetra nitrate ester PETN, which is an important explosive.

There appear to be no detailed kinetic investigation on acid catalyzed aldol condensation, but it is generally assumed that condensation proceeded by reaction between conjugate acid and the enol form of carbonyl compound, example for the formation of mesityl oxide (Scheme 108).

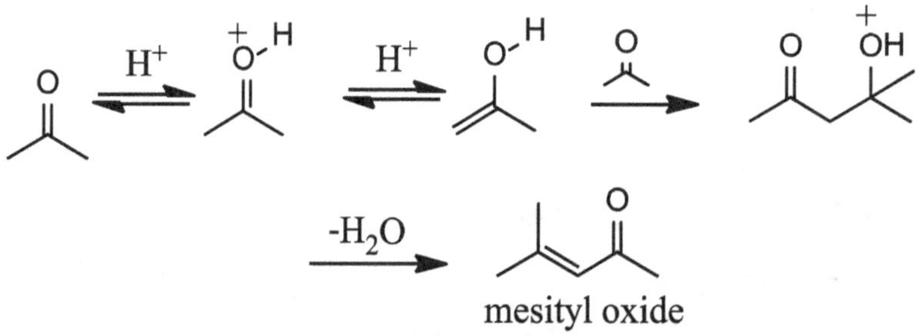

mesityl oxide

Scheme 108

Lithium enolates are usually made at low temperature in THF with a hindered lithium amide base (LDA) and are stable under these conditions because of the strong O-Li bond. The formation of the enolate begins with Li-O bond formation before the removal of the proton from the position by the basic nitrogen atom. This reaction happens very quickly that the partly formed enolate does not have a chance to react with unenolized carbonyl compound before proton removal is complete. Now if a second carbonyl compound is added, it too complexes with the same lithium atom. This allows the aldol reaction to take place by a cyclic mechanism in the coordination sphere of the lithium atom. The aldol step itself is now a very favourable intramolecular reaction with a six membered cyclic transtiton state. The product is initially the lithium alkoxide of the aldol, which gives the aldol on work up (Scheme 109).

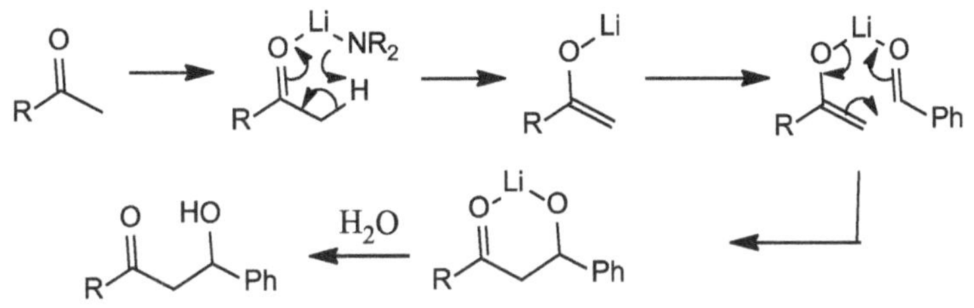

Scheme 109

2.7.13 Claisen–Schmidt condensation

When an enolate forms from an enolate will normally react with unreacted aldehyde to undergo the "aldol addition" or aldol condensation reaction. Since ketones are less reactive toward nucleophilic addition, the enolate formed from a ketone can be used to react with an aldehyde, a modification called Claisen-Schmidt reaction. The condensation of an aromatic aldehyde with an aliphatic aldehyde or ketone in the presence of a base or an acid to form an α,β-unsaturated aldehyde or ketone with high chemoselectivity is generally known as Claisen–Schmidt condensation. This reaction has been applied to the preparation of chalcone, flavanone, 1,3-diarylpropane derivatives, and a new family of macrocycles in a single step. This condensation is found to be accelerated by microwave activation either in a polar solvent (e.g., H_2O) or without solvent, starting from aldehydes or by means of acetals. The magnesium oxide crystal has been applied as catalyst. It has also been observed that the reactants with substituent groups in either of the two aromatic rings undergo the Claisen–Schmidt condensation at a slower rate than the unsubstituted reactants. This reaction is similar to the aldol condensation and is illustrated by the formation of cinnamaldehyde from benzaldehyde and acetaldehyde (Equation 52).

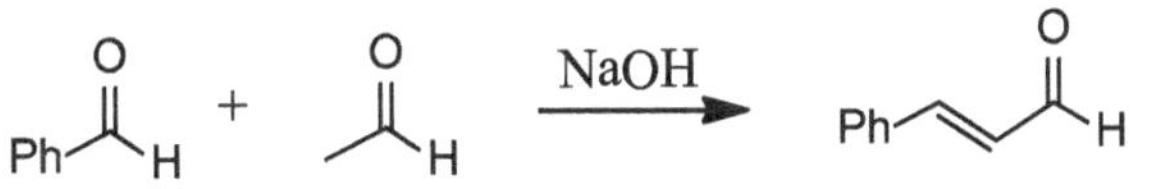

(Equation 52)

Because of the symmetry of acetone the reaction can now be repeated on the other side of the carbonyl leading to the final product, dibenzalacetone, a useful molecule which has been employed as an ultraviolet blocker in sunscreen preparations (Scheme 110).

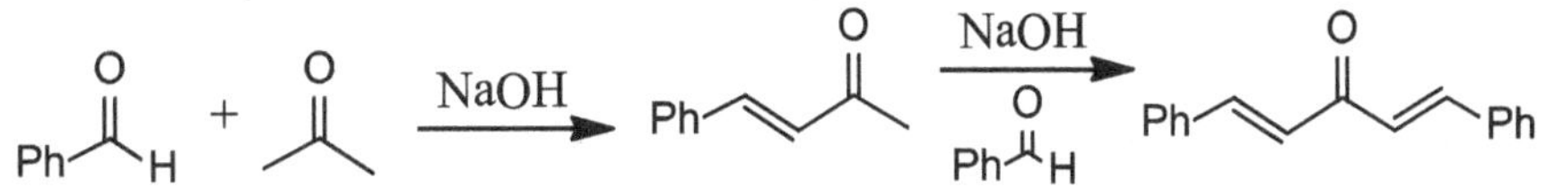

Scheme 110

2.7.14 Stobbe condensation

In 1893, H. Stobbe reported an unexpected reaction between acetone and diethyl succinate in the presence of a full equivalent of sodium ethoxide. Upon acidification of the reaction mixture, the major isolated product was found to be etaconic acid, an, β-unsaturated carboxylic acid, and its monoethyl ester. This result was surprising since the authors expected the formation of a 1, 3-diketone via a Claisen reaction. A subsequent extensive study by Stobbe and co-workers revealed that the transformation was general for esters of succinic acid with aldehydes and ketones. The formation of alkylidene succinic acids or their monoesters by the base-mediated condensation of ketones and aldehydes with dialkyl succinates is known as the Stobbe condensation (Equation 53).

$$\text{(Equation 53)}$$

The first step of the Stobbe condensation is the deprotonation of the succinate at the α-carbon to afford an ester enolate that in situ undergoes an aldol reaction with the carbonyl compound to form a β-alkoxy ester intermediate (Scheme 111). The following intramolecular acyl substitution gives rise to a γ-lactone intermediate which undergoes ring-opening and concomitant double bond formation upon deprotonation by the alkoxide ion. Isolation of the intermediate lactone-ester and the formation of an olefinic compound instead of an alcohol support the mechanism of the Stobbe condensation as written above.

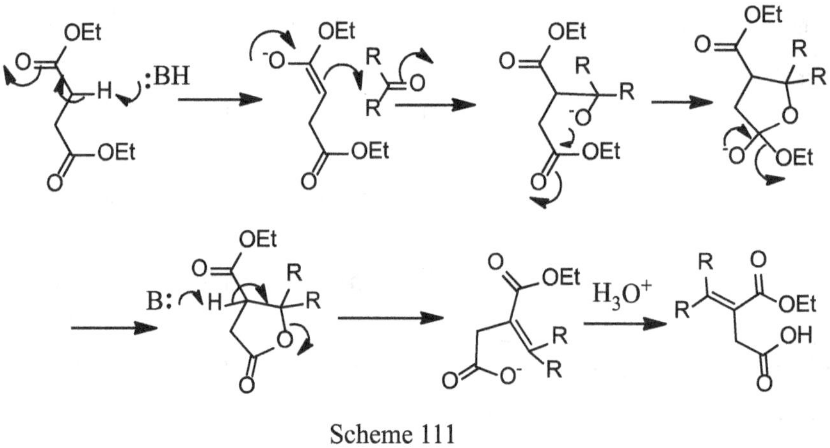

Scheme 111

Synthetic Applications: The asymmetric total synthesis of (+)-codeine, the unnatural enantiomer, was accomplished by J.D. White and coworkers using an intramolecular carbenoid insertion as the key step. The first stereogenic center that directed all subsequent stereochemical events was installed by the asymmetric hydrogenation of an alkylidene succinate that was obtained using the Stobbe condensation. Dimethyl succinate and isovanillin were reacted in the presence of excess sodium methoxide at reflux and the resulting reaction mixture was acidified to obtain the monomethyl ester.

The Stobbe condensation was applied to prepare the tetralin moiety of the target by reacting diethyl succinate in tert-butyl alcohol and using KOᵗBu as the base. The initially formed alkylidene compound was not purified but immediately subjected to in situ catalytic hydrogenation, and the resulting diacid was cyclized to afford a substituted tetralone, which was subsequently converted to the target. For instance, key step for the synthesis an anticancer agent, podophyllotoxin involves the Stobbe condensation of benzophenone with diethyl succinate in presence of potassium t-butoxide (Scheme 112).

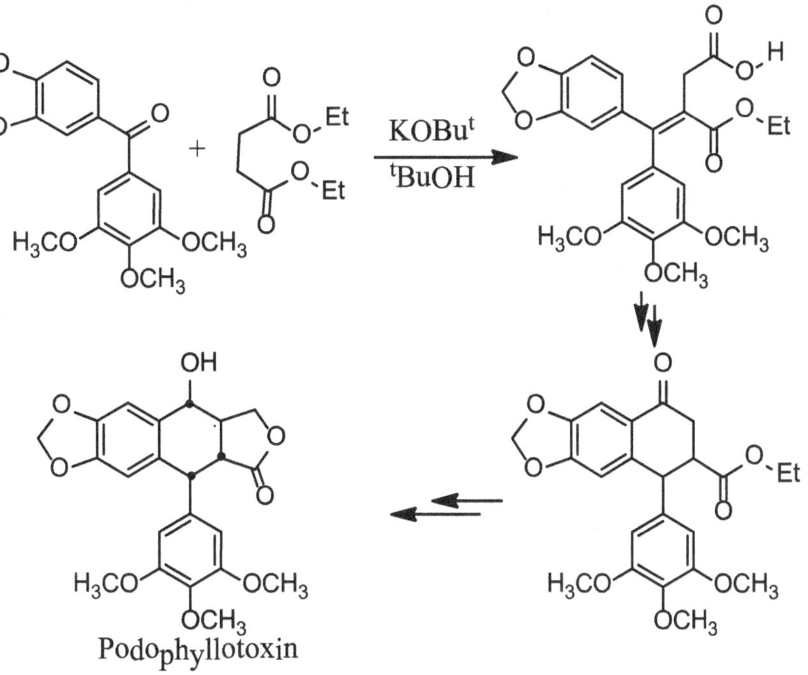

Scheme 112

2.7.15 Mukaiyama aldol reaction

The Mukaiyama aldol reaction is the nucleophilic addition of trimethylsilyl enol ether **31** to either an aldehyde **32** or a ketone in the presence of a Lewis acid ($TiCl_4$) to form a β-hydroxyketone [1913CL1011, 1974CL32] (Scheme 113). The silyl enol ether can be prepared from its parent compound by forming a small equilibrium concentration of enolate ion with weak base such as tertiary amine and trapping the enolate with the very efficient oxygen electrophile Me_3SiCl. The silyl enol ether is stable enough to be isolated but is usually used without storing.

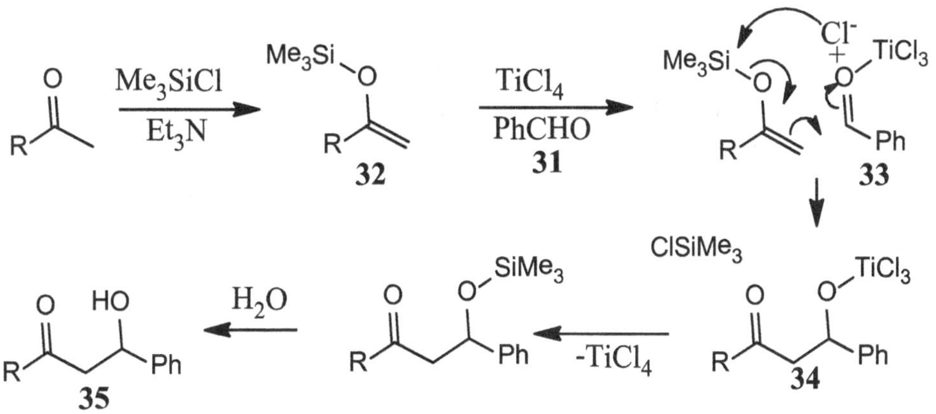

Scheme 113

Investigations concluded that when silyl enol ethers were reacted with ketones or aldehydes, the cross-aldol adducts was obtained in good yield with no self-addition or condensation products. A wide range of metal halide Lewis acids were utilized as mediators in the reaction of benzaldehyde with isopropenyl acetate demonstrating the reaction's tolerance to a wide range of reaction conditions. It was found that $TiCl_4$ was the most effective Lewis acid due to its ability to activate the carbonyl carbon thus making it susceptible to nucleophilic reactions.

The reaction begins with the coordination of a Lewis acid with aldehyde to form complex **33**. Due to its enhanced electrophilicity, complex **33** is attacked by the π-bond of the enol silane **31**, giving rise to the complex **34**. At this point, either intermolecular silyl cleavage upon hydrolysis or intramolecular silyl transfer to the product hydroxyl group occurs to give the product **35**.

The use of silyl enol ethers can be illustrated in a synthesis of manicone, a conjugated enone that ants use to leave a trail to a food source. It can be made by an aldol reaction between pentan-3-one and 2-methylbutanol (Scheme 114). Both partners are enolizable so we shall need to form a specific enol equivalent from the ketone. The product will be a mixture of diasteroisomers but it eliminates to give a single compound.

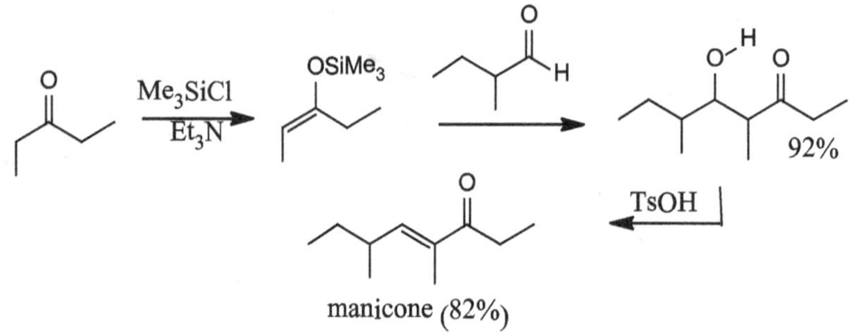

Scheme 114

2.7.16 Knoevenagel reaction

The Knoevenagel reaction is the condensation of an active methylene compound with an aldehyde or ketone, to give a, β-unsaturated dicarbonyl compound [1967OR204, 1991Misc341] (Equation 54).

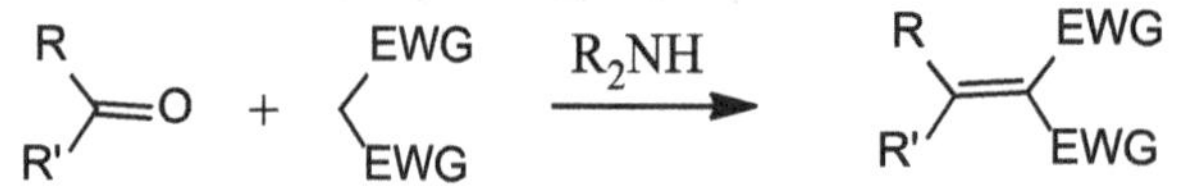

(Equation 54)

The Knoevenagel condensation is distinguished from the related aldol condensation in that the active methylene component (the nucleophile) must be doubly-activated by two electron-withdrawing groups (EWGs) such as NO_2, CN, COR, CO_2H, CO_2NR_2, SO_2OR, SO_2NR_2, SO_2R, SOR, SR, $PO(OR)_2$, aryl, and heteroaryl. The double activation of the active methylene unit allows the Knoevenagel condensation to take place under much milder conditions than the related aldol condensation.

The reaction may stop at step (i) or proceed to step (ii) via a Michael condensation involving the initial product (Scheme 115).

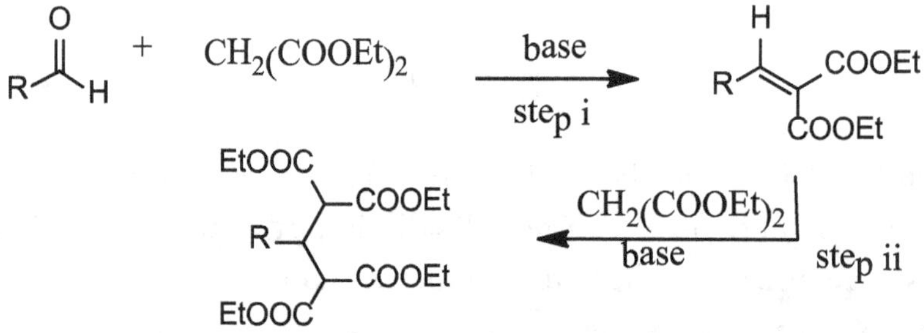

Scheme 115

Step (i) is favoured by using equivalent amount of aldehyde and ethyl malonate in the presence of pyridine. Step (ii) is favoured by using excess of malonic acid in the presence of pyridine and when the aldehyde is aliphatic. Furthermore it appears that the term Knovengol reaction is taken to mean the condensation when unsaturated compound are produced. Obviously then, to prepare α, β-unsaturated acids must be used followed by hydrolysis and heating (Scheme 116).

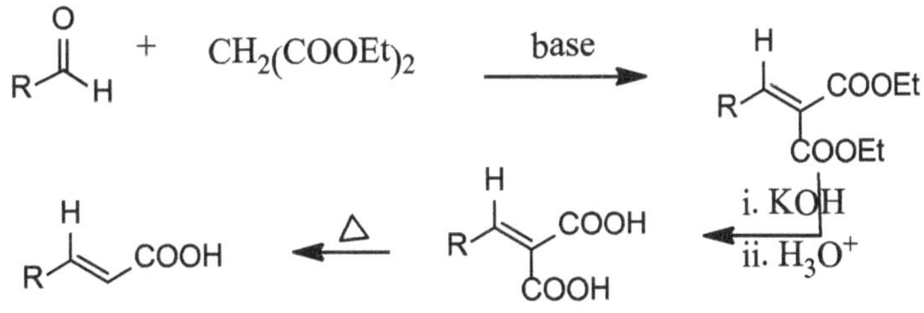

Scheme 116

In practice, it is usual to treat the aldehyde with malonic acid in the presence of pyridine (Equation 55).

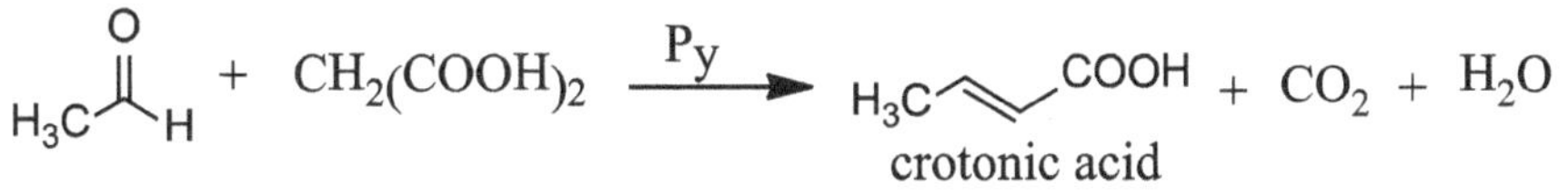

$$H_3C\overset{O}{\underset{H}{\parallel}}\ +\ CH_2(COOH)_2\ \xrightarrow{Py}\ H_3C\diagup\diagdown COOH\ +\ CO_2\ +\ H_2O\quad \text{(Equation 55)}$$
crotonic acid

The mechanism of this reaction has been the subject of mechanism discussion (Scheme 112). When pyridine is used as base and catalyst, then the mechanism is believed to be similar to that of aldol condensation.

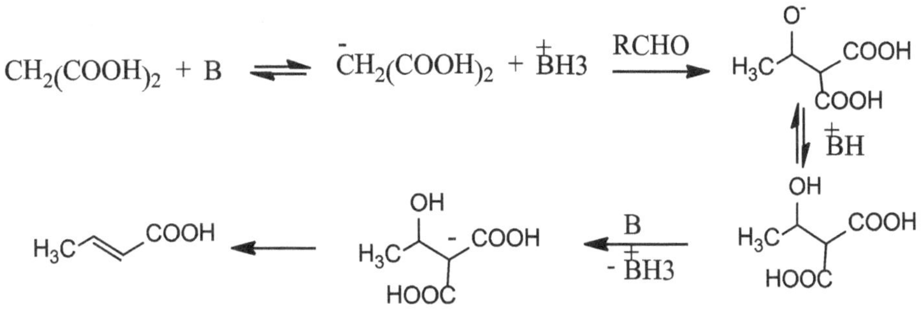

$$CH_2(COOH)_2\ +\ B\ \rightleftharpoons\ {}^-CH_2(COOH)_2\ +\ {}^+BH3\ \xrightarrow{RCHO}$$

Scheme 117

Hantzsch showed that the amine was an essential part of the condensation, as no reaction occurred between ethylacetoacetate and benzaldehyde in the absence of amine, but even a catalytic amount of amine was an effective promoter of this condensation. Hantzsch's synthesis of 1,5-dicarbonyl compound was probably the first reported instance of what was to become known as the Knoevenagel condensation (Scheme 118).

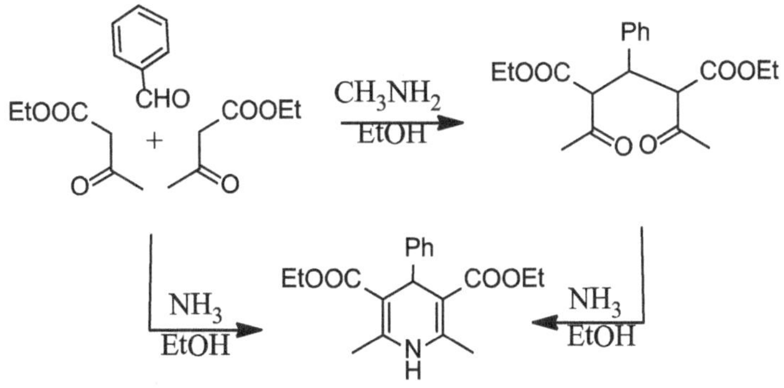

Scheme 118

2.7.17 Henry reaction

The Henry reaction, or nitroaldol reaction, is one of the classic carbon-chain formation methods utilized in organic synthesis. It involves the condensation of nitroalkanes with aldehydes or ketones in the presence of bases to afford the mixtures of diastereomeric 2-nitroalcohols (Equation 56), which in turn can be converted into other useful synthetic intermediates, such as 2-aminoalcohols, α-hydroxyketones, homologous ketones, and perhaps most importantly, nitroalkenes through various functional transformations [1991Misc321].

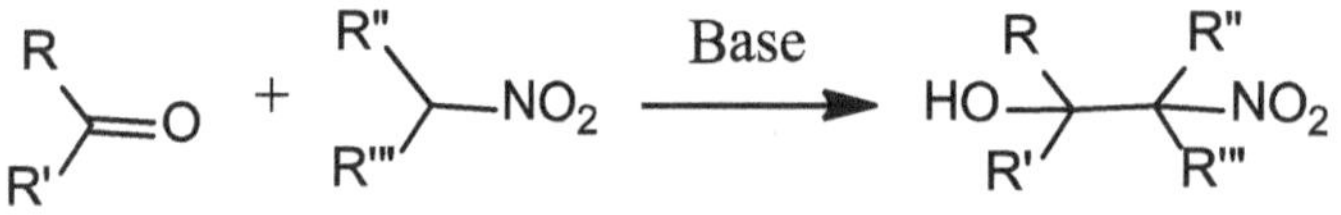

(Equation 56)

The vast majority of the Henry reactions involve the condensation of aliphatic and aromatic aldehydes with nitroalkanes. Nitroaldol condensation with aromatic aldehydes is especially prone to dehydration.

It is generally accepted that the condensation of nitroalkanes **35** with aldehydes proceeds with the nitronates **36** as the intermediates (Scheme 119). The role of the base is to shift the tautomeric equilibrium towards the formation of nitronic acids, or nitroalkanes **35**. Since 36 are much stronger acids (with a pKa range of 2-6) than 35 (with a pKa' range of 9-10), they are more readily deprotonated with the base. After the deprotonation, the formed nucleophilic nitronates 37 add to the carbonyl group of aldehydes or ketones (though less commonly seen) to form nitroalkoxy anions **38**. Subsequent protonation of **38** furnishes the final condensation products 2-nitroalcohols **39** and regenerates the base in the process to complete the catalytic cycle.

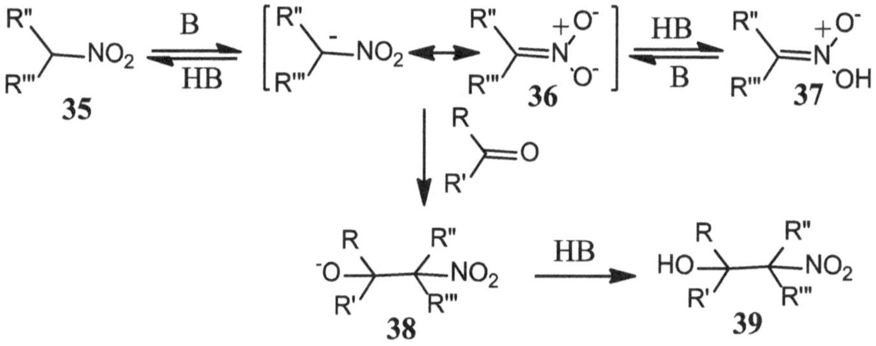

Scheme 119

Conjugated nitroalkenes are probably the most useful intermediates derived from the nitroaldol reaction products. Synthetic utility of the nucleophilic addition of nitroalkenes can be exemplified by transformation using enol silanes as nucleophiles to generate 1,4dicarbonyl compounds. For instance,

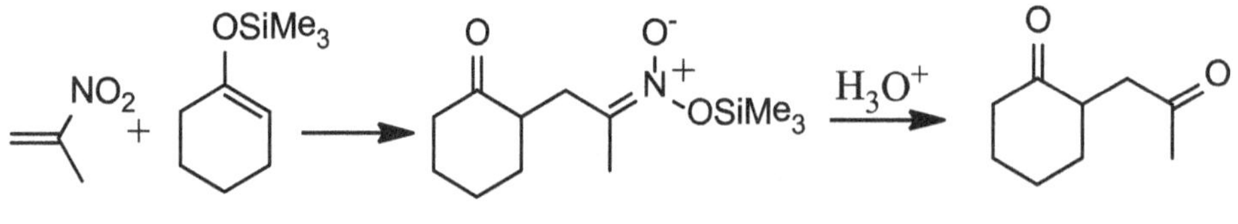

Scheme 120

Functionalized conjugated nitroalkenes, such as 2-nitro-2-propen-l-ol pivalate (NPP), deserve special considerations. They can serve as extremely versatile multiple coupling reagents by sequential coupling with various nucleophiles and electrophiles due to the unique feature of three reactive sites in one molecule.

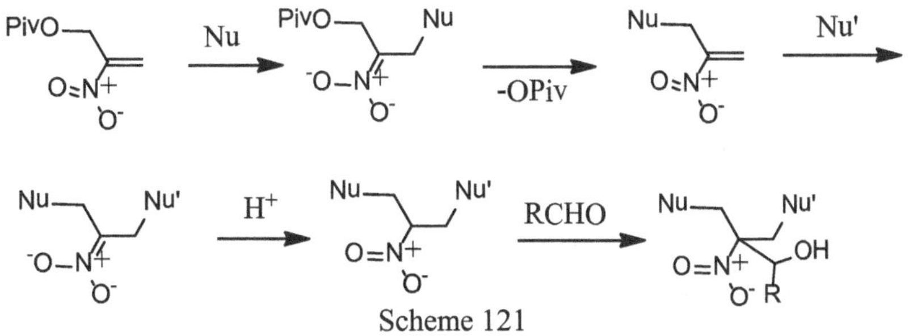

Scheme 121

2.7.18 Van Leusen reaction

The Van Leusen reaction involves the formation of nitrile from carbonyl compound by reaction with TosMIC in presence of tertiary butoxide (Equation 56). It was first described in 1977 by Van Leusen and co-workers [1977JOC3114, 2005OL3183]. When aldehydes are employed, the Van Leusen reaction is particularly useful to form oxazolesand imidazoles.

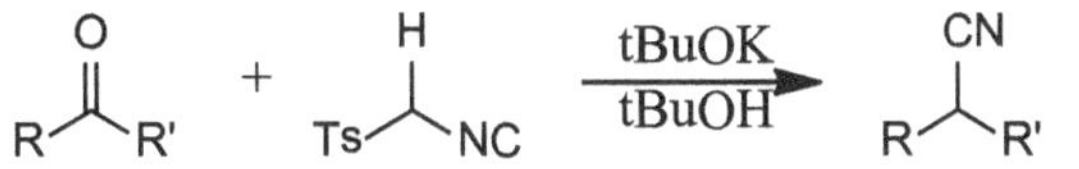

(Equation 56)

Probable mechanism for this conversion involves the initial deprotonation of TosMIC by tBuOK, which then attacks nucleophiically to the carbonyl compound to form alkoxide followed by cyclization to yield the oxazoline intermediate (Scheme 122). If the substrate is an aldehyde, then elimination of the tosyl group can occur readily to form oxazole derivatives.

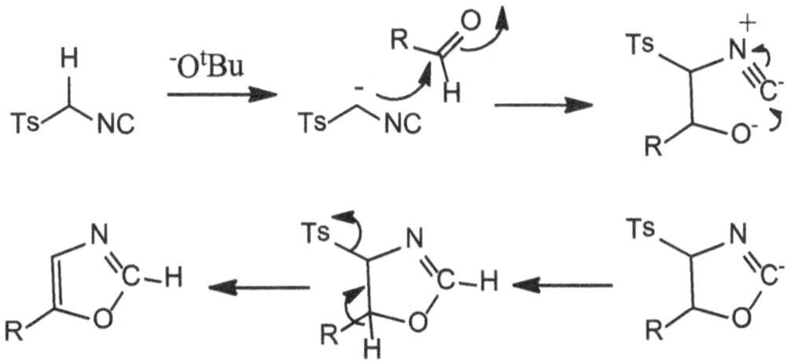

Scheme 122

When ketones are used instead, elimination cannot occur; rather, a tautomerization process gives an intermediate which after a ring opening process and elimination of the tosyl group forms an N-formylated alkeneimine. This is then solvolysed by an acidic alcohol solution to give the nitrile product (Scheme 123).

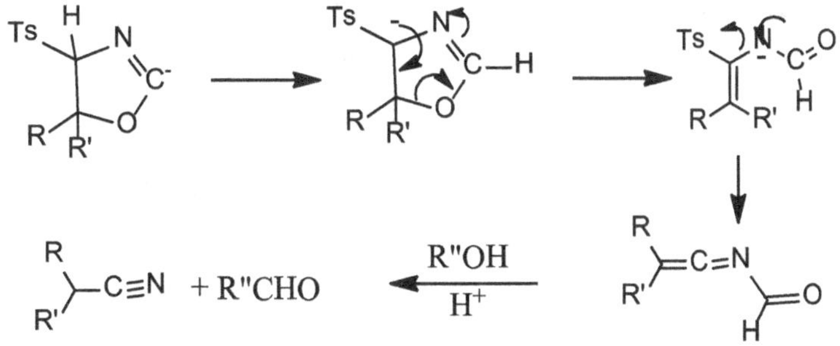

Scheme 123

2.7.19 The Prins reaction

The **Prins reaction** is an organic reaction consisting of an electrophilic addition of an aldehyde or ketone to an alkene or alkyne followed by capture of a nucleophile [1952CR505]. With water and a protic acid such as sulfuric acid as the reaction medium and formaldehyde the reaction product is a 1,3-diol (Equation 57). When water is absent, the cationic intermediate loses a proton to give an allylic alcohol. With an excess of formaldehyde and a low reaction temperature the reaction product is a dioxane. When water is replaced by acetic acid the corresponding esters are formed.

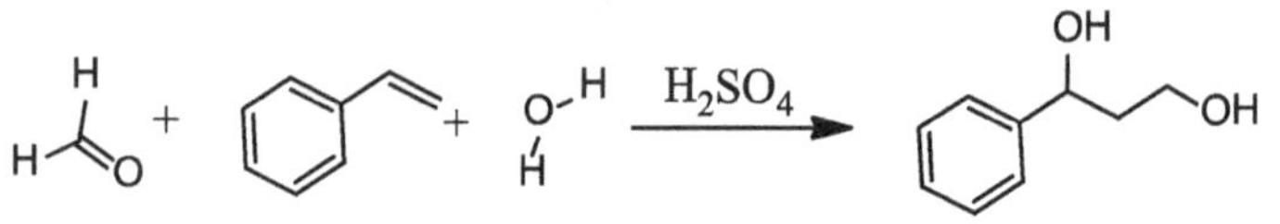

(Equation 57)

The probable mechanism for this reaction involves the protonation of carbonyl compound followed by the attack of alkene as nucleophile leads to the formation of cationic intermediate (Scheme 124). Capture of the carbocation by water followed by deprotonation leads to the formation of required 1,3-diol. In the absence of water, the carbocation loose proton to form the allyl alcohol. Capture of the carbocation by additional carbonyl reactant to produce oxonium ion which undergo ring closure via intramolecular attack of hydroxyl group leads to the formation of dioxane derivatives.

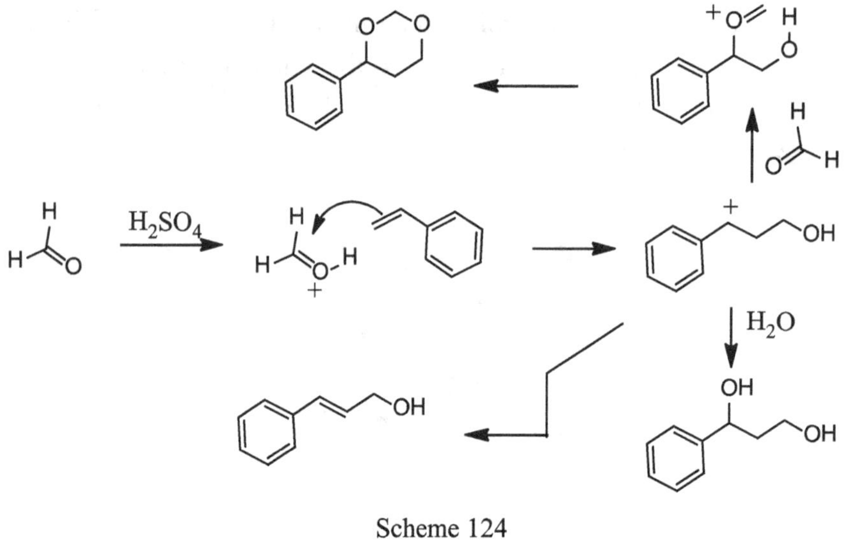

Scheme 124

EXERCISE

1. How bisulfite addition reaction is helpful in the synthesis of the synthesis of cyanohydrin derivatives?
2. Complete the following with suitable reagents and conditions:

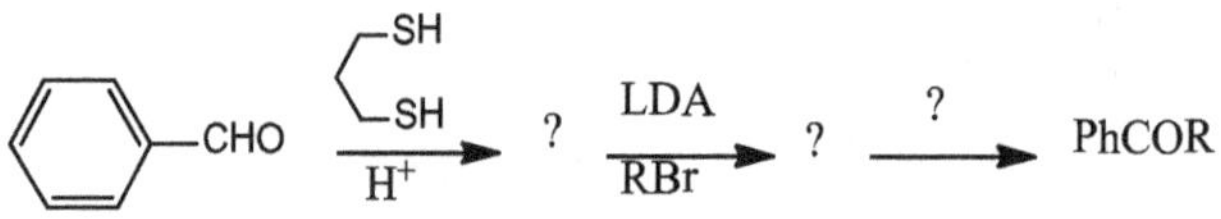

3. Predict the product for the following:

$$Me_3SiCl + AgCN \longrightarrow ? \xrightarrow{RCHO} ?$$

4. Predict the product with suitable mechanism: $RCHO + EtSH + HCl \longrightarrow ?$
5. Predict the product with suitable mechanism:.

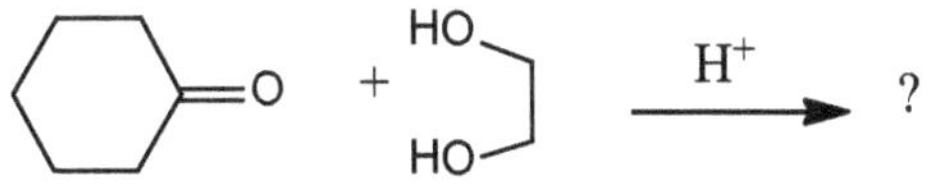

6. Mention the different nitrogen nucleophiles that can be added to carbonyl group and name the resultant product(s).
7. Complete the following:

8. Among 3-hexanone and cyclohexanone, which one is more reactive and why?
9. Protonation of carbonyl group will enhance its electrophilicity. Justify this statement.
10. How do you effect the following conversion? Write the mechanism.

11. Predict the products for the following

$$RCHO + NH_4Cl + NaCN \xrightarrow{NH_4OH} ?$$

12. Predict the products for the following with reasonable mechanism:

$$RCOR + HN_3 \longrightarrow ?$$

13. Predict the product for the following with reasonable mechanism:

$$RCHO + PhNHNH_2HCl \xrightarrow{AcONa} ?$$

14. Write the structure of hemithioacetal and acetal.
15. Complete the following:

$$RCHO \xrightarrow{Me_3SiCN} ? \xrightarrow[\text{ii. R'X}]{\text{i.LDA}} ? \xrightarrow{H^+} ?$$

16. Complete the following with reasonable reagents:

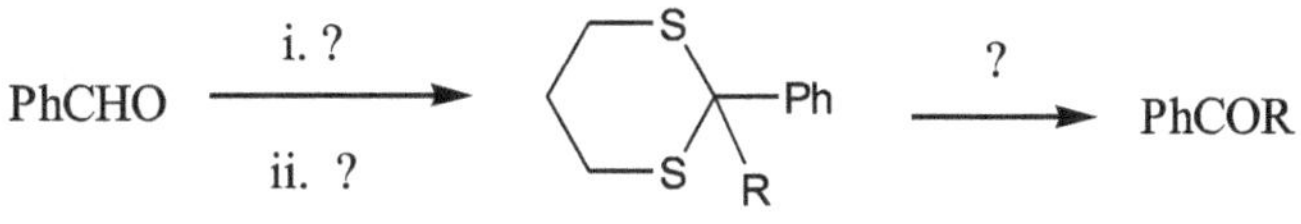

17. P-Aminobenzaldehyde doesnot undergo Benzoin condensation. Explain.

18. Predict the product for the following with reasonable mechanism:

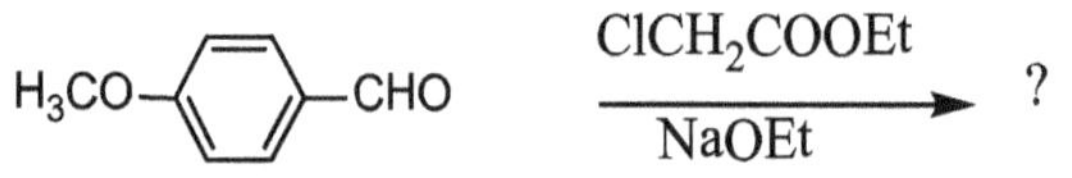

$$H_3CO-\text{⟨⟩}-CHO \quad \xrightarrow[\text{NaOEt}]{\text{ClCH}_2\text{COOEt}} \quad ?$$

19. How do you prove that Darzen's glysidic ester condensation undergoes via ionic mechanism.

20. Complete the following with reasonable reagents:

$$RCHO \quad \longrightarrow \quad RCOCH_2CH_2COR'$$

21. Write the reasonable mechanism for the following conversion:

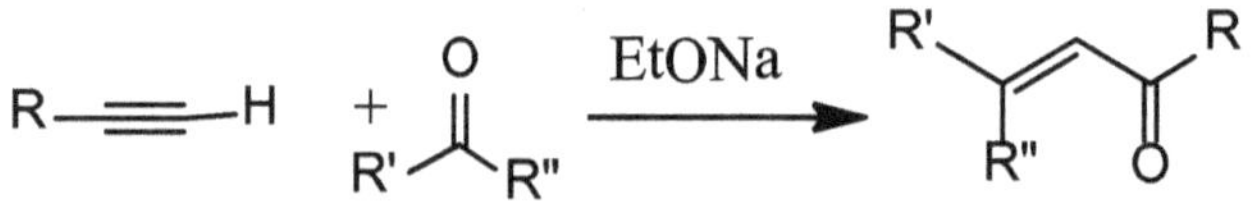

$$R-\!\!\equiv\!\!-H \;+\; \underset{R'}{\overset{O}{\underset{}{\|}}}\!\!-\!\!R'' \quad \xrightarrow{\text{EtONa}} \quad \underset{R''}{\overset{R'}{\diagdown}}\!\!\diagup\!\!\underset{O}{\overset{R}{\diagup}}$$

22. Acetal/ketals are acid labile while base stable. Justify this statement.

23. Dithiacetal/ketals are acid stable and base stable. Justify this statement.

24. Predict the product(s) for:

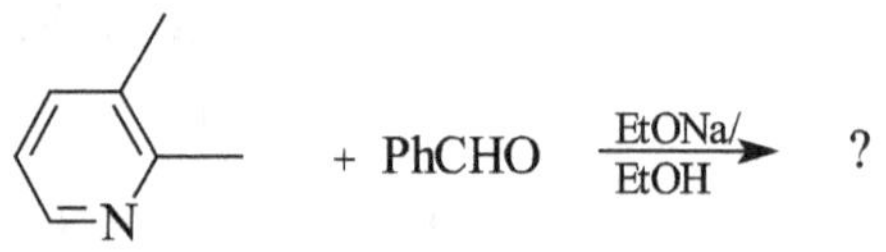

$$\text{⟨pyridine with CH}_3\text{⟩} \;+\; PhCHO \quad \xrightarrow[\text{EtOH}]{\text{EtONa/}} \quad ?$$

25. Write the structure of hemiacetal and dithioacetal.

26. Complete the following:

$$O=\text{⟨⟩}-CHO \quad \xrightarrow{?} \quad O=\text{⟨⟩}-CH_2OH$$

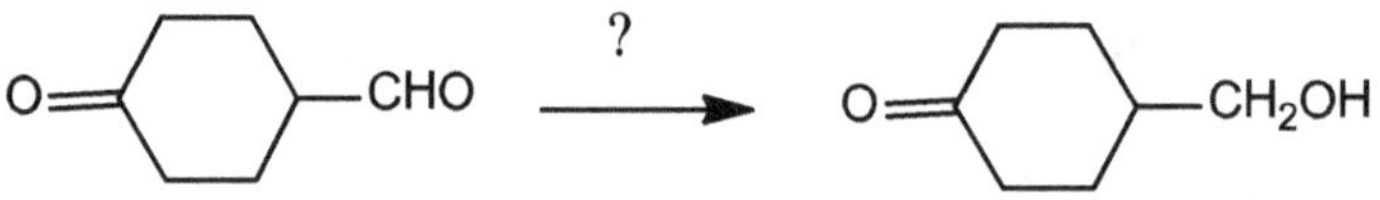

27. Complete the following:

$$O=\text{⟨⟩}-CHO \quad \xrightarrow{?} \quad HO-\text{⟨⟩}-CHO$$

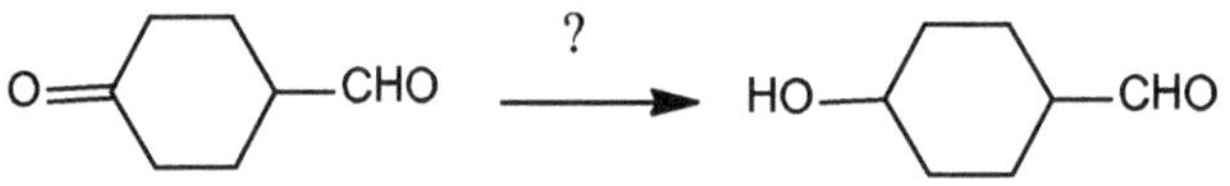

28. Complete the following:

$$O=\text{⟨⟩}-CHO \quad \xrightarrow{?} \quad \text{⟨⟩}-CHO$$

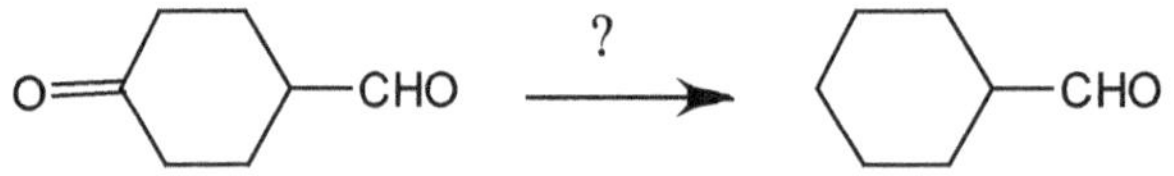

29. Complete the following:

$$O=\text{⟨⟩}-CHO \quad \xrightarrow{?} \quad O=\text{⟨⟩}-$$

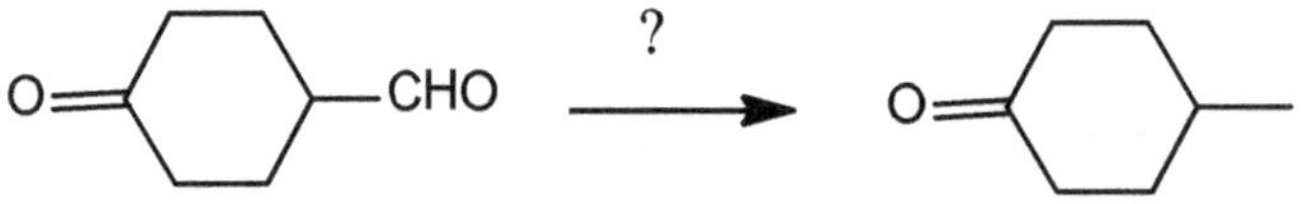

30. Write notes on addition of bisulfite and amines to the carbonyl compounds.

31. What happens when benzaldehyde is treated with: i. NaOH & ii. NaCN.
32. Complete the following with suitable mechanism:

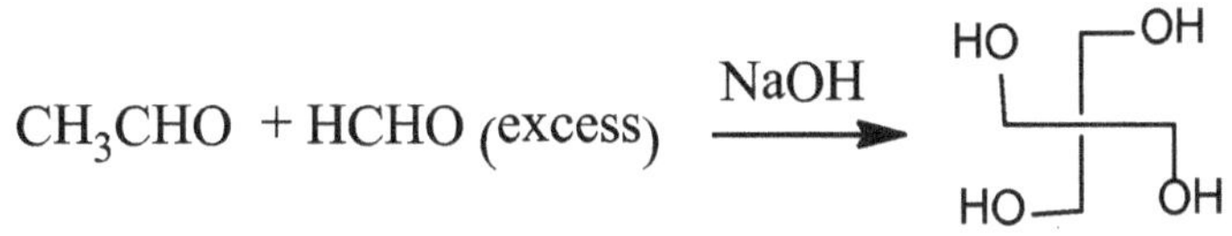

CH_3CHO + HCHO (excess)
NaOH
HO
OH
HO
OH

33. Suggest suitable regents for the following conversions.

Acrolein ⟶ Glyceraldehydes.

34. Complete the following with reasonable reagents:

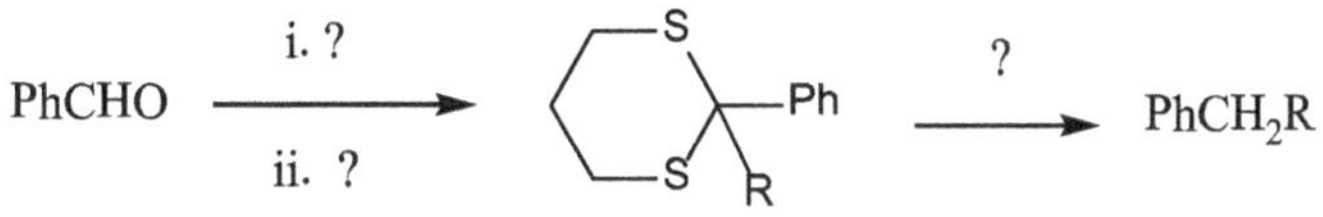

PhCHO
i. ?
ii. ?
S
S
Ph
R
?
$PhCH_2R$

35. How Grignard reagent is useful for the synthesis of aldehydes starting from formic acid?
36. Acid catalyzed addition of water to carbonyl group is reversible and the backward reaction is predominant. Why?
37. Lithium salts of 2-Alkyl-1,3-dithiane acts as acyl anion equivalent. Justify this statement with illustrations.
38. Write the reasonable mechanism for the following conversion:

Ph
Ph
O
i. diethyl succinate
ii. ^{t}BuOK, ^{t}BuOH
Ph
Ph
COOK
COOEt

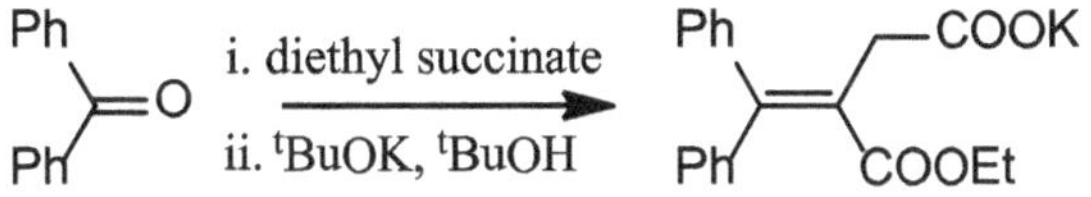

39. Write a reasonable mechanism for the condensation of benzaldehyde with nitromethane.
40. Outline the synthesis of coumarin through Perkin reaction.
41. Using Wittig reaction as synthetic tool, how do you convert RCHO to RCH_2CHO.
42. Complete the following with reasonable mechanism:

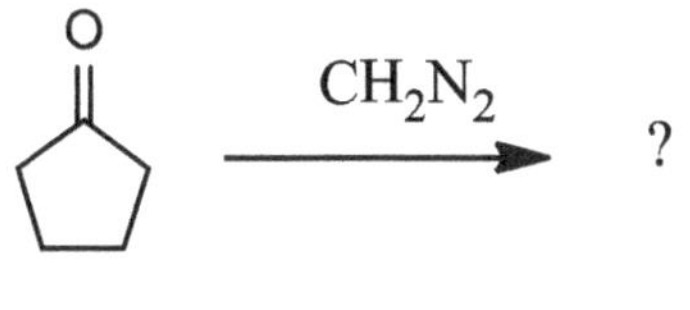

O
CH_2N_2
?

43. Complete the following with suitable steps.

O
O
O

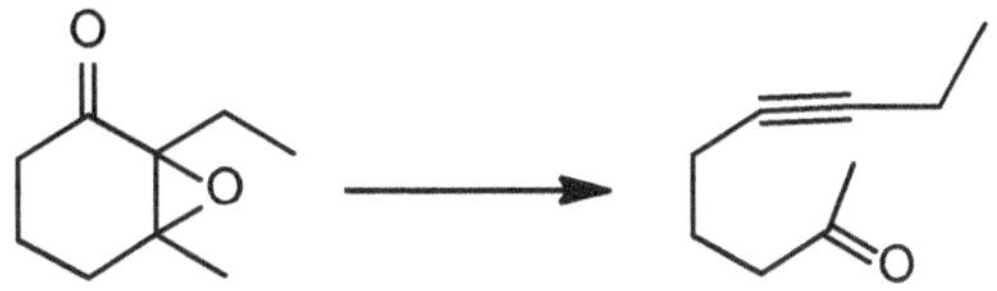

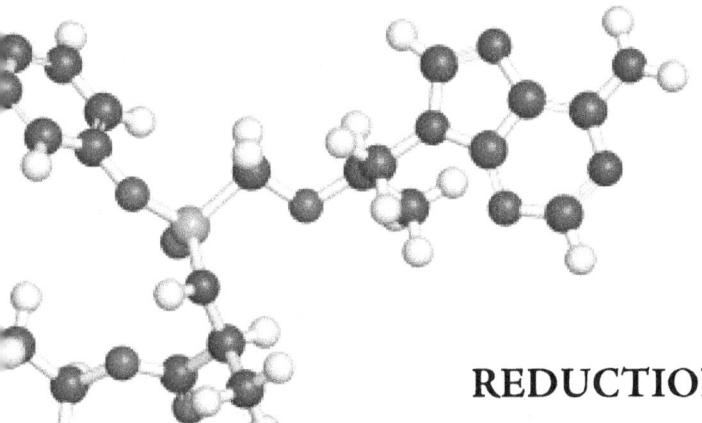

REDUCTION AND OXIDATION REACTIONS OF CARBONYL COMPOUNDS

3.1 REDUCTION OF CARBONYL COMPOUNDS

3.1.1 Catalytic hydrogenation

The easiest large scale reduction method for the conversion of aldehydes and ketones to alcohols is by catalytic hydrogenation.

$$\text{(cyclopentanone)} \quad \xrightarrow[\text{1000psi}]{\text{H}_2 + \text{Ni}} \quad \text{(cyclopentanol)} \qquad \text{(Equation 58)}$$

The advantage over most chemical reduction schemes is that usually the product can be obtained simply by filtration from the catalyst followed by distillation. The usual catalysts are nickel, palladium, copper chromate or platinum promoted with ferrous ion. Hydrogenation of aldehydes and ketone carbonyl groups is much slower than C-C double bonds and rather rigorous conditions are required. It follows that it is generally not possible to reduce a carbonyl group in the presence of a double bond.

$$\xrightarrow{\text{Ni}} \quad + \quad \Delta\text{H} = 3.0 \text{ Kcal}$$

$$\xrightarrow{\text{Ni}} \quad + \quad \Delta\text{H} = 3.0 \text{ Kcal}$$

Aldehydes and ketones are reduced to alcohols by catalytic hydrogenation. It is usually possible to reduce an alkene double bond selectively in the presence of a carbonyl group; the carbonyl group does not have to be protected. Palladium catalysts are particularly effective in this reduction.

$$\xrightarrow[\text{H}_2]{\text{Pd/C}} \quad \underset{\text{20\% yield}}{} + \underset{\text{73\% yield}}{} \qquad \text{(Equation 59)}$$

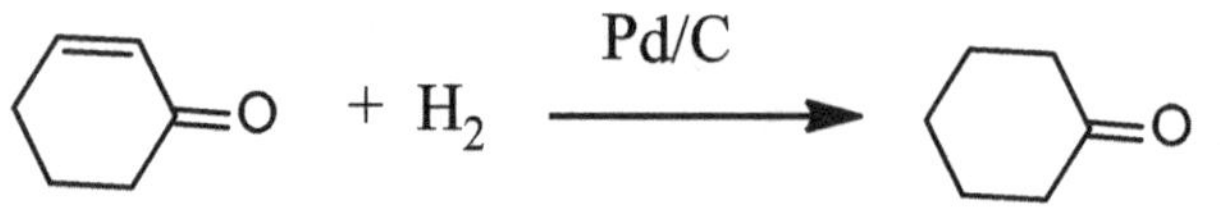

(Equation 60)

3.1.2   The Nystrom and Brown reduction by LiAlH$_4$

In recent years inorganic hydrides such as LiAlH$_4$ or NaBH$_4$ have become extremely important as reducing agents of carbonyl compounds since it is a good hydride ion donor and compounds containing >=O moiety are good hydride acceptors. For this reason the latter compounds are reduced very often by LAH or its derivatives. The polar nature of >C=O moiety makes the C atom electrophilic towards the nucleophilic hydride ion and enables the compound containing >C=O moiety to be reduced by LAH. Thus aldehydes, ketones, carboxylic acids, esters, amides, acid chlorides, nitriles etc., get reduced on treatment with LAH in ether medium.

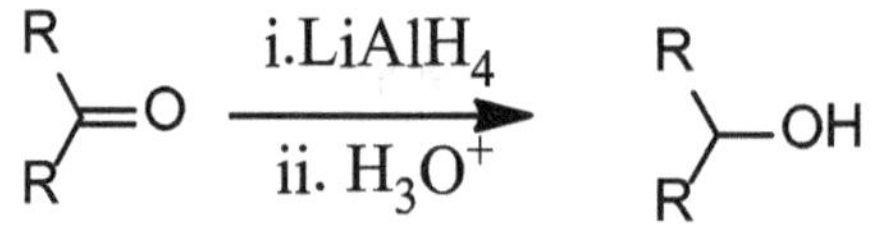

LAH being soluble in ether, the reaction is usually carried out in ether (THF etc.,). At first a lithium aluminium alkoxide type complex is formed which on acidification liberates the reduction product. Water is then added to the reaction mixture to protonate alkoxide. Because LAH react vigorously with protic solvents and violently with water, it cannot be used with these solvents.

The mode of addition of LAH to the carbonyl group involves a prior complexing or perhaps simply an electrostatic attraction between the oxygen atom of the carbonyl group and the lithium cation. Since the aluminium atom in $^-$AlH$_4$ has its octet of electrons and a negative charge; it is unlikely that this species enter into prior complex formation with the oxygen atom. A hydride is then transformed by the $^-$AlH$_4$ to the carbon atom of the carbonyl group. A single mole of LAH is able to reduce four moles of carbonyl compounds (Scheme 125). LAH usually reduces carbonyl groups without affecting carbon-carbon double bonds. All alkoxide species are transformed into alcohols when water is subsequently added to the reaction mixture.

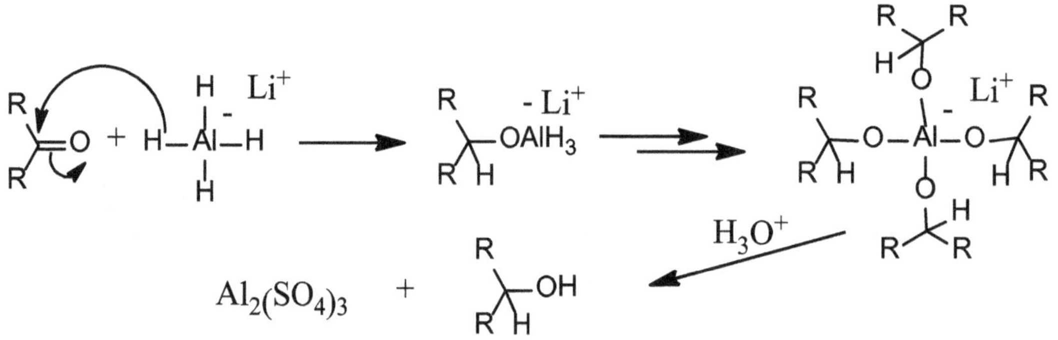

Scheme 125

When a nucleophile adds to cyclohexanone, the nucleophile can adopt either of the two conformations shown below, with Nu axial or equatorial, depending on the relative size of the nucleophile and OH.

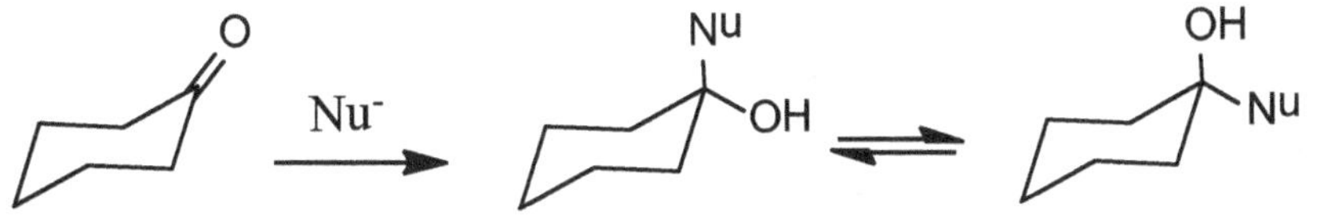

Now think of a nucleophile attacking 4-t-butylcyclohexanone wherein t-butyl group attain equatorial position. Here, attack on the same face as the t-butyl group leaves the nucleophile axial and the hydroxyl group equatorial while attack on the opposite face leaves the nucleophile equatorial and the hydroxyl group axial.

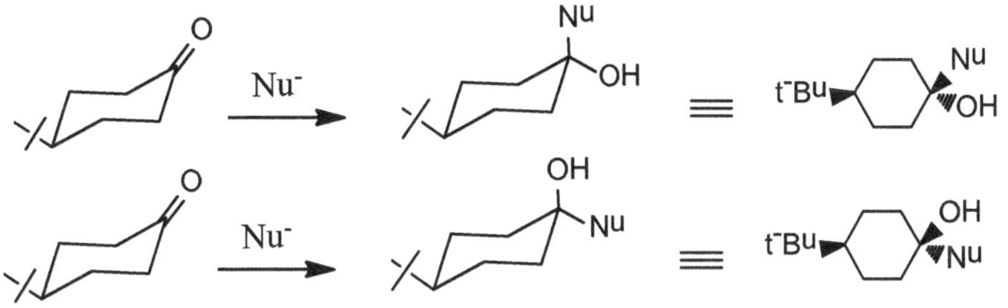

For instance, reduction of 4-t-butylcyclohexanone with LAH in Et$_2$O gives trans alcohol (90% of the hydride has added axially, AlH$_4^-$ is quite small as nucleophile go) while with L-selectride (lithium tri-sec-butyl borohydride) gives cis alcohol to make more of the cis alcohol (hydride has added equatorially, selectride is quite large as nucleophile go) (Scheme 126).

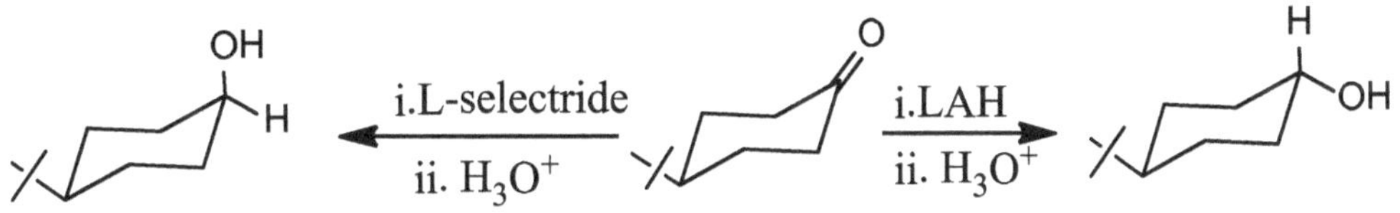

Scheme 126

LAH of course does not usually reduce C-C double or triple bonds. However, a C-C multiple bond in conjugation with an aromatic system on one side and a carbonyl group on the side gets reduced by LAH (Equation 61).

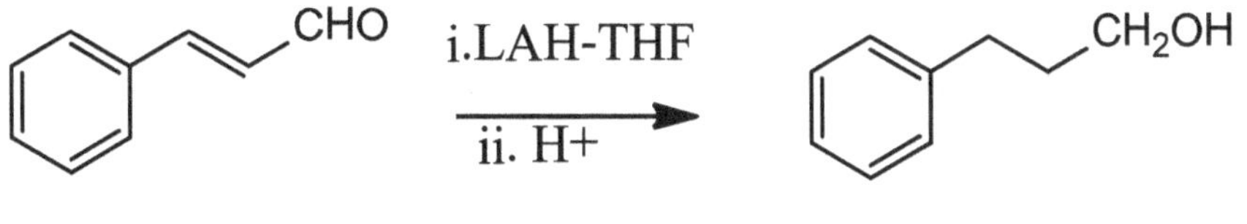

(Equation 61)

A C-C multiple bond is an electron dense moiety where electrophilic attack occurs. A hydride ion is an electron dense species and thus behaves as a strong nucleophile. So, a hydride ion experiences repulsion when it comes close to a C-C multiple bonds and thus fails to reduce it usually. On the other hand, electron withdrawing effect of an aromatic system on one side and that of the carbonyl group on the other side make the C-C multiple bonds less electron dense and enable to hydride ion attack the C-C moiety. For this reason such a C-C multiple bonds gets reduced by LAH.

Alkynyl ketone is reduced with good eneantioselectivity derived from reaction of LAH, N-methylephedrine and 3,5-dimethylphenol (Equation 62) [1980CC1026] while reduction of α,β-unsaturated compounds with high enanatiomeric excess by LAH in presence of N-ethylaniline (Equation 63) [1991PAC307].

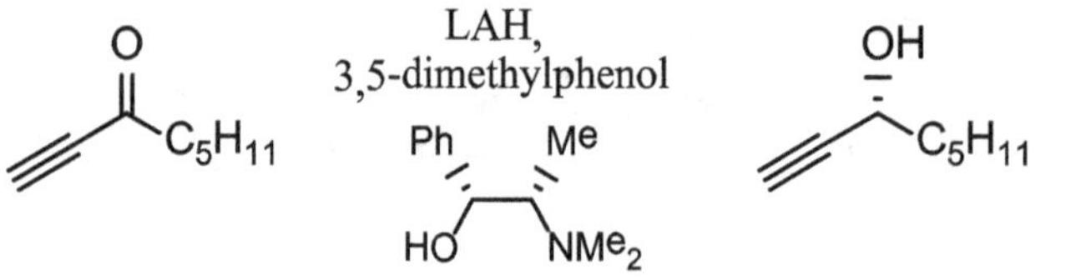

(Equation 62)

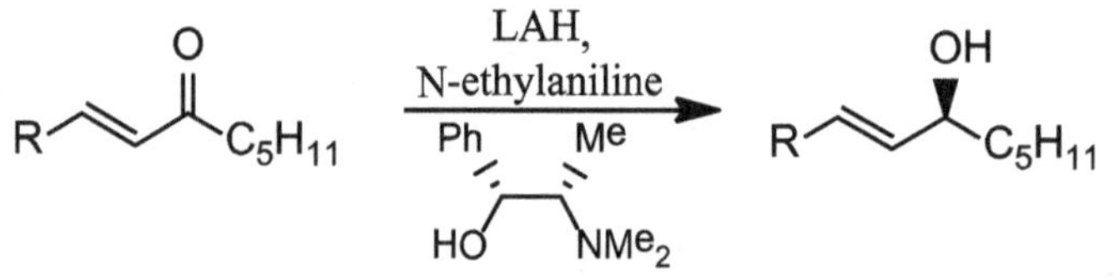

(Equation 63)

3.1.3   Reduction by NaBH$_4$

The reaction of NaBH$_4$ with aldehydes and ketones is consequently similar to that of LAH. The sodium ion however does not coordinate to the carbonyl group as well as the lithium ion. For this reason, reactions of NaBH$_4$ with carbonyl compounds are often carried out in protic solvents such as alcohols. Hydrogen bonding between the alcohol solvent and the carbonyl group serves as a weak acid catalysis that activates the carbonyl group. NaBH$_4$ reacts only slowly with alcohols and can even be used in water if the solution is not acidic. The sodium methoxyborohydride formed after the first reduction can continue as a reducing agent until all equivalents of hydride are used.

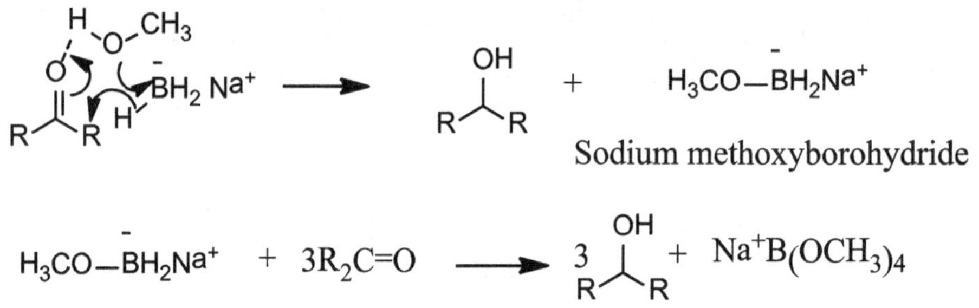

Scheme 127

Because NaBH$_4$ and LAH are hydride donors, reductions by these and related reagents are generally referred to as **hydride reductions**.

Both NaBH$_4$ and LAH are highly useful in the reduction of aldehydes and ketone. NaBH$_4$ is however a milder reducing agent than LAH but more selective reagent. It can be used to reduce aldehyde and ketones but not acids or esters. It can be used to reduce carbonyl groups in the presence of certain other functional groups that are sensitive to LAH. For instance,

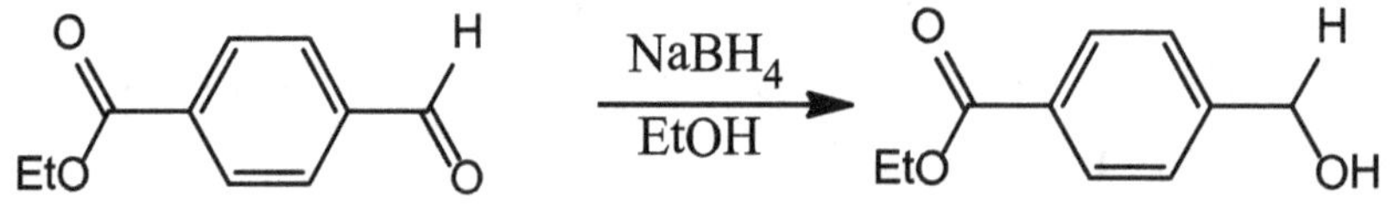

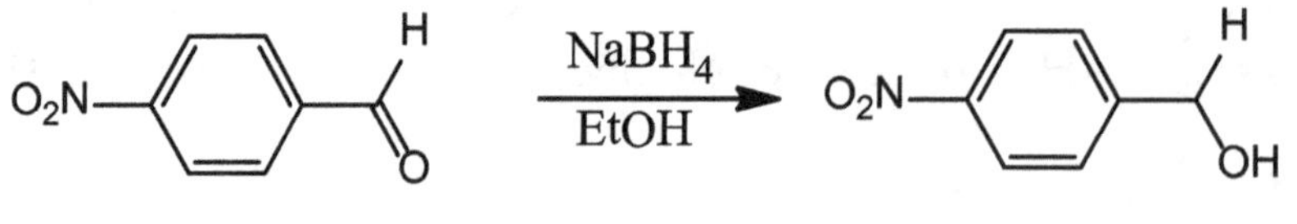

3.1.4 The Meerwein-Pondorf-Verly reduction

The conversion of a carbonyl compound into an alcohol by the action of aluminium isopropoxide in isopropyl alcohol medium is called the Meerwein-Pondorf-Verly reduction. At first, aluminium alkoxide of the newly alcohol is formed which gives the alcohol by the reaction with isopropyl alcohol. However, the alcohol is now generated by the reaction of dilute sulphuric acid on the aluminium alkoxide. The process is reversible, an alcohol can be converted to the corresponding ketone by the same material and the oxidation is called Oppenauer oxidation.

$$R_2CO + [(CH_3)_2CHO]_3Al \rightleftharpoons (CH_3)_2CO + (R_2CHO)_3Al \longrightarrow R_2CHOH$$

This reducing agent is specific for the carbonyl group, and so may be used for reducing aldehydes and ketones containing some other functional group that is reducible (a double bond or a nitro group).

The mechanism of the Meerwein-Pondorf-Verley reduction is considered to involve the direct hydride transfer from the α-carbon of the aluminium isopropoxide to the carbonyl carbon of the carbonyl compound reversibly through a cyclic six membered transition state. The products are the aluminium salt of the alcohol and acetone. The reaction is reversible, however in an inert solvent at moderate temperature, the acetone distills out as it is formed and the equilibrium is shifted towards the products. The free alcohol is thus obtained by hydrolyzing the aluminium salt with acid. When isopropanol is present, the isopropyl alcohol enters into equilibrium with aluminium salt of the desired alcohol. The free alcohol is generated and the chromium salt of isopropyl alcohol is formed instead.

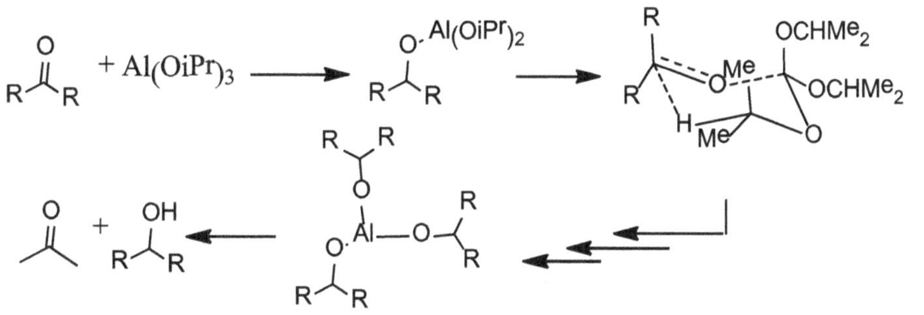

Scheme 128

When the reaction is carried out with α-deutero-aluminium isopropoxide, the product is found to contain deuterium at the carbonyl carbon atom of the starting carbonyl compound. This is evidence in favour of the direct hydride transfer from the α-C atom of the aluminium isopropoxide to the carbonyl carbon.

$$CH_3CH_2COCH_3 + [(CH_3)_2CDO]_3Al \xrightarrow[]{Me_2CHOH} CH_3CH_2CD(OH)CH_3 + (CH_3)_2CO$$

Since an excess of isopropyl alcohol is present, the equilibrium is also shifted towards product. The aluminium salt of isopropyl alcohol is just the starting reagent aluminium isopropoxide, which has been regenerated. Of all the metal alkoxide that may be used for the reaction viz., Na, K, etc., aluminium alkoxide is the best probably because of its great ability to form coordination compounds. Aluminium alkoxide are much less polar than alkali metal alkoxide, the aluminium-oxygen bonds being almost covalent and having little tendency to dissociate to give free alkoxide ions. Aluminium isopropoxide is the reagent of choice, is a low melting solid (118°C) while distills at 140-50°C (12mm).

3.1.5   Yeast reduction

A great deal of work has been carried out on the enanatioselective reduction of ketones with fermenting yeast [1991CR49]. If the two ketone substituents are classified as "large" (R_L) and "small" (R_S), yeast reduction often gives the product of reduction from the lower face of the carbonyl group as shown below.

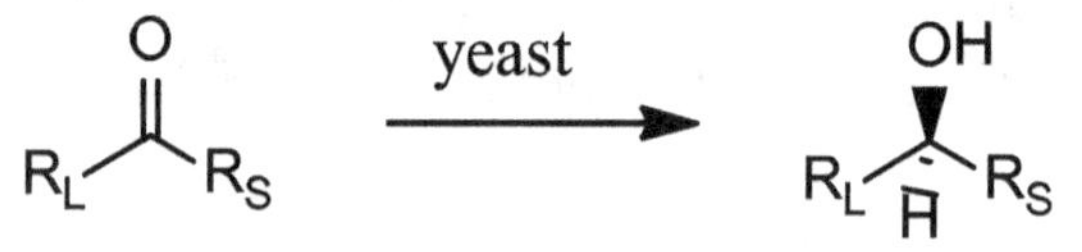

(Equation 64)

Represenatative examples of alcohols which have been prepared by this type of yeast reduction are shown below and all conform to the Prelog'rule [1988JOC5215].

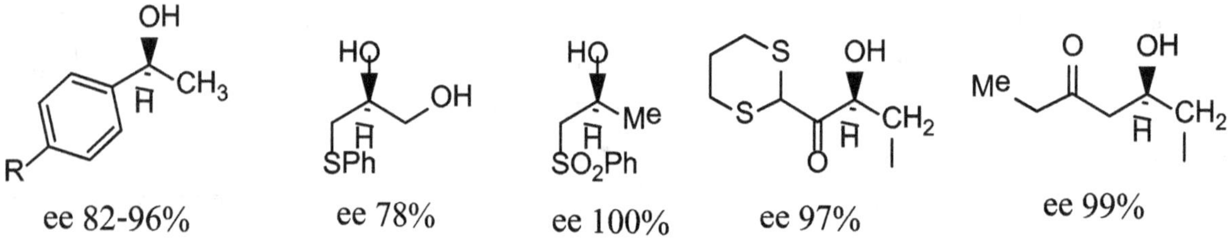

ee 82-96%   ee 78%   ee 100%   ee 97%   ee 99%

The reduction of β-keto esters with yeast has received much attention [1991CR49]. Although Prelog's rule often applies, the absolute configuration and eneantiomeric excess of the product can depend on the size and nature of the ketone substituents, and on the ester used. Instances are also known in which variables such as the glucose concentration, the P^H, and the physiological state of the cells influence the reduction stereoselectivity. The origin of these effects and the effect of additives on the stereoselectivity, can be traced to the presence in the yeast cells of several oxidoreductases, which possess different enantiselecivity and the substrate selectivity.

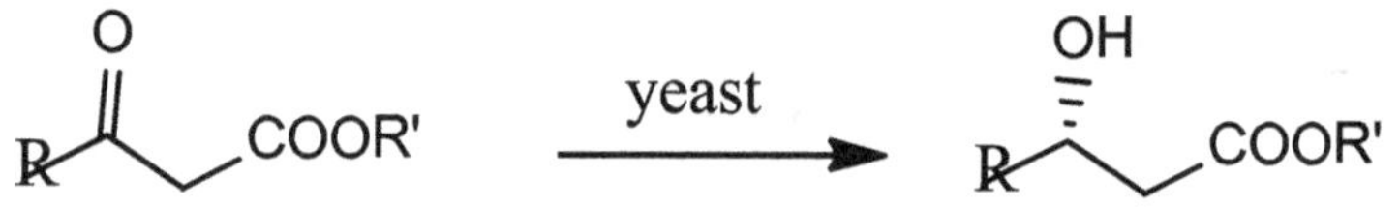

(Equation 65)

Table: Yeast reduction of β-keto esters

R	R'	Yield	e.e (%)
Me	Bu^t	61	85(S)
Ph	Et	70	100(S)
Et	C_8H_{17}	75	95(R)
$(CH_2)_2CH=CH_2$	H	35	99(R)
$(CH_2)_2CH=CH_2$	Me	30	92(R)
EtS	Me	55	70(R)
CH_2Br	Bn	40-50	100(S)
CH_2N_2	Bn	70-80	95(R)
CH_2CH_2Bn	CH_2Bu^t	35	96(S)

3.1.6 Clemmensen reduction

Clemmensen reported that simple ketones and aldehydes were converted to the corresponding alkanes upon refluxing for several hours with 40% aqueous hydrochloric acid, amalgamated Zinc (Zn/Hg) and a hydrophobic organic co-solvent as toluene. The method of converting a carbonyl group to the corresponding methylene group is known as Clemmensen reduction [1942OR155].

The mechanism of the Clemmensen reduction is not well understood. The lack of a unifying mechanism can be explained by the fact that the products formed in the various reductions are different when the reaction conditions (e.g., concentration of the acid, concentration of zinc in the amalgam) are changed. The early mechanistic papers came to established that simple aliphatic alcohols are not intermediates of these reductions, since they do not give alkanes under the same usual Clemmensen reductions. Currently, there are two proposed mechanisms for the Clemmensen reduction and they are somewhat contradictory. In one of the mechanisms the rate determining step involves the attack of zinc and chloride ion on the carbonyl group and the key intermediates are carbanions, whereas in the other heterogeneous process, the formation of a radical intermediate and then a zinc carbenoid species is proposed.

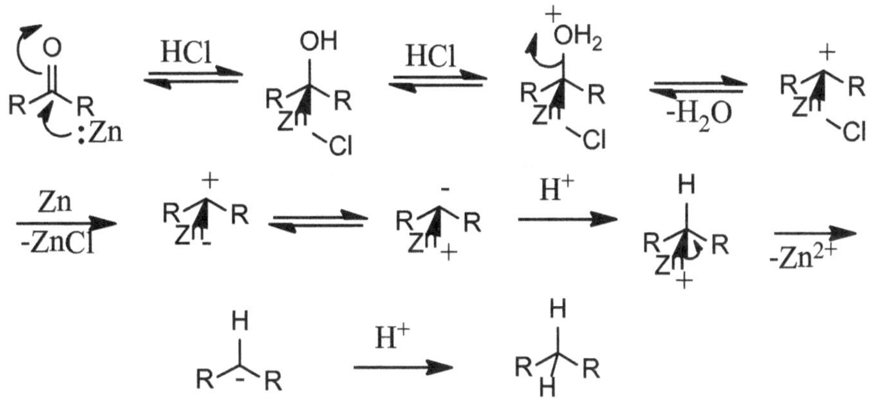

Scheme 129

3.1.7 Wolf-Kishner reduction

When an aldehyde or ketone is treated with hydrazine and strong base under forcing conditions (heat), the carbonyl group is completely reduced to methylene (-CH_2-) group. This reaction is called the Wolf-Kishner reduction [1948OR378; 1991Misc327], typically utilizes glycol or similar compounds as co-solvents. The boiling point of these compounds ensures the high temperature required for the reaction to take place at a reasonable rate.

The Wolf-Kishner reduction involves formation of a hydrazine, a type of Schiff base. The first step in Wolf-Kishner reduction reaction is the formation of a hydrazone anion by deprotonation of the terminal nitrogen by KOH. A range of mechanistic data suggests that the rate-determining step involves formation of a new carbon–hydrogen bond at the carbon terminal in the delocalized hydrazone anion. This proton capture takes place in a concerted fashion with a solvent-induced abstraction of the second proton at the nitrogen terminal. Several molecules of solvent have to be involved in this process in order to allow for a concerted process. The final step of the Wolff–Kishner reduction is the collapse of the diimide anion in the presence of a proton source to give the hydrocarbon via loss of dinitrogen (Scheme 130).

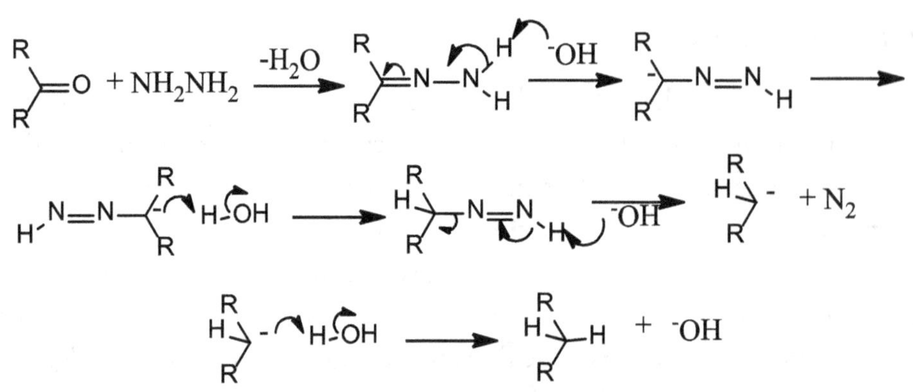

Scheme 130

A detailed Hammett analysis of aryl aldehydes [1964JACS2909], methyl aryl ketones and diaryl ketones showed a non-linear relationship which the authors attribute to the complexity of the rate-determining step. Mildly electron-withdrawing substituents favor carbon-hydrogen bond formation, but highly electron-withdrawing substituents will decrease the negative charge at the terminal nitrogen and in turn favor a bigger and harder solvation shell that will render breaking of the N-H bond more difficult. The exceptionally high negative entropy of activation values observed can be explained by the high degree of organization in the proposed transition state.

Kishner noted that in some instances, α-substitution of a carbonyl group can lead to elimination affording unsaturated hydrocarbons under typical reaction conditions [1955JACS3272]. He found that the amount of elimination increases with increasing steric bulk of the leaving group.

3.1.8 Dissolving Metal Reductions of carbonyl compounds

There are a number of metals at the zero or low oxidation state that readily donate one or more electrons to molecules with an accessible LUMO orbital (Li°, Na°, K°, Zn°, Mg°, Ca°, Cr^{+2}, Ti°, Sm^{+2}, etc). These function primarily as electron transfer reagents, although the transition metals and lanthanides also have a strong component of sigma bonding between cations and intermediate anions.

The reduction of aldehydes, ketones and esters by metallic sodium in ethanol or n-butanol to the corresponding alcohol is known as the Bouveault-Blanc reduction [2009Misc493]. The probable mechanism for the reduction involves the formation of a radical anion by electron transfer from the metal to the low lying antibonding orbitals of the carbonyl π-bond. The radical anion formed by the addition of an electron to a ketone is known as ketyl. The single electron is in the π* orbital, so we can represent a ketyl with the radical on oxygen or on carbon and the anion on the other atom.

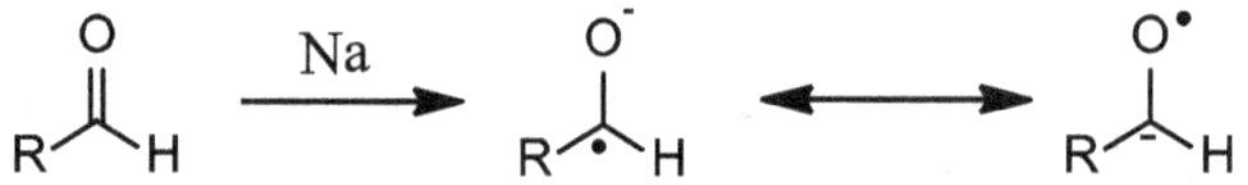

Ketyl behaves in a manner that depends on the solvent that they are in. In protic solvents ethanol), the ketyl gets protonated to form stable radical, which then can accept electron from the metal and gets protonated by the solvent to afford the alcohol as shown in the scheme 131. Notice that this is a reaction using sodium in ethanol and not sodium ethoxide, which is the basic product that forms once sodium has dissolved in ethanol. It is important that the sodium is dissolving as the reaction takes place, since only then is the free electron available.

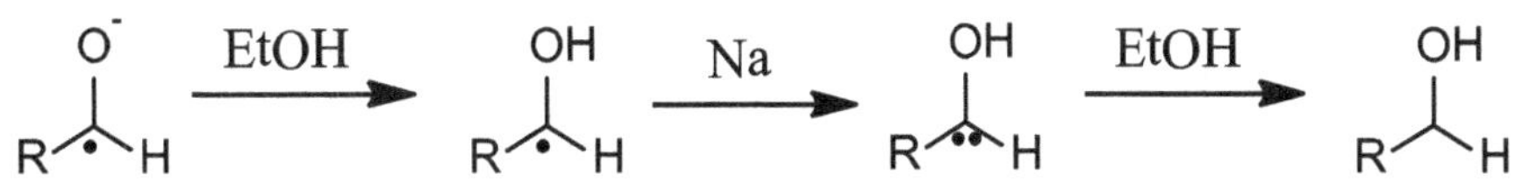

Scheme 131

In aprotic solvents, such as benzene or ether, no protons are available so the concentration of ketyl radical builds up significantly and the ketyl radical anion starts to dimerize. The key to success is to use a metal such as magnesium or aluminium, which forms strong covalent metal-oxygen bonds and can coordinate to more than one ketyl at once. Once two ketyls are coordinated to the same metal atom, they react rapidly. The salt on treatment with acid gives 1,2-diol which is popularly known as pinacol (Scheme 132).

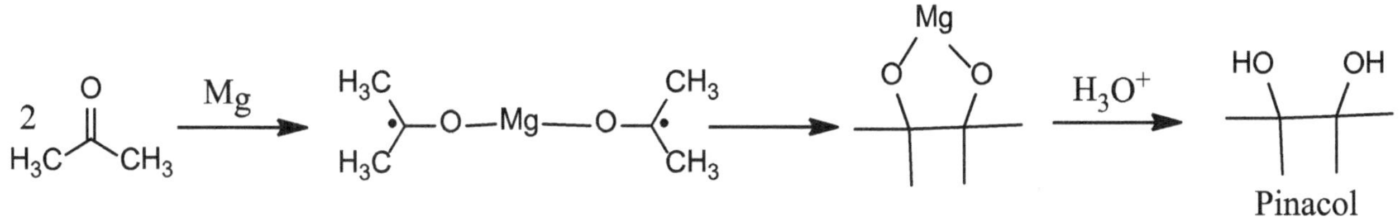

Scheme 132

McMurry [1989TL1173] developed a reduction procedure that is more reliable than the Mg pinacol conditions using the much more oxophilic metal titanium. The reagent, thought to be a mixture of Ti(0) and Ti(II) species, is usually formed by reduction of $TiCl_3$ with Zn/Cu, although other reluctant such as $LiAlH_4$, potassium-graphite (C_8K), Zn, and Mg can be used. The reagent is most useful for intramolecular reductions to form cyclic glycols and alkenes, but can also perform high-yield cross-coupling of aldehydes and ketones if one reactant can be present in excess.

Tetrahdrofuran (THF) is an important organic solvent in which many low-temperature inert atmosphere reactions are conducted. It has a drawback, however, it is quite hygroscopic and often the reactions for which it is used as a solvent must be kept absolutely free of water, it is therefore always distilled immediately before use from sodium metal, which reacts with any traces of water in the THF. However, it is necessary to have an indicator to show that the THF is dry that the sodium has done its job. The indicator used is a ketone, benzophenone.

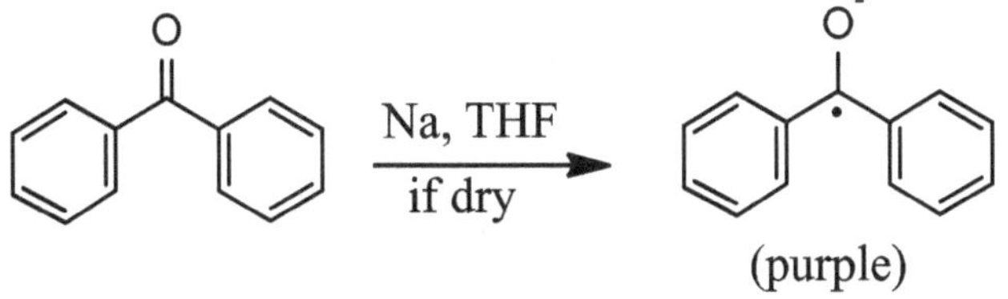

(Equation 66)

(purple)

When the THF is dry, the distilling liquid containing the benzophenone becomes purple colour. This colour is due to the ketyl of benzophenone. It should also come as no surprise that this ketyl being stabilized by conjugation and quite hindered, is persistent (long lived), it does not undergo pinacol dimerization. If water is present, the ketyl is rapidly quenched in the manner of the reduction to give the colourless alkoxde anion. [If ether is used as solvent, then it gives pale blue colour].

The key step for the synthesis of anticancer compound, taxol involves the intramolecular pincol reaction of the compound containing two carbonyl groups using titanium (produced during the reaction of $TiCl_3$ using a zinc-copper mixture) as the source of electron (Scheme 133).

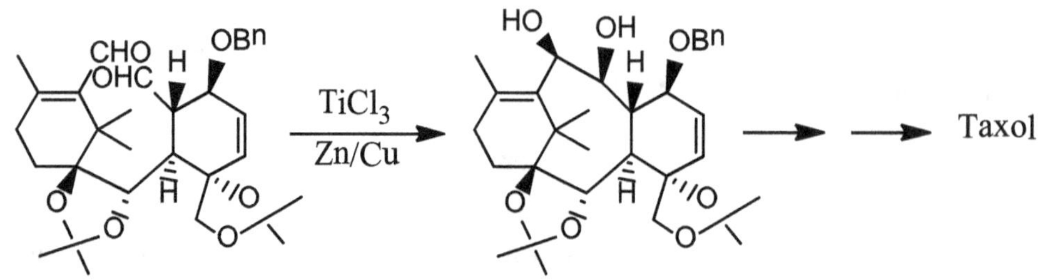

Scheme 133

Phosphines have also been found to reduce activated carbonyl groups for instance the reduction of an α-keto ester to an α-hydroxy ester in scheme 134 [2006CC1218]. In the proposed reaction mechanism, the first proton is on loan from the methyl group in trimethylphosphine (triphenylphosphine does not react).

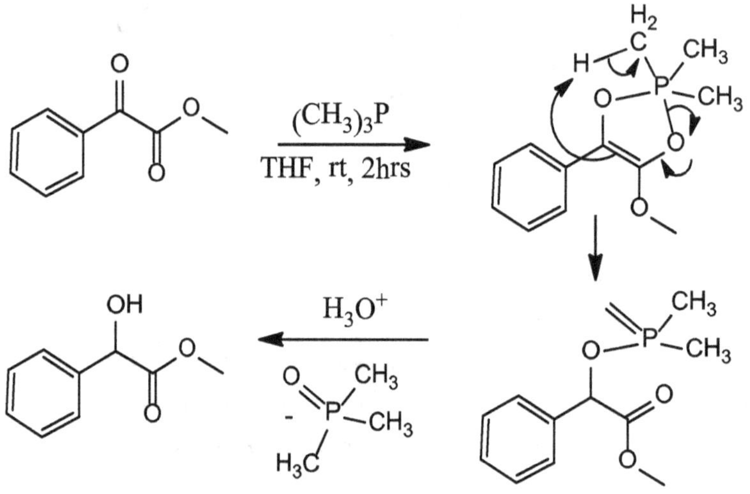

Scheme 134

3.1.8.1   Acyloin condensation

When an ester is refluxed with sodium in an inert solvent such as ether, benzene, toluene, xylenes etc. is in the absence of any free alcohol and then the product of reflux is treated with an acid or alcohol, a dimeric α-hydroxyketone, called acyloin is obtained and the reaction is called acyloin synthesis or condensation (Scheme 135).

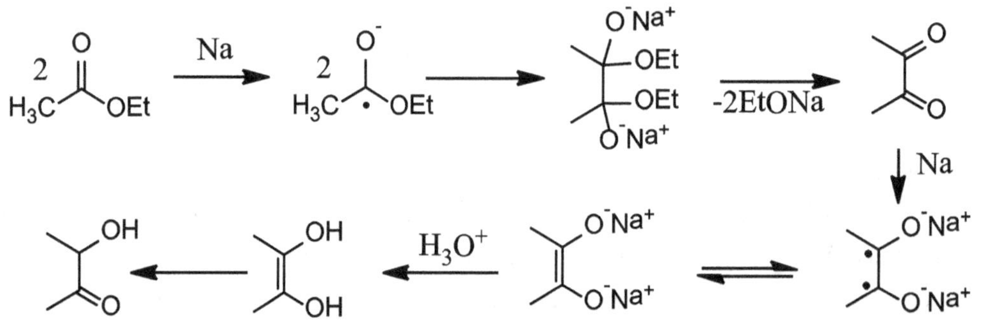

Scheme 135

Probable mechanism for the formation of acyloin involves the formation of ketyl radical anion from ketones as in the case of ketone followed by dimerization. The product of the dimerization looks very much like tetrahedral intermediate in a carbonyl addition-elimination reaction and it collapses to give a diketone. 1,2-Diketones are more reactive towards electrophiles and reducing agents than ketones because their π^* is lower in energy and straight away two electrons transfer takes place to form a molecule called enediolate. On quenching the reaction with acid, this dianion is protonated twice to give the enol of a α-hydroxyketone that is the final product of the acyloin reaction.

It is a good means for preparing long chain acyloins containing 18 to 20 carbon atoms. Ten to 20 membered cyclic acyloins have been prepared from diesters of dicarboxylic acid using high dilution technique. The reaction is best done in the presence of trimethylsilyl chloride, which traps the intermediate enolates. The reaction is especially useful for the formation of rings.

For instance,

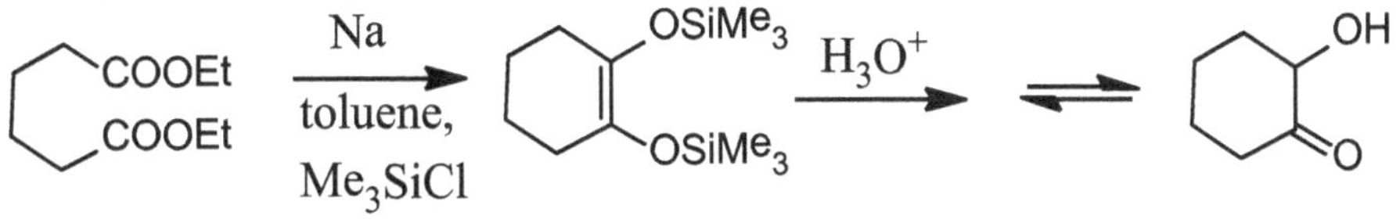

Scheme 136

3.1.8.2 McMurry reaction

The McMurry reaction is an organic reaction in which two ketone or aldehyde groups are coupled to an alkene using titanium chloride compound such as titanium(III) chloride and a reducing agent (viz., potassium, zinc, and magnesium [1974JACS4708]. This reaction is related to the pinacol coupling reaction which also proceeds by reductive coupling of carbonyl compounds. This reductive coupling can be viewed as involving two steps. First is the formation of a pinacolate (1,2-diolate) complex, a step which is equivalent to the pinacol coupling reaction. The second step is the deoxygenation of the pinacolate which yields the alkene, this second step exploits the oxophilicity of titanium (Scheme 137).

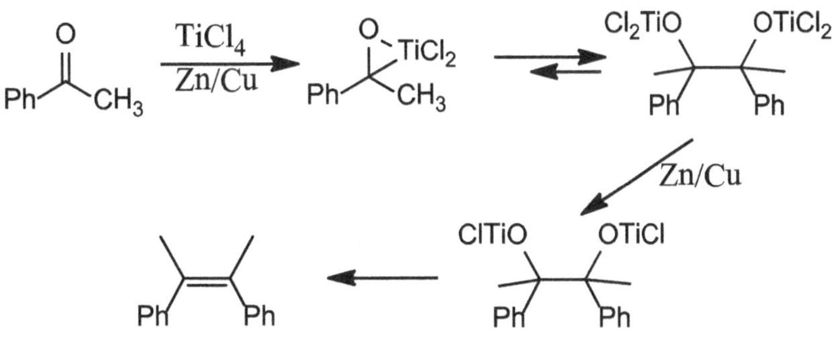

Scheme 137

Low-valent titanium species induce coupling of the carbonyls by single electron transfer to the carbonyl groups. The required low-valent titanium species are generated via reduction, usually with zinc powder. This reaction is often performed in THF because it solubilizes intermediate complexes, facilitates the electron transfer steps, and is not reduced under the reaction conditions. The nature of low-valent titanium species formed is varied as the products formed by reduction of the precursor titanium halide complex will naturally depend upon both the solvent (most commonly THF or DME) and the reducing agent employed: typically, lithium aluminum hydride, zinc-copper couple, zinc dust, magnesium-mercury amalgam, magnesium, or alkali metals.

3.1.9 Trimethyl silane as reducing agent

Reductions with hydrosilanes are chemical reactions that involve the combination of an organosilane (R_3SiH) with an organic substrate containing unsaturated or electron-withdrawing functionality. Products in which the electron-withdrawing group has been replaced by hydrogen or the unsaturated group has been hydrogenated result.

Silicon (1.90) is less electronegative than hydrogen (2.20); as a result, the silicon-hydrogen bond possesses some hydridic character. In the presence of a strong electrophile, organosilanes containing an Si-H bond (hydrosilanes) can serve as hydride donors to highly electrophilic organic substrates. Alcohols, alkyl halides, acetals, orthoesters, alkenes, aldehydes, ketones, and carboxylic acid derivatives may be reduced in good yield using hydrosilanes in conjunction with either a Brønsted or Lewis acid or an activating nucleophile. Because only reactive electrophiles undergo reduction, selectivity is possible in reactions of substrates with multiple reducible functional groups. Chiral Lewis acids and metal complexes may be used for the enantioselective reduction of ketones with hydrosilanes.

Hydrosilanes are not intrinsically nucleophilic; thus, they react only with highly electrophilic functional groups that have carbocationic character. However, many resonance-stabilized carbocations are insufficiently electrophilic to react with hydrosilanes. Upon the generation of a carbocation, rate-determining hydride transfer from the organosilane occurs to yield a reduced product and a neutral silane in which the counterion of the carbocation has replaced hydrogen. Prior to hydride transfer, the carbocation intermediate is liable to undergo Wagner-Meerwein rearrangements.

Trimethylsilane in aqueous sulfuric acid or hydrochloric acid give quantitatively yields of alcohols from ketones and aldehydes at room temperature while in presence of alcohols give high yields of ethers (Scheme 138).

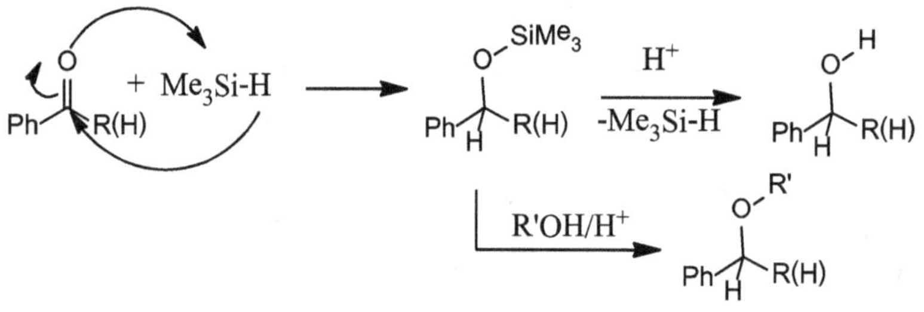

Scheme 138

Reduction of carbonyls to methylene in trifuroacetic acid with hydrosilanes (excess) is superior to the Clemmenson and Wolf-Kishner reactions and the reaction of acid chlorides with hydrosilanes presents greater ease of handling compared with the Rosenmund reaction (Scheme 139).

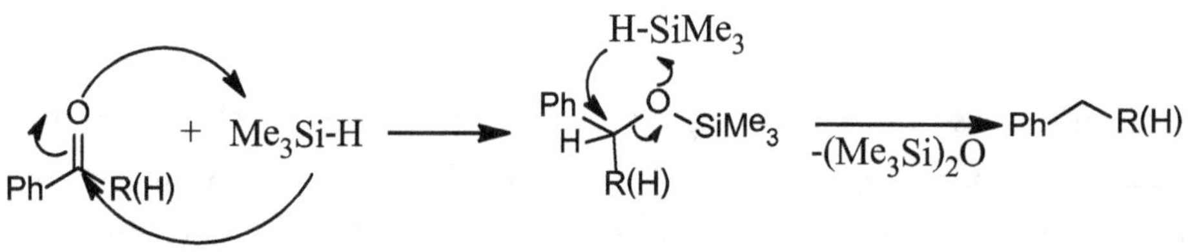

Scheme 139

Aromatic carboxylic acids are reduced to methyl compounds by reacting with trimethyl silane in acetonitrile (Scheme 140).

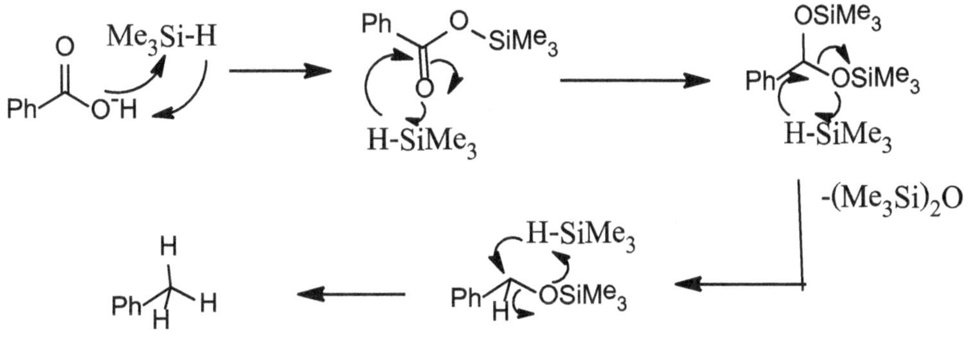

Scheme 140

Lactones are reduced to tetrahdrofurans by reacting with trimethyl silane in acetonitrile (Scheme 140a).

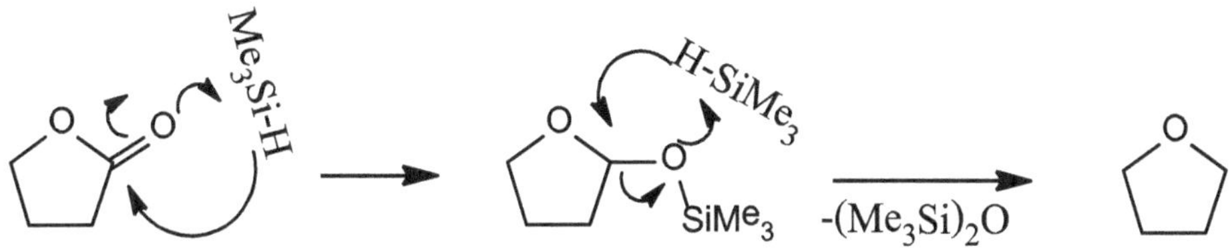

Scheme 140a

Acid chlorides can be reduced to aldehydes at room temperature by trimethyl silane in the presence of palladium catalyst in yields of 50-70%. Compared to the use of Li(t-BuO)$_3$AlH yields are higher from aliphatic substrates and lower with aromatic acid chlorides. The use of trimethylsilane is advantageous when considering ease of operation although yields are a little less than those using the Rosenmund reduction.

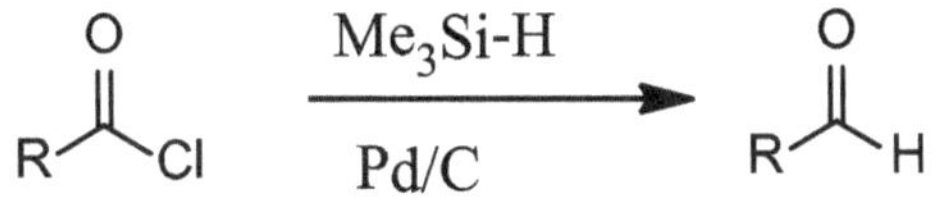

(Equation 67)

3.1.10 Samarium(II) iodide as reducing agent

Samarium diiodide is prepared by the oxidation of samarium metal with diiodomethane or with iodine (Equation 68) [1987CL501]. Deep blue solution of SmI2 (0.1M in THF) are generated in quantative yields by these process and can be stored for long periods. For synthetic purposes, SmI2 is typically generated and utilized in situ.

$$2\ Sm\ +\ 2\ ICH_2I \xrightarrow[0°C,\ 1hr]{THF} SmI_2\ +\ H_2C{=}CH_2 \qquad \text{(Equation 67)}$$

The mechanism of reductions of aldehydes and ketones by samarium iodide is based primarily on mechanisms elucidated for similar one-electron reducing agents [1983ACR399]. Upon single-electron transfer, a ketyl dimer **43** forms. In the absence of protic solvent, this dimer collapses to form 1,2-diols. In the presence of a proton source, however, the dimer may undergo either disproportionation to form a samarium alkoxide and carbonyl compound, or protonation to form a carbinol radical **42** followed by a second reduction and protonation, yielding an alcohol.

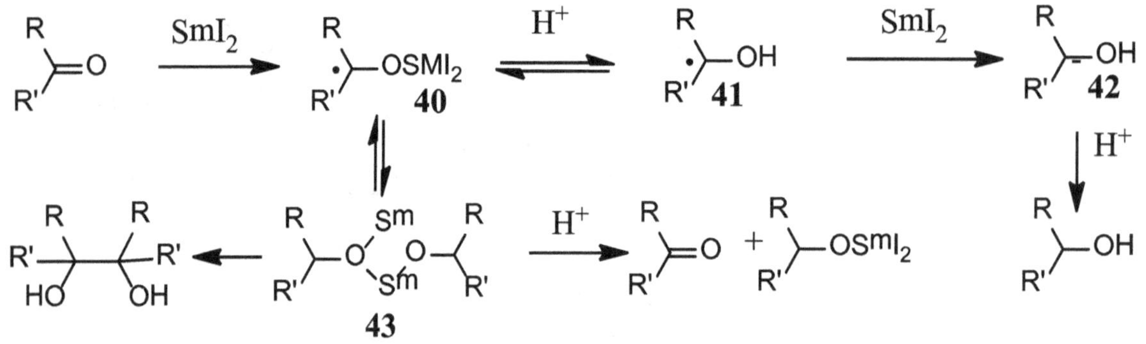

Scheme 141

α-Functionalized carbonyl compounds are reduced to the corresponding unfunctionalized carbonyl compounds in the presence of samarium iodide. This process may be initiated by initial electron transfer to either the substituent at the α position or the carbonyl moiety, depending on the relative electron affinity of the functional groups. A second reduction immediately follows, after which either protonation or elimination-tautomerization affords the product (Scheme 142) [1986JOC1135].

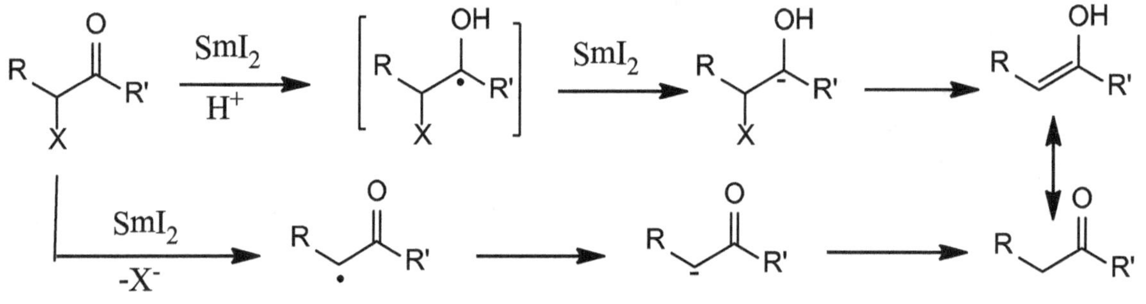

Scheme 142

Isolated cyclopentanols can be synthesized with considerable diastereodelectivity when appropriately substituted ω-iodoalkyl ketones are treated with SmI$_2$ in THF at -78°C and allowed to warm to room temperature (Equation 68) [1987SC901].

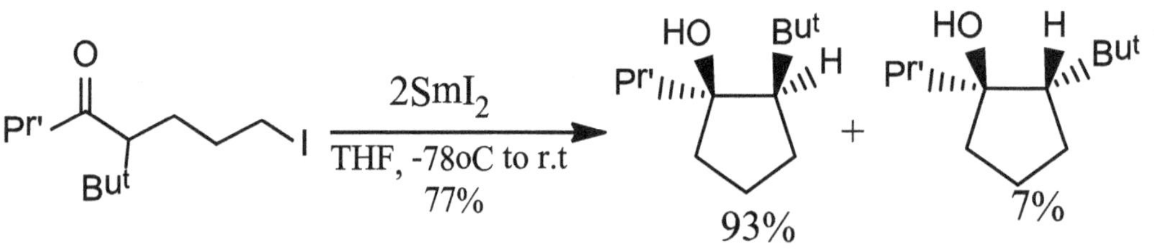

(Equation 68)

Treatment of an iodomethylsamarium alkoxide (generated in situ by reaction of aldehydes or ketones with SmI$_2$/CH$_2$I$_2$) with SmI2/HMPA and N,N-dimethylaminoethanol (DMAE) induces a reductive elimination, resulting in the generation of the corresponding methylated material (Scheme 143) [1987CL2101].

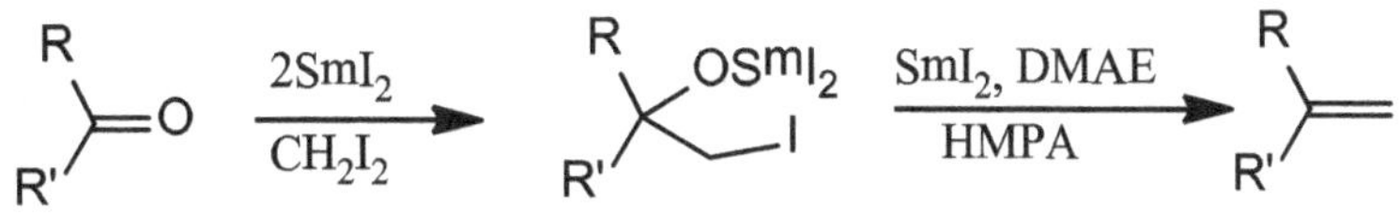

Scheme 143

3.2   OXIDATION OF CARBONYL COMPOUNDS

Aldehydes, RCHO, can be oxidized to carboxylic acids, RCO_2H by chromic acid, permanganate etc. The oxidation with dichromate is believed to proceed through the hydrate by a mechanism similar to that of primary and secondary alcohol (Scheme 144).

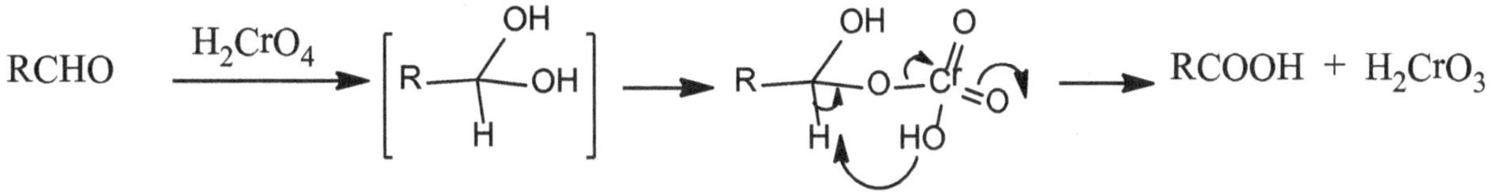

Scheme 144

Ketones are not oxidized under these conditions as they lack the critical H for the elimination to occur. Ketones are resistant to the conditions typically used for the oxidation of alcoholic and aldehydes provided that the conditions are not too severe. Thus, acetone can be employed as an inert solvent for the chromium(VI) oxidation of alcohols to aldehydes.

Strong oxidizing agents, e.g. acid dichromate, permanganate, nitric acid etc., oxidize keones, only carbon atom adjacent to the carbonyl group being attacked and the carbon atom joined to the smaller number of hydrogen atoms is preferentially oxidized.

$$CH_3COCH_2CH_3 \xrightarrow{(O)} 2CH_3COOH$$

Probable mechanism for the oxidation is:

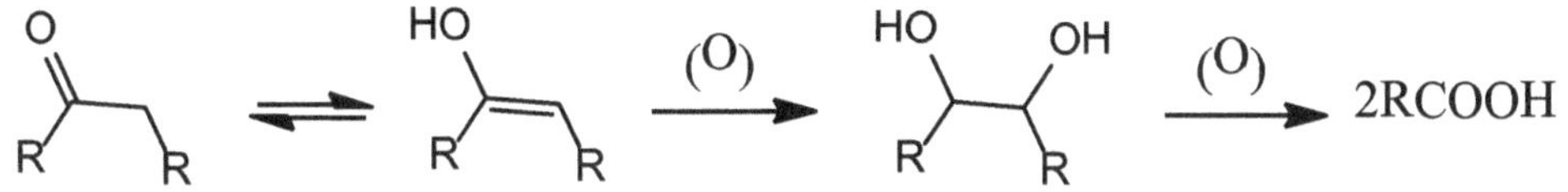

Scheme 145

When ketones are treated with nitrous acid, the half oxime of the α-dicarbonyl compound is formed, e.g. acetone gives oxaminoacetone (Equation 69).

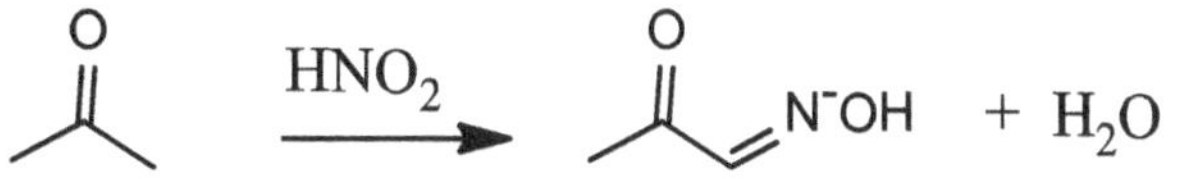

(Equation 69)

All compounds containing the –CH_2CO- group form the oximino derivatives with nitrous acid and this has been used to detect the presence of –CH_2CO- group. The mechanism of this reaction is uncertain. A possibility is via attack by the nitrosonium ion to the enol to form nitroso compound which then rearranges to the oxime (Scheme 146).

$$HO\text{-}NO \ + \ H^+ \ \rightleftharpoons \ H_2\overset{+}{O}\text{-}NO \ \rightleftharpoons \ H_2O \ + \ \overset{+}{N}O$$

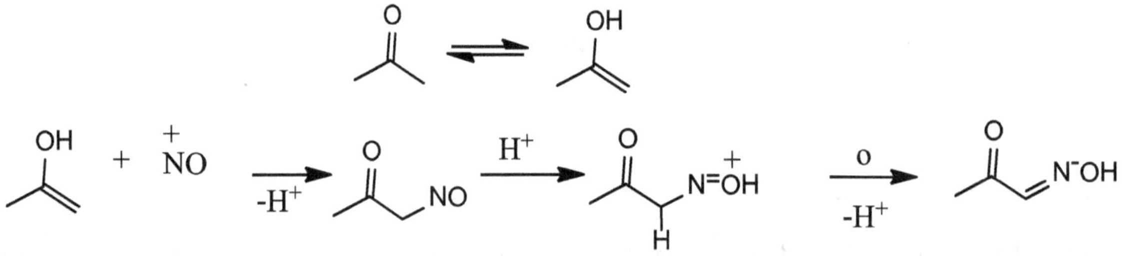

Scheme 146

Nitroso compounds exist as such only when the nitroso group is attached to a tertiary carbon atom. If the nitroso group is attached to a primary or secondary carbon atom, the nitroso compound is generally unstable, because it can tautomerize with the transfer of a proton from carbon to the oxygen of the nitroso group. This process is exactly like enolization but with N=O in the place of C=O group. It gives an oxime as the stable "enol" which can form an intramolecular hydrogen bond with the ketone carbonyl group. Hydrolysis of the oxime forms the second ketone (Scheme 147).

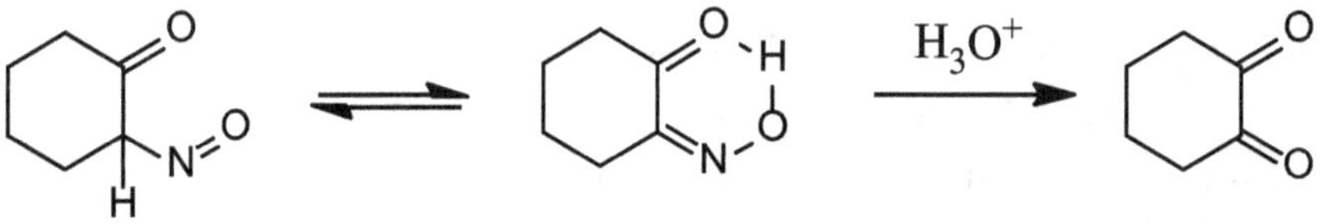

blue colour (unstable)   colourles (stable)
                                      solid

Scheme 147

If the ketone is unsymmetrical, this reaction will occur on the more substituted side, for the same reason that acid catalyzed enol bromination gives the more substituted α-bromocarbonyl compound.

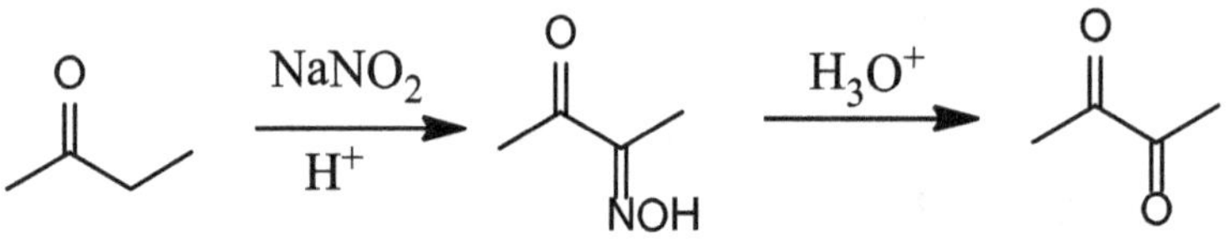

Scheme 148

This reaction is very useful in predicting the reactive methylene group in the unknown molecule possessing carbonyl group. For instance, the nitrosation of camphor gave positive test with the formation of α-keto oxime clearly indicates that in camphor adjacent to C=O, CH_2 group is present (Equation 70).

(Equation 70)

Aldehyde and ketones with a methyl or methylene group adjacent to the carbonyl group are oxidized by selenium dioxide at room temperature to dicarbonyl group.

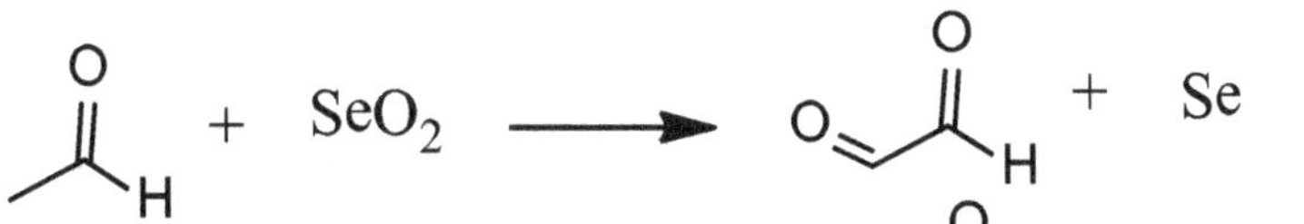

$$\text{(Equation 71)}$$

$$\text{(Equation 72)}$$

The reaction is usually carried out in acetic acid and the actual reagent is selenous acid as shown in the scheme 149.

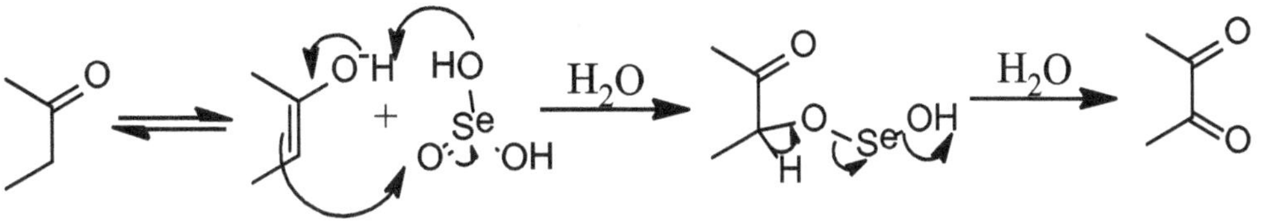

Scheme 149

3.2.1 Fehling's test

Fehling's solution is prepared by combining two separate solutions, known as Fehling's A and Fehling's B. Fehling's A is aqueous solution of copper(II) sulfate, which is deep blue. Fehling's B is a colorless solution of aqueous potassium sodium tartrate (also known as Rochelle salt) made in a strong alkali, commonly with sodium hydroxide. Typically, the L-tartrate salt is used. The copper(II) complex (**44**) in Fehling's solution is an oxidizing agent and the active reagent in the test [1849Misc106]. The deep blue active ingredient in Fehling's solution is the bis(tartrate) complex of Cu^{2+}. The tartrate tetraanions serve as bidentate alkoxide ligands [2016EJIC1798].

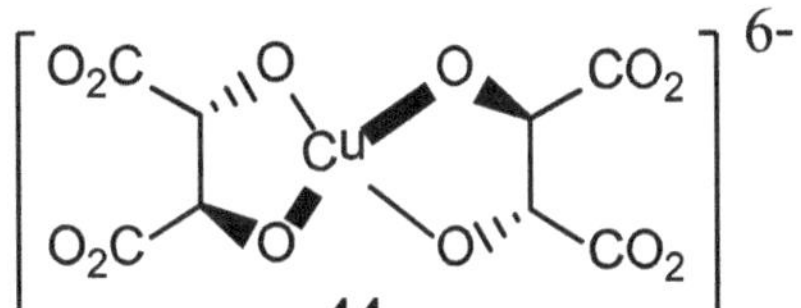

44

Fehling's can be used to distinguish aldehyde vs. ketone functional groups. The compound to be tested is added to the Fehling's solution and the mixture is heated. Aldehydes are oxidized, giving a positive result, but ketones do not react, unless they are alpha-hydroxy-ketones. The bistartratocuprate(II) complex oxidizes the aldehyde to a carboxylate anion, and in the process the copper(II) ions of the complex are reduced to copper(I) ions. Red copper(I) oxide then precipitates out of the reaction mixture, which indicates a positive result i.e. that redox has taken place (this is the same positive result as with Benedict's solution). A negative result is the absence of the red precipitate; it is important to note that Fehling's will not work with aromatic aldehydes; in this case Tollens' reagent should be used.

$$\text{+ 2CuO + 5 }^-\text{OH} \longrightarrow \text{+ Cu}_2\text{O + 3H}_2\text{O} \quad \text{(Equation 73)}$$

3.2.2 Tollen's reaction

Tollens' reagent is a chemical reagent used to determine the presence of an aldehyde, aromatic aldehyde and α-hydroxy ketone functional groups. The reagent consists of a solution of silver nitrate and ammonia. It was named after its discoverer, the German chemist Bernhard Tollen's [1882Misc1635]. A positive test with Tollens' reagent is indicated by the precipitation of elemental silver, often producing a characteristic "silver mirror" on the inner surface of the reaction vessel.

This reagent is not commercially available due to its short shelf life, so it must be freshly prepared in the laboratory. One common preparation involves two steps. First a few drops of dilute sodium hydroxide are added to some aqueous silver nitrate. The OH^- ions convert the silver aqua complex form into silver oxide, Ag_2O, which precipitate from the solution as a brown solid:

$$2\,AgNO_3 + 2\,NaOH \rightarrow Ag_2O\,(s) + 2\,NaNO_3 + H_2O$$

In the next step, sufficient aqueous ammonia is added to dissolve the brown silver(I) oxide. The resulting solution contains the $[Ag(NH_3)_2]^+$ complexes in the mixture, which is the main component of Tollen's reagent.

$$Ag_2O\,(s) + 4\,NH_3 + 2\,NaNO_3 + H_2O \rightarrow 2\,[Ag(NH_3)_2]NO_3 + 2\,NaOH$$

Represenatative examples of alcohols which have been prepared by this type of yeast reduction are shown below and all conform to the Prelog'rule [1988JOC5215].

The carbonyl group is oxidized in the process and the Ag+ in AgX is reduced. The resultant oxidized aldehyde (now a radical cation) reacts with hydroxide to form a tetrahedral intermediate (Scheme 150). A gem-diol like intermediate is formed via a hydrogen shift, which then continues on to the final carboxylate anion.

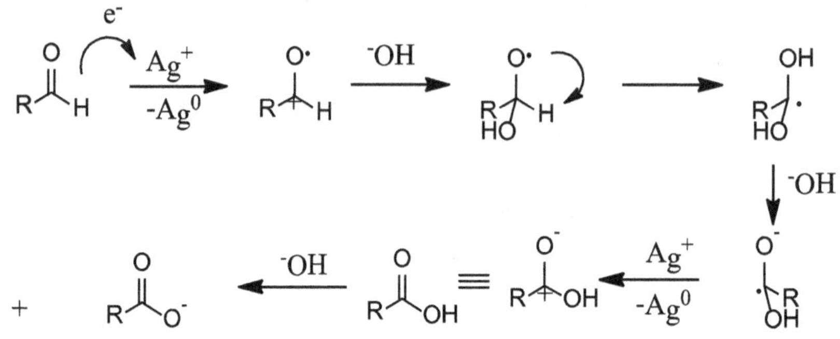

Scheme 150

3.2.3 Baeyer-Villiger Reaction

The Baeyer-Villiger Oxidation is the oxidative cleavage of a carbon-carbon bond adjacent to a carbonyl, which converts ketones to esters and cyclic ketones to lactones. The Baeyer-Villiger can be carried out with peracids, such as MCBPA, or with hydrogen peroxide and a Lewis acid.

The regiospecificity of the reaction depends on the relative migratory ability of the substituents attached to the carbonyl. Substituents which are able to stabilize a positive charge migrate more readily, so that the order of preference is: tert. alkyl > cyclohexyl > sec. alkyl > phenyl > prim. alkyl > CH_3. In some cases, stereoelectronic or ring strain factors also affect the regiochemical outcome.

In the first steps of the reaction mechanism, the peroxy acid protonates the oxygen of the carbonyl group. This makes the carbonyl group more susceptible to attack by the peroxy acid. In the next step of the reaction mechanism, the peroxy acid attacks the carbon of the carbonyl group forming what is known as the Criegee intermediate. Through a concerted mechanism, one of the substituents on the ketone migrates to the oxygen of the peroxide group while a carboxylic acid leaves. This migration step is thought to be the rate determining step. Finally, deprotonation of the oxygen of the carbonyl group produces the ester (Scheme 151).

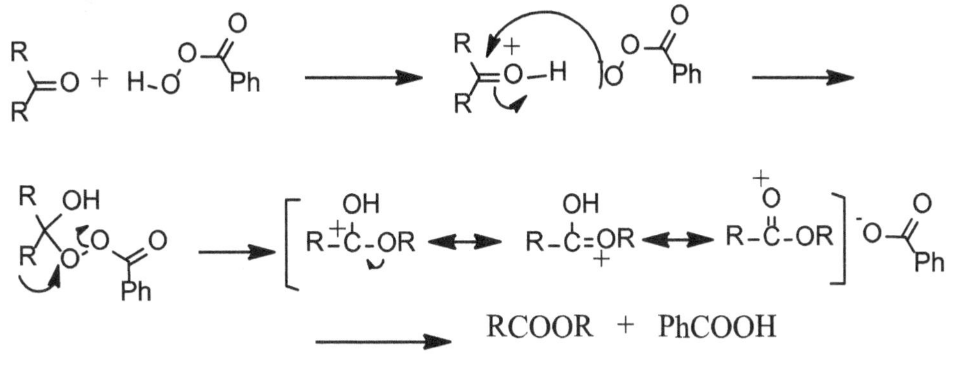

Scheme 151

The products of the Baeyer-Villiger oxidation are believed to be controlled through both primary and secondary stereoelectronic effects. The primary stereoelectronic effect in the Baeyer-Villiger oxidation refers to the necessity of the oxygen-oxygen bond in the peroxide group to be antiperiplanar to the group that migrates. This orientation facilitates optimum overlap of the σ orbital of the migrating group to the σ^* orbital of the peroxide group. The secondary stereoelectronic effect refers to the necessity of the lone pair on the oxygen of the hydroxyl group to be antiperiplanar to the migrating group. This allows for optimum overlap of the oxygen nonbonding orbital with the σ^* orbital of the migrating group. This migration step is also (at least in silico) assisted by two or three peroxyacid units enabling the hydroxyl proton to shuttle to its new position.

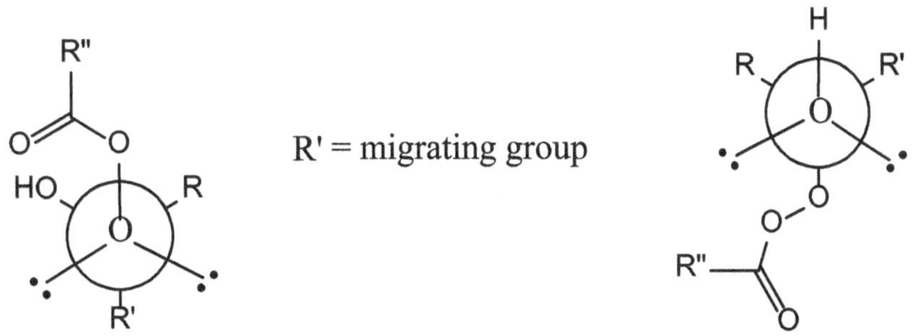

R' = migrating group

Primary stereoelectronic effect   Secondary stereoelectronic effect

The migratory ability is ranked tertiary > secondary > phenyl > primary. Allylic groups also migrate better than primary groups but not as well as secondary groups. If there is an electron withdrawing group on the substituent, then it decreases the rate of migration. There are two explanations for this trend in migration ability. One explanation relies on the carbocation resonance structure of the Criegee intermediate. Keeping this structure in mind, it makes sense that the substituent that can maintain positive charge the best would be most likely to migrate. Tertiary groups are more stable carbocations than secondary groups, and secondary groups are more stable than primary. Therefore, the tertiary > secondary > primary trend is observed.

Another explanation uses stereoelectronic effects and steric bulk to explain the trend. As mentioned, the substituent that is antiperiplanar to the peroxide group in the transition state will be the group that migrates. This transition state has a gauche interaction between the peroxyacid and the non-migrating substituent. If the bulkier group is placed antiperiplanar to the peroxide group, the gauche interaction between the substituent on the forming ester and the carbonyl group of the peroxyacid will be reduced. Thus, it is the bulkier group that ends up antiperiplanar to the peroxide group making it the group that migrates (Scheme 152). This explains the trend of tertiary > secondary > primary because tertiary groups are generally bulkier than secondary and primary groups.

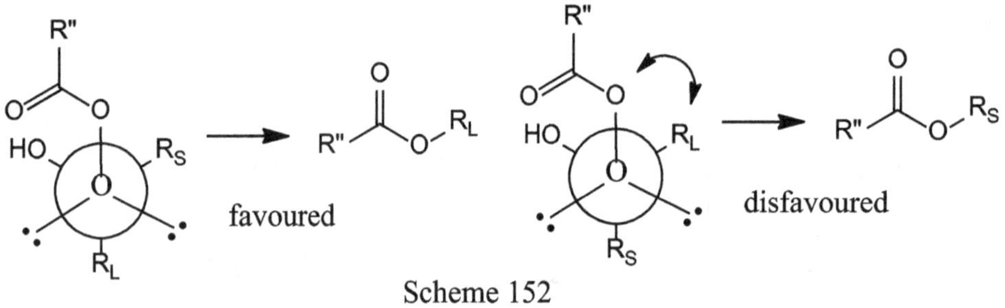

Scheme 152

3.2.4   Oxidative esterification of aldehydes

Aldehydes in the prescenc of methanol undergo oxidative transformation to corresponding ester under treatment with catalytic amounts of vanadium pentoxide in combination with oxidant hydrogen peroxide [2000OL577]. It is most probable that aldehydes under acidic condition react with alcohol to form hemiacetal (**45**). Initially formed hemiacetal react with peracid formed by the reaction of V_2O_5 and H_2O_2 to form a vanadium hemiacetal type intermediate (**46**) (Scheme 153). The conjugate base of the peracid **46** is an excellent leaving group for nucleophilic displacement. Subsequent elimination produced the desired product and releases the catalyst.

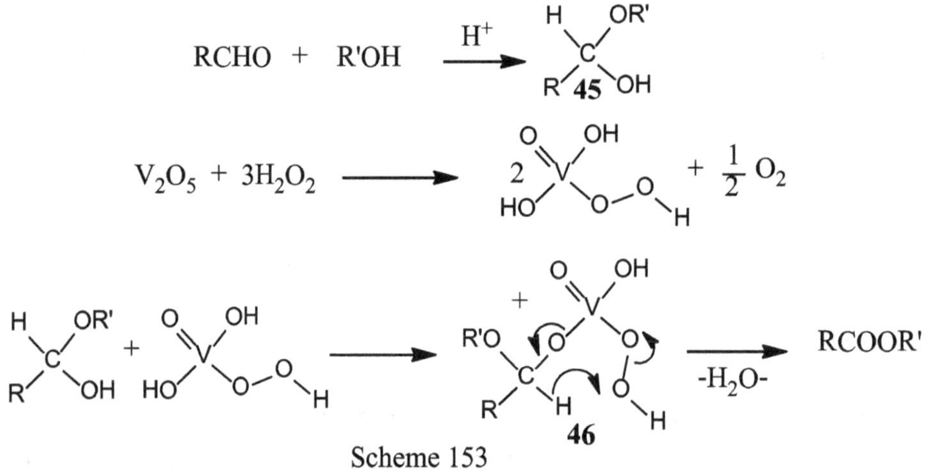

Scheme 153

Mori and Togo reported that oxidation of a mixture of unbranched aliphatic aldehydes and primary alcohols in the presence of stoichiometric amounts of iodine affords the corresponding esters in up to 91% yield (Scheme 154)[2005TL5915], The reaction proceeds at room temperature and involves oxidation of an intermediate hemiacetal hypoiodite **47**.

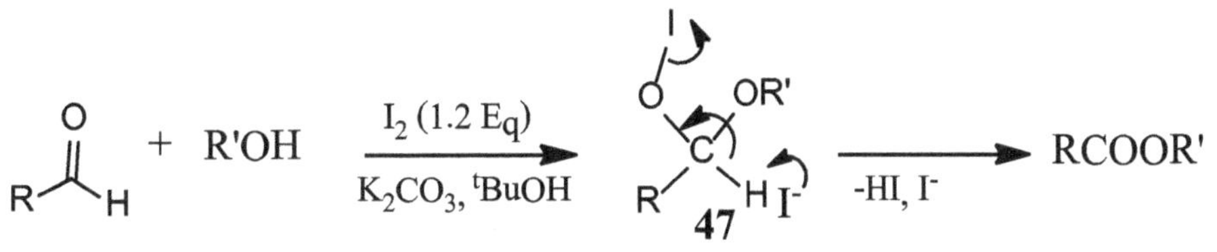

Scheme 154

Evans and others [1990JACS6447] have utilized the Tishchenko reaction for diastereoselective samarium-catalyzed preparation of anti-1,3-diols from b-hydroxy ketones and aldehydes (Scheme 155). The remarkable asymmetric induction was attributed to efficient chelation control in an activated samarium-derived complex **48** involving coordination of both the aldehyde and the hemiacetal oxygen atoms. The reaction furnishes a series of anti-1,3-diol monoesters that proved to be useful precursors of polyketide-derived natural products.

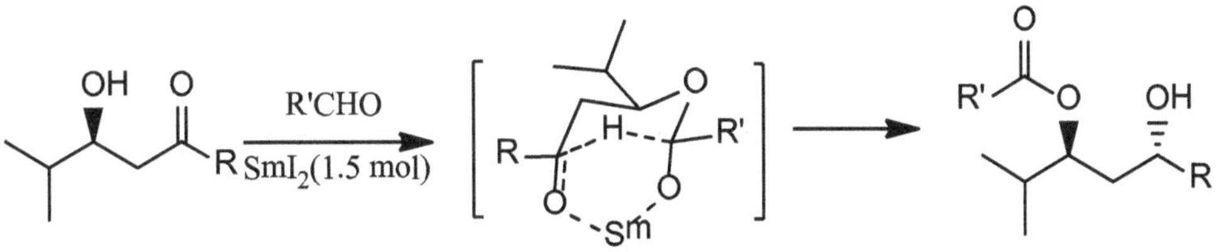

R = C_6H_{13}- ; R' = Me 96% anti/syn>99:1

R = C_6H_{13}- ; R' = iPr 95% anti/syn>99:1

R = C_6H_{13}- ; R' = Ph 94% anti/syn>99:1

R = iPr- ; R' = Ph 99% anti/syn>99:1

R = iPr- ; R' = Me 85% anti/syn>99:1

Scheme 155

3.2.5 Oxidative amidation of aldehydes

Wolf et al [2008EJC6302, 2007OL3429] reported a metal-free oxidative amidation of various aldehydes with secondary amines in the presence of tert-butyl hydroperoxide (TBHP). In analogy to the oxidative esterification mechanism involving formation of an intermediate hemiacetal discussed above, this reaction probably proceeds via a carbinolamine **48**, which is oxidized by TBHP (Scheme 156).

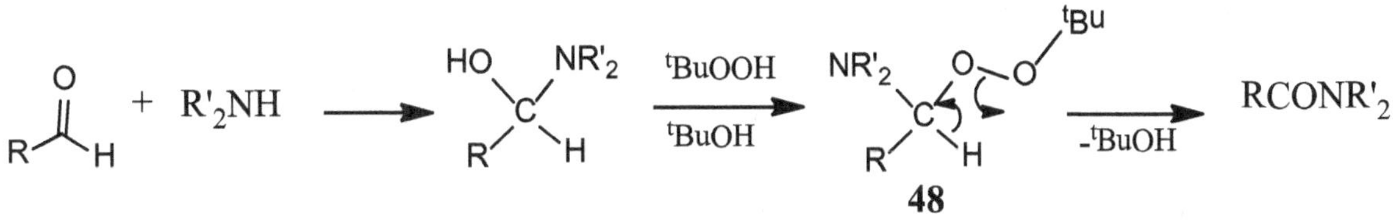

Scheme 156

EXERCISE

1. Write the steps involved in the reduction of carbonyl compounds to alcohols using $LiAlH_4$ as reducing agent

2. Predict the products for the following and explain with suitable mechanism. Cyclohexan-1,3-dione + $Br_2/NaOH \longrightarrow$?

3. Predict the product with suitable mechanism:

$$RCOR' \; + \; PhCOOOH \; \xrightarrow{\;\; H^+ \;\;} \; ?$$

4. Write the steps involved in the reduction of carbonyl compounds to alcohols using $NaBH_4$ as reducing agent.

5. Complete the following:

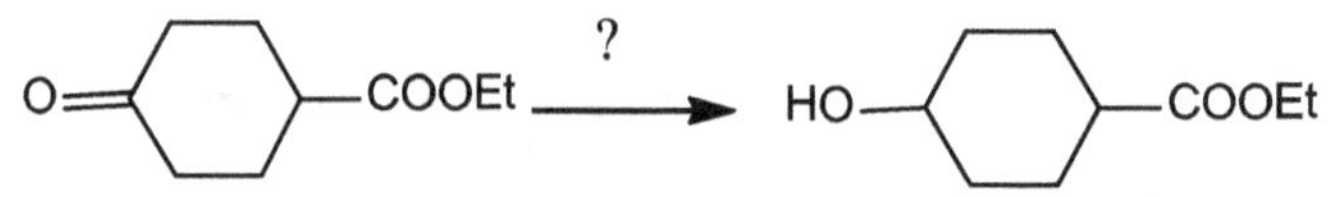

6. Discus the mechanism for the synthesis of pinacol from acetone using sodium in ethanol as reducing agent.

7. Mention the different functional groups that can be reduced by i. $LiAlH_4$ & ii. $NaBH_4$. Comapre their reactivity.

8. How Tollen's reagent oxidizes aldehydes.

9. Write a reasonable mechanism for the following conversion:

$$RCOCH_3 \; + \; Me_3SiH \; \longrightarrow \; RCH_2R$$

10. Formulate the steps involved in the reduction of carbonyl compounds to alkanes employing Clemenson reduction.

11. Formulate the steps involved in the reduction of carbonyl compounds to alkanes employing neutral reduction.

12. Predict the product with suitable mechanism:

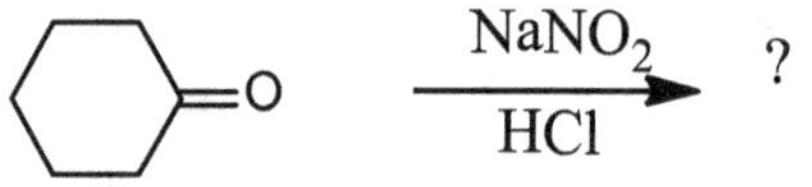

13. Write the major products for the following conversions
    i.   Cycohexanone + $LiAlH_4 \longrightarrow H_3O \longrightarrow$?

14. Write a reasonable mechanism for the following conversions:
    $RCOCH_2R' + SeO_2 \longrightarrow RCOCOR'$

15. Predict the product with suitable mechanism:

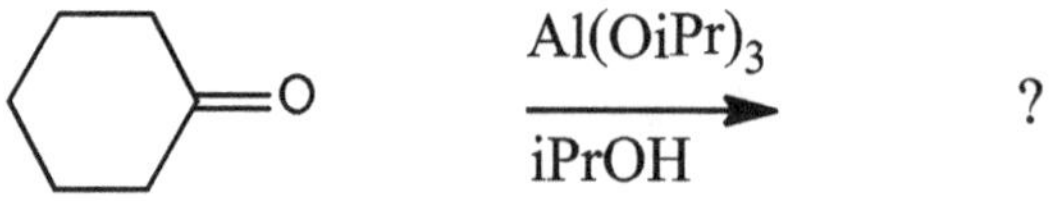

16. Explain why PCC is a better oxidizing agent than CrO_3 for conversion of a primary alcohol to an aldehyde?

REACTIONS INVOLVING THE SUBSTITUENT GROUPS

4.1   ENOLIZATION

The phenomenon of enolized is closely bound up with carbaranion chemistry. A carbonyl compounds (aldehyde or ketone, ester or even a carboxylic anhydride) which has at least one hydrogen atoms on an α-carbon may undergo an internal proton transfer which converts it into a tautomer, the enol form. The equilibrium between keto and enol tautomers usually lies heavily on the side of the keto but the reactivity of the enol is often far greater than that of the keto form and its formation as an intermediate is frequently necessary for reaction to occur. The conversion of keto to enol is both general acid & general base catalyzed. Intermediate in the tautomerization are respectively the protonated carbonyl compound and the enolate ion. The enolate ion is a resonance stabilized carbanion. Many of the moderately strong carbon acids dissociate to form enolate ions, e.g. dimethyl malonate.

Scheme 157

The equilibrium proportion of enol in pure sample of aldehydes or ketones in neutral solution is very small (and those esters even smaller). 1, 3-Dicarbonyl compounds on the other hand frequently contain a high proportion of enol which is stabilized by internal hydrogen bonding, e.g., in acetyl acetone.

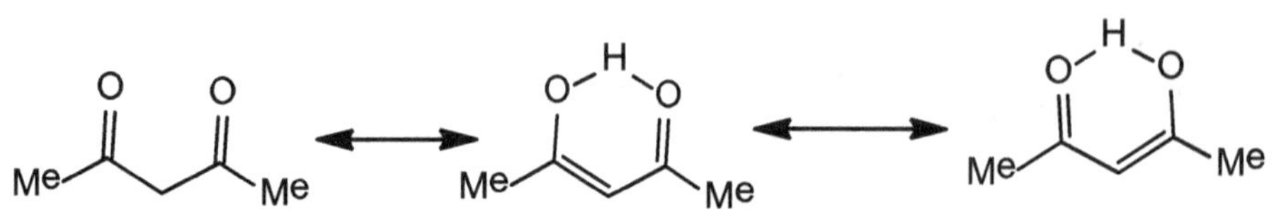

Compound	% enol	Compound	% enol
Acetone	2.5×10^{-4}	Cyclopentanone	4.8×10^{-3}
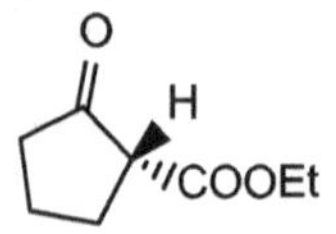	5.0	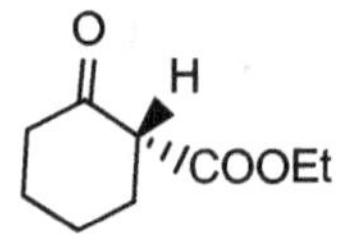	18.0
Ethyl acetoacetate	7.5	Acetyl acetone	78
			> 95

The determination of these ratios much makes use of IR spectroscopy if the tautomer is present in comparable amounts. Distinct carbonyl stretching frequencies are shown by the two forms. If the enol is present in very low concentrations, it may be determined chemically by a rapid bromine titration at low temperature. Only the enol form reacts and the titration may be carried out at before a significant shift of the equilibrium has occurred (Scheme 158).

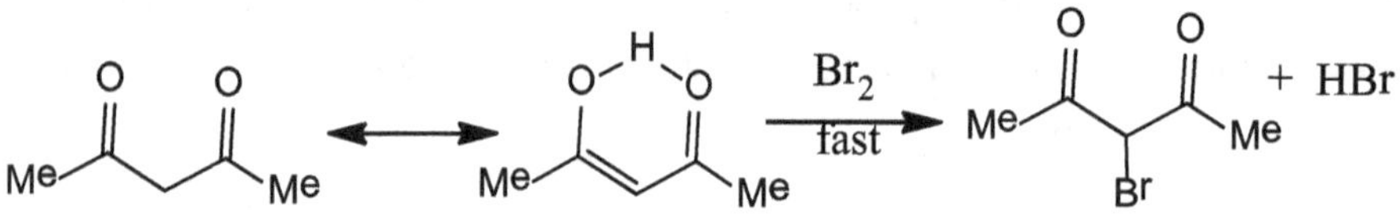

Scheme 158

It was first demonstrated by Lapworth (around 1912) that substitution reaction at carbon is proportional to a carbonyl group proceeded via the enol (or enolate). One of the interesting pieces of evidence was the identical rates measured for different substitution reaction such as bromination deuterium exchange, iodination, racemization, it was postulated that all those reactions required the same step, the conversion of keto to enol (Scheme 159). The reactions of optically active isobuterophenone illustrate these points.

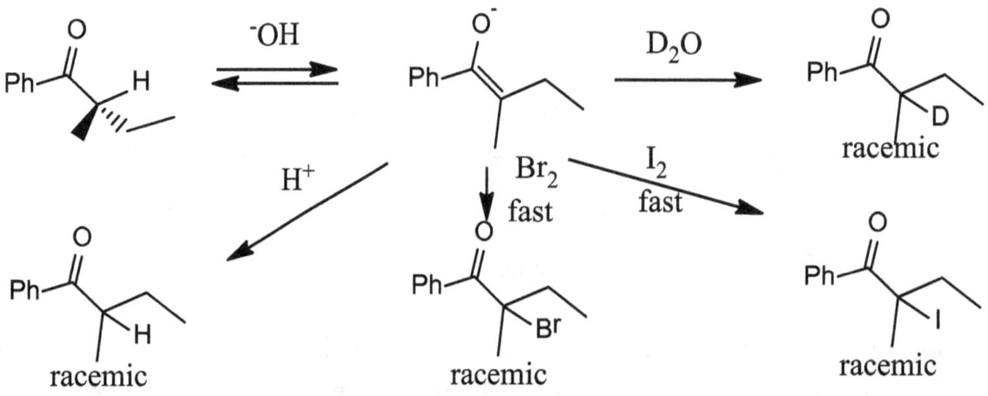

Scheme 159

The intermediate is likely to be the enol is acid-catalyzed reaction and the enolate ion under base-catalyzed conditions, both of which have a plan of symmetry. The rate determining loss of the α-proton is confirmed by the large primary isotopic effects observed. They are also finding in substitution of aliphatic nitro compounds for which a similar enolization mechanism occurs (Scheme 160).

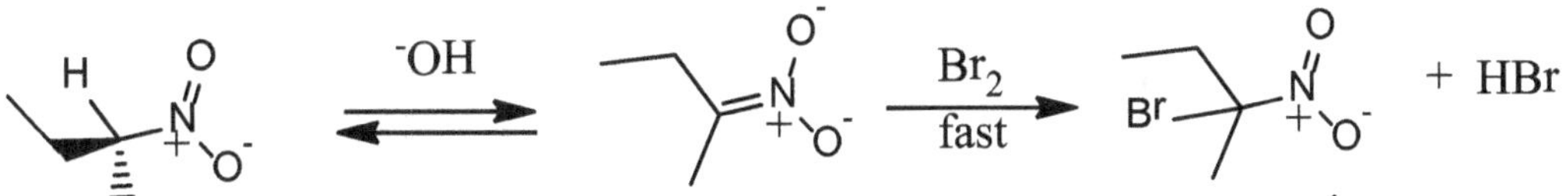

Scheme 160

Deuterium isotopic effects for reactions occurring via enolization were studied and observed that the rate determining step involves the breakage of C-H bond.

For racemization of [structure: Ph, H, p-Tol substituted acetic acid] k_H/k_D 7.7

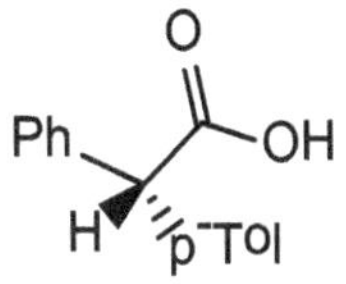

Acetone + Br_2 ⟶ $BrCH_2COCH_3$ 4.5-6.5

CH_3NO_2 + Br_2 ⟶ $BrCH_2NO_2$ 4.6

4.2 α-HALOGENATION OF CARBONYL COMPOUNDS

4.2.1 Halogenation of saturated aldehydes and ketones

Halogenation of saturated aldehydes and ketones usually occurs exclusively by replacement of hydrogen α to the carbonyl groups.

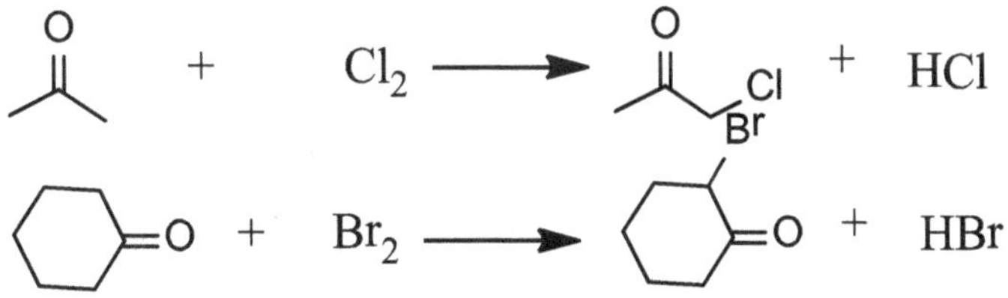

The characteristics of such reactions are very different from those of the halogenation of alkanes. An important feature of the reaction is that acetone reactions at the same rate with each halogen. Indeed, the rate of formation of halogenated acetone is independent of the concentration of halogens, even at quite at low halogen concentrations. Furthermore halogenation of acetone is catalyzed by both acids and bases. The rate expression for the formation of halogenated acetone in acid/base conditions are -

rate = k[CH3COCH3][-OH] at moderate concentrations of -OH.

rate = k'[CH3COCH3][H⁺] at moderate concentrations of H⁺.

The ratio of k to k' is 12,000, which means that hydroxide ion is much more effective catalyst than hydrogen ion.

To account for the role of the catalysts and the independence of the rate on the halogen concentration, the acetone must necessarily by slowly converted by the catalysts to an intermediate that reacts rapidly with halogen to the products. The intermediate is α-methyl vinyl alcohol, which is the unstable enol form of acetone (Scheme 161).

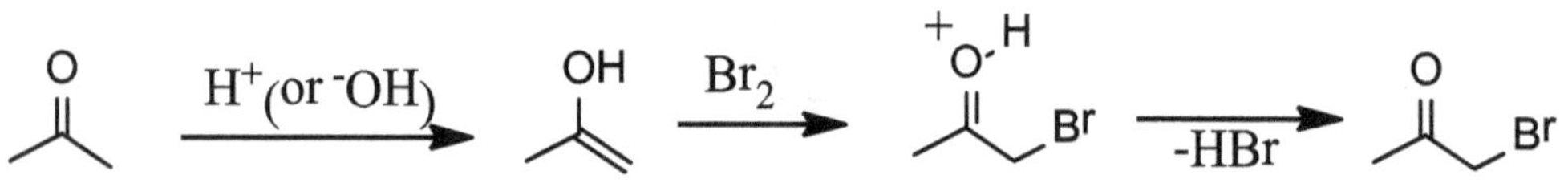

Scheme 161

As long as the first step is slow compared with the second and third steps, the rate will be independent of both the concentration of halogen and its nature, whether chlorine, bromine or iodine.

Base catalyzed bromination:

Normally, C-H bonds are highly resistant to attack by basic reagents, but removal of a hydrogen, α- to a carbonyl group, results in the formation of considerably stabilized anion with a substantial proportion of a negative charge on oxygen. As a result, hydrogen, α- to carbonyl groups have acidic character and can be removed as proton. In contrast to the dissociation of many weak acids, the acidic protons on carbon are removed slowly and equilibrium between the ketone and its enolate anion (i) is not established rapidly. This means of course that if the proton is removed slowly from carbon, the reverse reaction must also be slow. As a result, the enolate anion has ample time to add a proton to oxygen to form the enol.

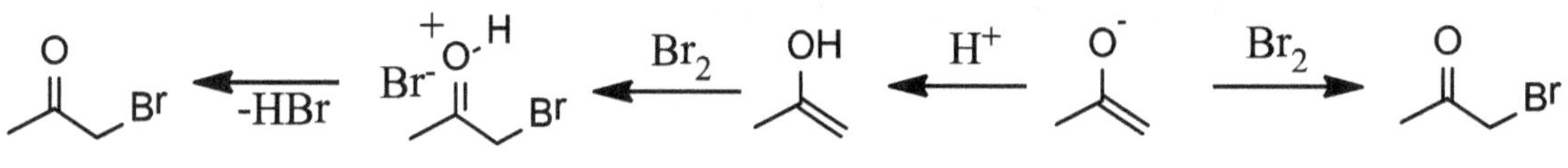

Scheme 162

Represenatative examples of alcohols which have been prepared by this type of yeast reduction are shown below and all conform to the Prelog'rule [1988JOC5215].

Either the enol or the enolate anion can combine rapidly with halogen to give the α-haloketone. The slowest step in the whole sequence is the formation of the enolate anion, and the overall rate is thus independent of the halogen concentration.

Acid catalyzed bromination:

In the acid catalyzed bromination of acetone, it is possible to isolate the mono, di and tribromo derivatives. Hence in contrast with base catalyzed reaction, the introduction of a second and third bromine atom is less rapid than the first.

4.2.2   Free radical halogenation

Halogenation of carbonyl compounds at the α-position may also be carried out under free-radical conditions. Since a free radical alpha to a carbonyl group, like an allylic or benzyllic free radical is resonance stabilized, bromination at this position is effected by the free radical brominating agent N-bromosuccinimide (NBS) (Eqaution 74).

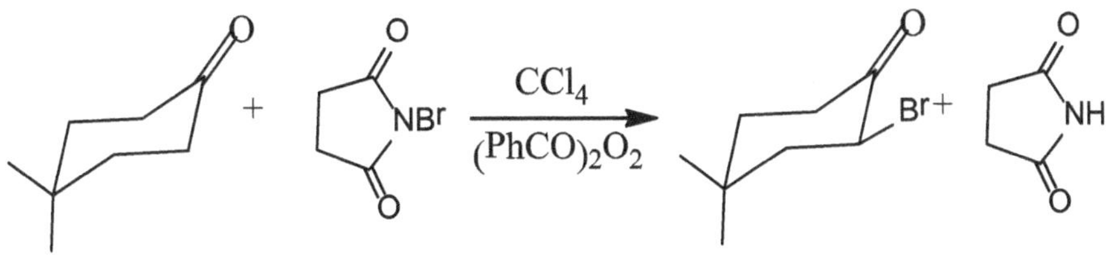

(Equation 74)

Sulfuryl chloride (SO_2Cl_2) is a free radical chlorinating agent. Under mild conditions it selectively chlorinates at allylic and benzyllic positions as well as at the α-carbon of carbonyl compounds (Equation 75).

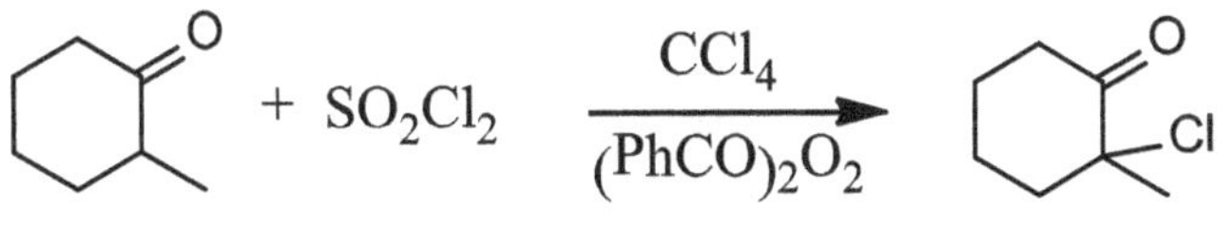

(Equation 75)

Most α-halocarbonyl compounds are very reactive in S_N2 displacement reactions with nucleophiles that are not basic enough to remove α-protons. The explanation for the enhanced reactivity is probably similar to that for the increased reactivity of allylic alkyl halides in S_N2 displacements.

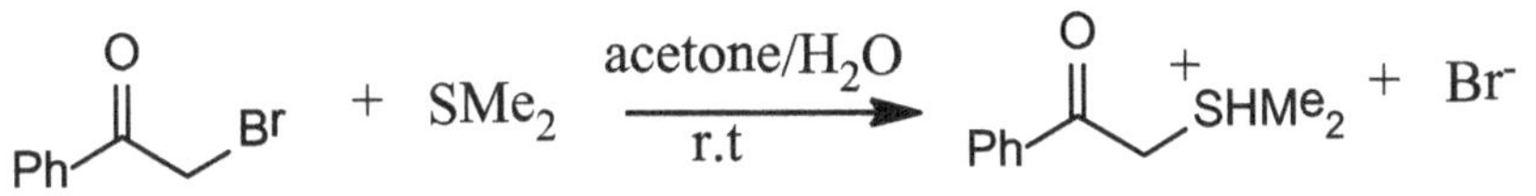

(Equation 77)

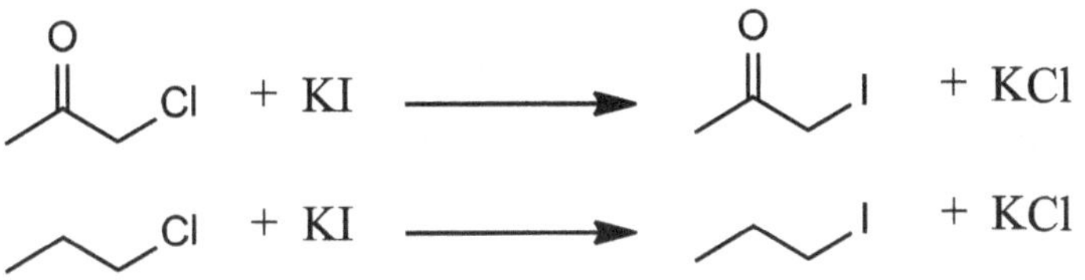

35,000 (rel. rate)

1 (rel. rate)

4.2.3 Haloform reaction

The bromination of acetone in basic medium called bromoform reaction corresponds to the following stoichiometric equation (Equation 78).

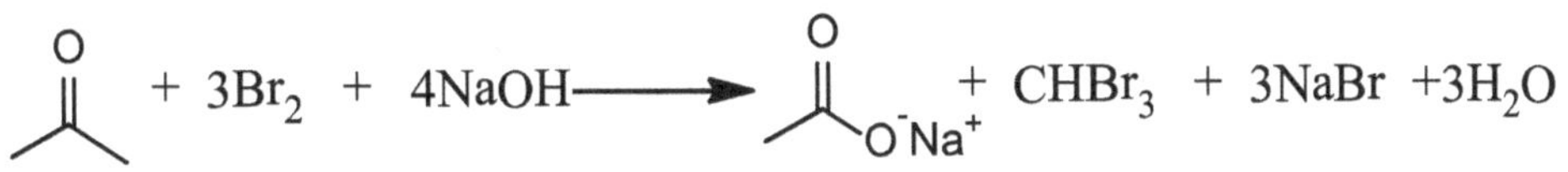

(Equation 78)

The base promoted halogenation involves the following steps; first step involves the removal of α-hydrogen to form a resonance stabilized carbanion as an intermediate (Scheme 163).

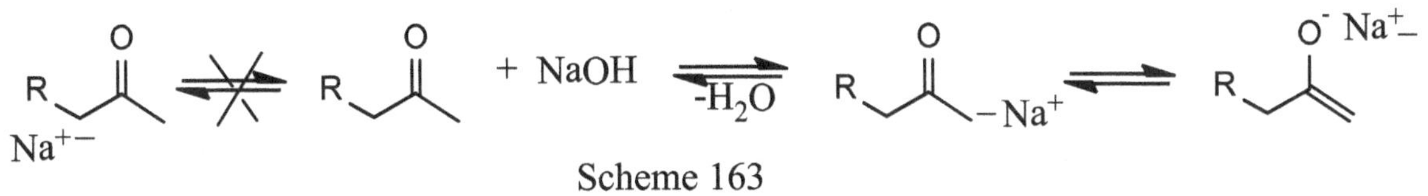

Scheme 163

In this step $RCH_2COCH_2^-$ forms preferentially. This is because the H atom of the RCH_2- group is less acidic than H of the $-CH_3$ group; the +I effect of the R group decreases the acidity of the H of the methylene group whereas there is no such effect on CH_3 group. However, this is the slow step and this step determines the rate of reaction. This is further confirmed by studying the kinetic effect. This reaction exerts primary kinetic isotopic effect with k_H/k_D value approximately equals to 7.

The second step involves the electrophilic attack by the halonium ion on the nucleophilic carbanion. This is a fast step:

The formation of monohalo derivative follows the second order kinetics: it is first order with respect to the substrate and first order with respect to the base;

The rate = k[substrate][⁻OH]

This reaction may be designated as S_E1 reaction though the rate-determining step involves the concentrations of the substrate and the base (but it does not involve the halogen). Though the stoichiometric concentrations of the reagents predict a much higher order it is only first order with respect to the concentrations each of acetone and hydroxide ion. Obviously stoichiometry and the rate law have no relationship.

The following are the evidences in favour of the mechanism written above:

i.   The reversible formation of the carbanion is supported by the fact that when the reaction is carried out with ⁻OD in D_2O, exchange of deuterium occurs.

ii.   The kinetic study shows that the rate of the reaction is independent of the concentration of the halogen but it depends on the concentrations of the substrate as well as that of the base. This supports the rate equation as predicted above.

iii. The base promoted halogenation of acetone shows that the rates of bromination and iodination of acetone are one and the same. This also indicates that the rate does not depend on the concentration of halogen.

The rate of base promoted racemization of optically active phenyl sec-butyl ketone is the same as that of the halogenation. The formation of the optically inactive phenyl sec-butyl ketone from the +) or (-)-phenyl sec-butyl ketone indicates that a bond of chiral C must have broken and the formation of the carbanion fits this requirement. In the absence of halogen, the carbanion takes up a proton from the conjugate acid; the protonation occurs from the two sides equally, from above the plane and form below the plane of the carbanion which leads to racemization.

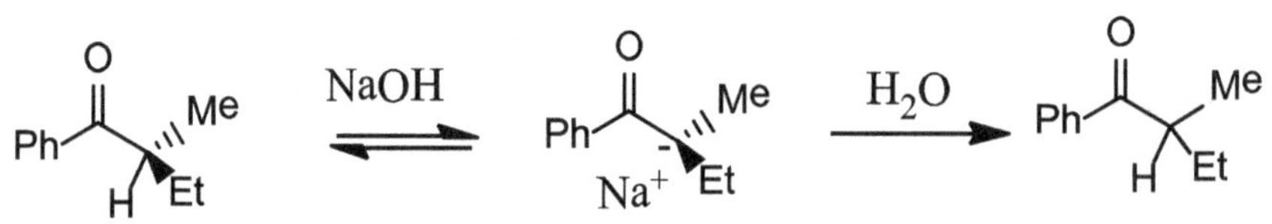

Scheme 164

However, the reaction does not stop at the monohalogenation stage; usually a trihalogenated product is formed (Scheme 152). Since the carbonyl group and the X atom of CH_2X group draw electron towards themselves, the α-C atom becomes more electron-deficient than the nonhalogenated CH_3 group. So abstraction of proton from the CH_2X group takes place and then the electrophilic attack by X⁺ gives rise to

$RCH_2COCH_2X_2$. $RCH_2COCH_2X_2$ still contains an acidic H atom which gets replaced by X atom in the same manner as stated above. Here it is important to note that the H atoms of RCH_2 are much less acidic than the H atoms of CH_3, CH_2X and CHX_2 groups because of the +I effect of the R group and –I effect of the X atoms.

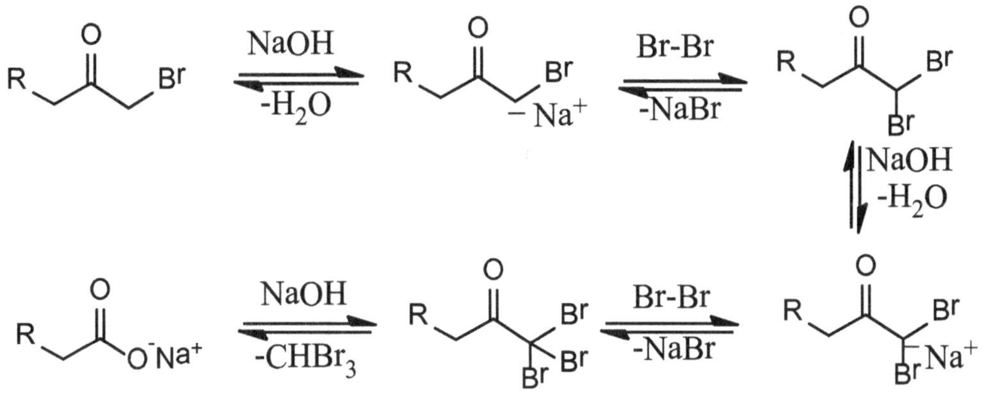

Scheme 165

If there is an excess of base, the base now attacks the highly electron deficient carbonyl C atom and develops the species C. The carbonyl C now bears high steric strain owing to the presence of the RCH_2, OH, CBr_3 groups and the bulky anionic O atom. In order to get rid of steric strain $C-CX_3$ bond breaks and trihalomehyl cabanion forms, the latter is rather a stable species since the three halogen atoms draw the negative charge of carbanion towards themselves by the –I effect and by $p\pi$-$d\pi$ conjugation. Thus the steric and the stability of trihalomethide ion help to break the $C-CX_3$ bond as shown below. The trihalomethide ion then takes up the proton to give haloform and a carboxylate ion (Scheme 166).

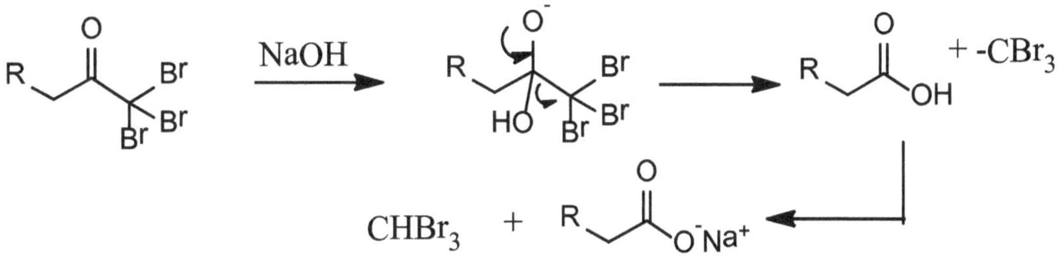

Scheme 166

The haloform reaction is a useful method for detection of methyl ketones, particularly when iodine is used because iodoform is highly insoluble, bright yellow solid. The reaction is also useful for the synthesis of carboxylic acids when the methyl ketone is more available than the corresponding acid.

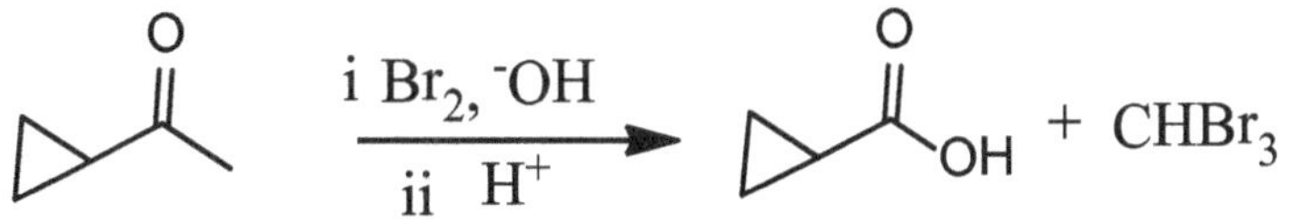

(Equation 79)

Since the haloform reaction is fast, it can in some cases be used to prepare unsaturated acids from unsaturated ketones without serious complication caused by addition of halogen to the double bond.

$$\text{(structure)} \quad \xrightarrow[\text{ii} \; H^+]{\text{i } Cl_2, \; ^-OH} \quad \text{(structure)} \; 50\% \quad + \; CHCl_3 \qquad \text{(Equation 80)}$$

There are compounds which give haloform against our expectation. Isopropyl amine and compounds other type $RCOCH_2COR$, $RCHOHCH_2CHOHR$, $CH_3(NHCOR)COR$ give haloform on treatment with alkaline halogen solutions. However the reaction takes place merely on mixing a methyl ketone with an alkaline solution of hypohalite.

The haloform reaction is useful in organic degradation. The reaction may be used to distinguish between different possible structures since methyl ketones give the positive test. For example, between $CH_3COCH_2CH_2CH_3$ and $CH_3CH_2COCH_2CH_3$ only the former gives the haloform test and not the latter since the former one only contains a keto ethyl group. So by the haloform reaction one can determine the structure of a compound. In practice iodoform test is performed instead of bromoform or chloroform test since iodoform is a yellow crystalline solid insoluble in water and can be readily identified its melting point. The iodoform reaction is best carried out by taking the solution of the substance in dioxane and treating first with dilute NaOH solution and then with a slight excess of iodine in potassium iodide solution, warming the mixture and finally adding water to it. If the compound under consideration contains a – $COCH_3$ group or CH_3CHOH- group or if it is acetaldehye or ethyl alcohol, yellow precipitate of iodoform will be obtained.

How is it possible for C-H bound to be broken so readily? This has explained as follows. Owing to inductive effect of the carbonyl group, the electrons of the C-H bonds on the α-carbon atom are displaced towards the carbon atoms, thus facilitating the release of proton from Cα. At the same time, the anion produced has greater resonance stabilization than the parent carbonyl compound in the conjugate base is stabilized with respect to its acid.

The inductive effect is very much weaker on a β-carbon since I effect falls rapidly from the source. Hence proton release is far less easy than on α-carbon. Also if a proton were released from the β-carbon, this negative atom would be insulated from the carbonyl group by the intervening saturated carbon atom and consequently such a, these groups in anion would not have increases resonance stabilization. Thus because of increased activity of hydrogen in a methylene group ($-CH_2-$) or in a methylene group when adjacent to a carbonyl group or other strongly electron attracting group, these groups are referred to as the active methylene group.

4.3   MANNICH REACTION

The Mannich reaction is an organic reaction which consists of an amino alkylation of an acidic proton placed next to a carbonyl functional group by formaldehyde and a primary or secondary amine or ammonia [1912ADP647, 1940OR0001, 2002JACS1866]. The final product is a β-amino-carbonyl compound also known as a Mannich base. Reactions between aldimines and α-methylene carbonyls are also considered Mannich reactions because these imines form between amines and aldehydes. The reaction is named after chemist Carl Mannich. The Mannich reaction is an example of nucleophilic addition of an amine to a carbonyl group followed by dehydration to the Schiff base (Equation 81). The Schiff base is an electrophile which reacts in the second step in an electrophilic addition with a compound containing an acidic proton (which is, or had become an enol). The Mannich reaction is also considered a condensation reaction.

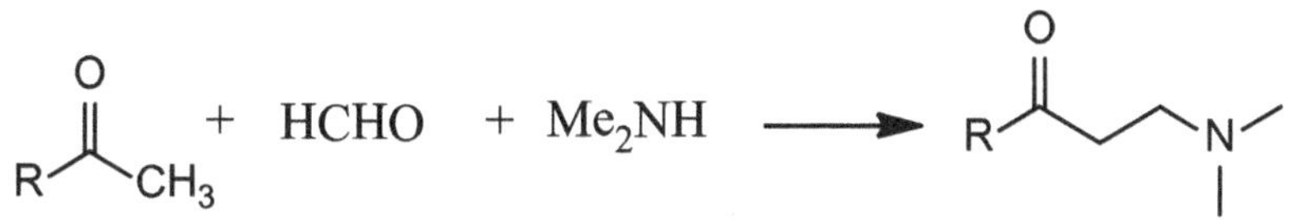

(Equation 81)

In the Mannich reaction, primary or secondary amines or ammonia, are employed for the activation of formaldehyde. Tertiary amines lack an N–H proton to form the intermediate enamine. α-CH-acidic compounds (nucleophiles) include carbonyl compounds, nitriles, acetylenes, aliphatic nitro compounds, α-alkyl pyridines or imines. It is also possible to use activated phenyl groups and electron-rich heterocycles such as furan, pyrrole, and thiophene.

The mechanism involves the preliminary formation of an imine salt from the amine and formaldehyde (Scheme 167). The amine is nucleophilic and attacks the more electrophilic of the two carbonyl compounds available (HCHO). No acid is needed for this addition step, but acid-catalyzed dehydration of the addition product gives the imine salt. In the normal Mannich reaction, this is an intermediate but it is quite stable and the corresponding iodide is sold as Eschenmoser's salt for use in Mannich reactions. The compound with the carbonyl functional group (in this case a ketone) can tautomerize to the enol form, after which it can attack the iminium ion.

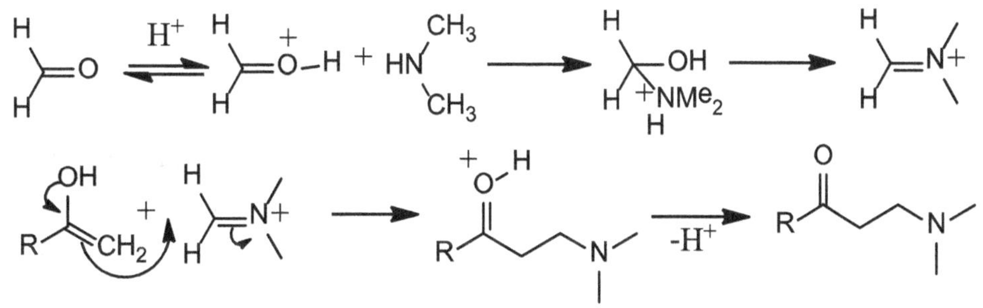

Scheme 167

The Mannich products can be converted to enones. The most reliable method for making the enone is to alkylate the amine product of the Mannich reaction with methyl iodide and then treat the ammonium salt with base. Enolate ion formation leads to an E1cB reaction rather like the dehydration of aldols, but with the better leaving group (Scheme 168).

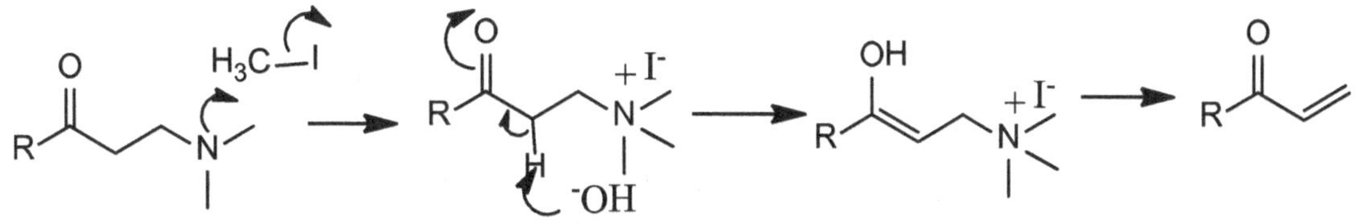

Scheme 168

4.4   STORK-ENAMINE REACTION

In 1954 Stork et al [56JACS5129] reported that the alkylation of the pyrrolidine enamine of cyclohexanone with CH_3I followed by acid hydrolysis led to monomethylated ketone. It was thus obvious that the enamine (**49**) derived by the loss of proton form the intermediate methylated iminium cation failed to undergo any further alkylation (Scheme 169).

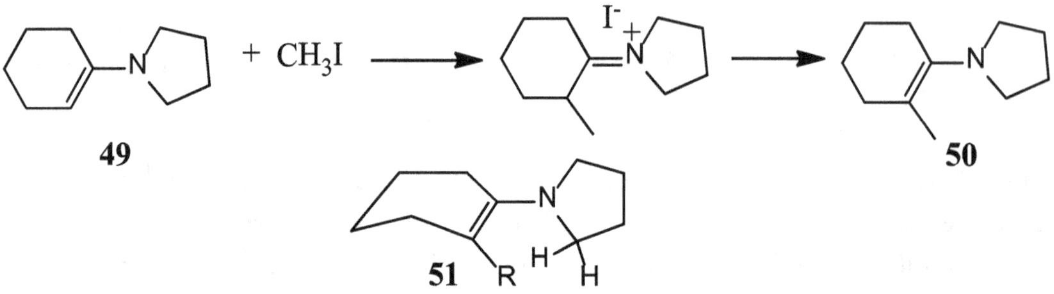

49   **50**

51 R H H

Scheme 169

The pyrrolidine enamine of α-methyl cyclohexanone (**50**) was in fact found to be quite inert toward further alkylation and was shown to consists only of the tetra substituted isomer on the basis of NMR. The formation of tetra substituted isomer (**51**) since it would involve a severe interaction between the methyl group and the methylene group adjacent to the nitrogen if an overlap between the electron pair on the nitrogen atom and the double bond were to be maintained. Kuehne has reported that the pyrrolidine enamine of 2-phenyl cyclohexamine failed to show any styrene type absorption in the UV which would have been exhibited by the tetra substituted isomer.

A rationale for the inert behavior of the enamine **50** toward further alkylation was put forward by Williamson who argued that the methyl group in **50** should assume an axial orientation as shown in conformation (**52**) if the overlap between the electron pair in the nitrogen atom and the double bond is to be maintained, since in alternate conformation (**53**) the equatorial methyl group is pretty much coplanar with the methylene group adjacent to the nitrogen atom and thus interfere severely with it. The alkylation being subject to stereoelectronic control would, i.e., involve a severe 1,3-diaxial alkyl-alkyl interaction, thus increasing the energy of the transition state.

52   **53**

Johnson and et al found that careful hydrolysis of pyrrolidine enamine of the conformationally more stable system, in 2-methyl-4-t-butyl cyclohexanone led to a 1:4 mixture of cis and trans isomers of the ketone, showing that the methyl group in isomer shows an extremely weak band at $1675 cm^{-1}$, which is undoubtedly a reflection of the low degree of the electronic overlap in the latter case.

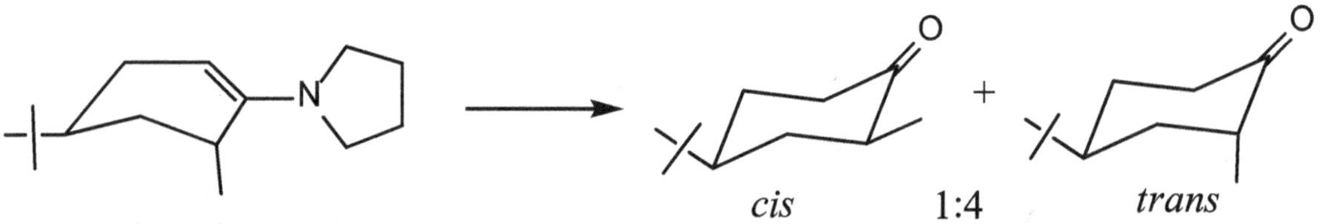

cis   **1:4**   *trans*

Unlike the pyrrolidine enamine of 2-methyl cyclohexanone which consists predominantly of the tetra substituted isomer, the morpholine and piperidine enamines of 2-methylcyclohexanone were shown by Gurowitz and Joseoph as an almost 1:1 mixture of tri and tetra substituted isomers by NMR.

Amine	% tri substituted	% tetra substituted
Pyrrolidine	90	10
Morpholine	52	48
Piperidne	46	54
Diethylamine	24	75
N-methylaniline	0	100

In N-methylaniline enamine of 2-methyl cyclohexanone, electron pair on the nitrogen atom could overlap predominantly with the phenyl group and not with the enamine double bond, thus minimizing the steric interference between the C-2 methyl group and the substituted attached to nitrogen.

The increase in the proportion of the tetra substituted isomer in the cases of the morpholine and piperidine enamine of the 2-methylcyclohexanone has been ascribed to both steric and the electronic factors. The authors proposed that the overlap of the electron pair on the nitrogen atom and the π-electrons of the double bond is much more important in the case of pyrrolidine enamines and less with the others. The greater amount of overlap or electron delocalizations, in the case of pyrrolidine enamines, is in accord with the postulate of Brown et al that the double bond exo to the five membered rings is more favoured than the double bond into the double bond exo to the six membered rings.

Synthetic applications:

Alkylation of **49** with acrolein gives the bicyclic ketones **55**, which can be converted to 4-cyclooctene-1-carboxylic acid by the action of base on its methiodide (Scheme 170) [56JACS5129]. The formation of **54** must have taken place via the normal alkylation product, which undergoes hydrolysis with water followed by pyrrolidine with the more reactive aldehyde group to give an intermediate which can then cyclize to give the observed product.

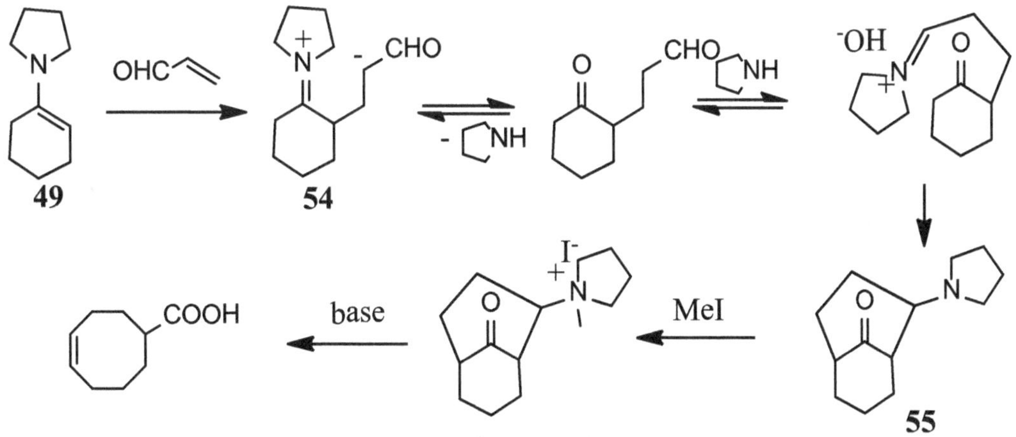

Scheme 170

Key step for the synthesis of 1,7-dimethylindole involve the alkylation of 1-methyl-2-ethylidine purrolidine to acrolein (Scheme 171) [58CI979].

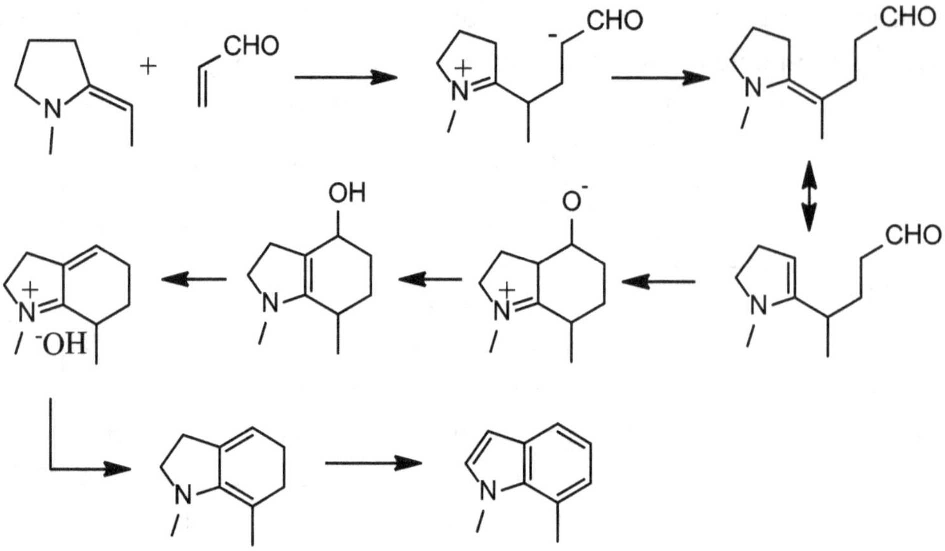

Scheme 171

A variety of cyclic ketones have been allowed to react in this way. For instance enamine derived from dihydro-3-(2H)-furanone react with dimethyl acetylene dicarboxylate followed by ring expansion leads to the formation of 2,7-dihydro-3-pyrolidienyl-4,5-oxepinedicarboxylate at room temperature in ether as solvent (Scheme 172) [1963JOC3134].

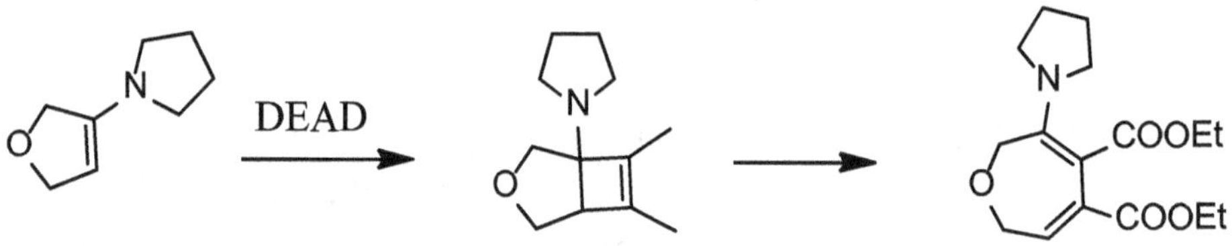

Scheme 172

Under identical conditions, enamine derived from acetyl acetone undergoes cycloaddition with diethyl acetylene dicaroboxylate followed by cyclization yields diethyl-3-pyrrolidineo-5-methyl phthalate (Scheme 173).

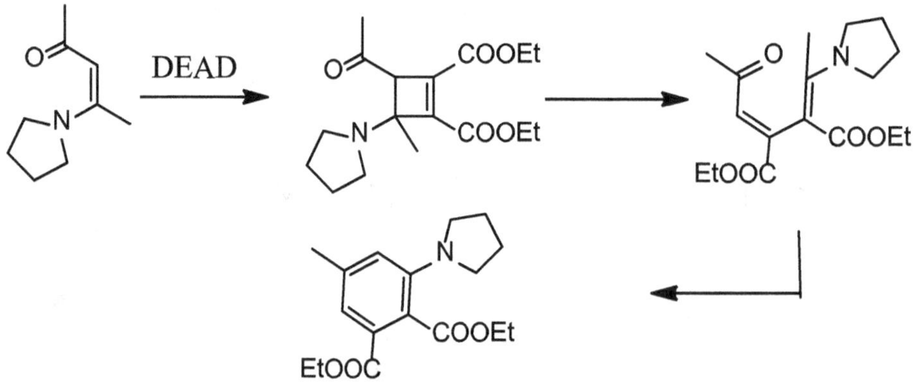

Scheme 173

Ninitzescu [1929BSC37] synthesized 1-phenyl-2-methyl-3-carbethoxy-5-hydroxyindole by condensing ethyl-β-anilnocronoate with p-benzoqunione (Scheme 174)

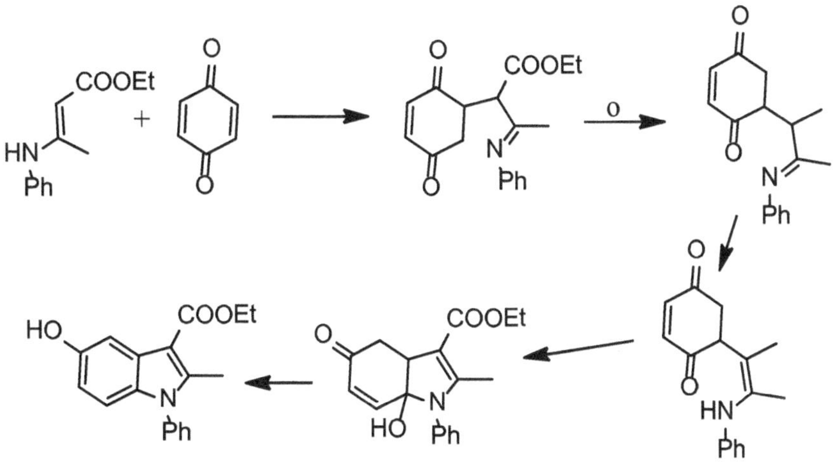

Scheme 174

The total synthesis of the phenolic sesquiterpene (±)-parviflorine was accomplished by L.A. Maldonado and coworkers (Scheme 175) [1998JOC2918]. The key step in the synthetic sequence was the reaction of an enamine with acrolein to form a bicyclic intermediate, which was subjected to a Grob fragmentation to afford the eight-membered ring of the natural product. The bicyclic ketone substrate was refluxed in benzene using a Dean-Stark trap and the resulting enamine was taken to the next step as crude material.

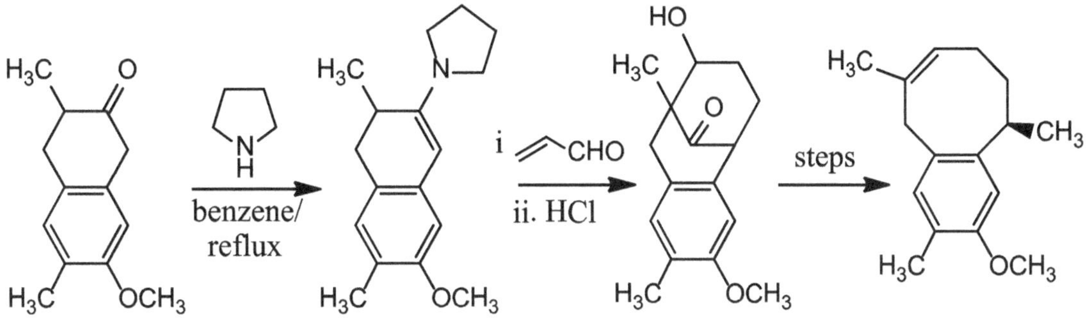

Scheme 175

An intramolecular variant of the Stork enamine synthesis was utilized during the asymmetric total synthesis of (−)-8-aza-12-oxo-17-desoxoestrone by A.I. Meyers (Scheme 176) [1992JOC4732].

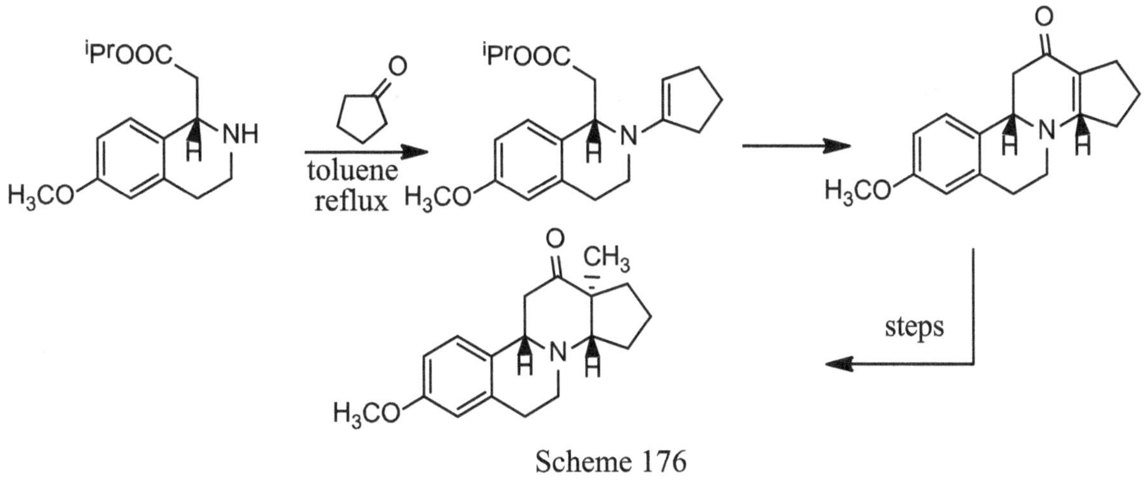

Scheme 176

4.5 FAVORSKII REARRANGEMENT

When an α-haloketones is treated with alkoxide, we expect three types of reaction:

i. Substitution reaction:

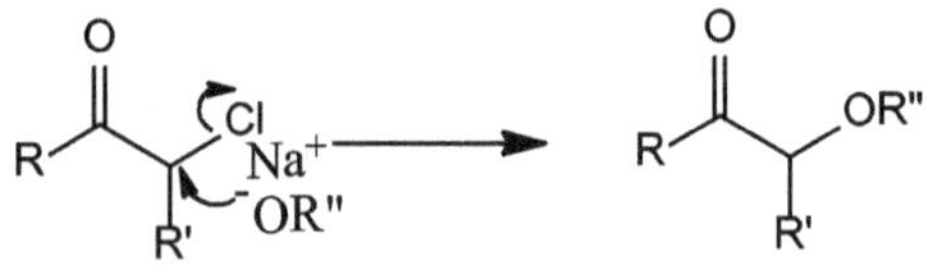

(Equation 82)

ii. Condensation reaction:

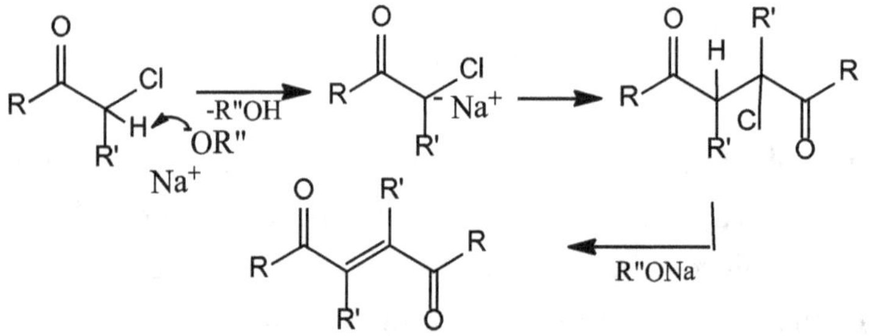

Scheme 177

iii. Epoxidation:

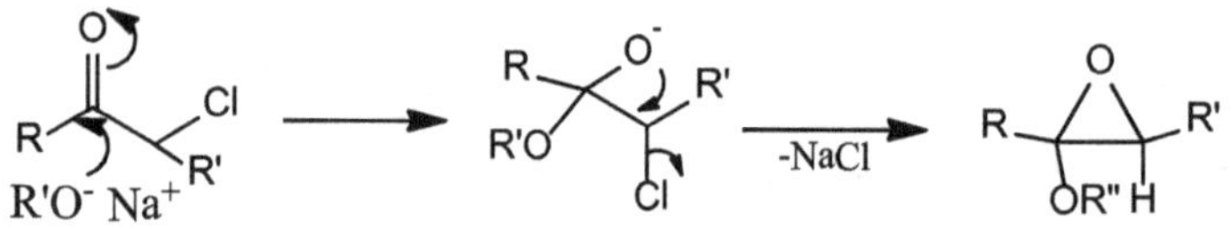

Scheme 178

But Favorokii in 1894 observed a different product being formed when an α-haloketone bearing a hydrogen atom on the cholrine of the carbon atom adjacent to carbonyl group, when treated with alkoxide. The compound obtained was an ester.

$$PhCH_2COCH_2Cl + NaOEt \longrightarrow PhCH_2CH_2COOEt + NaCl \quad (Equation\ 83)$$

Thus the reaction of the above type in which α-halo carbonyl compounds on treatment with alkoxide yield rearranged product (ester) with same number of carbon atom is called Favorskii rearrangement. Depending on whether the base is hydroxide ion or alkoxide ion or an amine, the produt is an acid or an ester or an amide respectively. In case of cyclic ketones, ring contraction tkes place (Scheme 179).

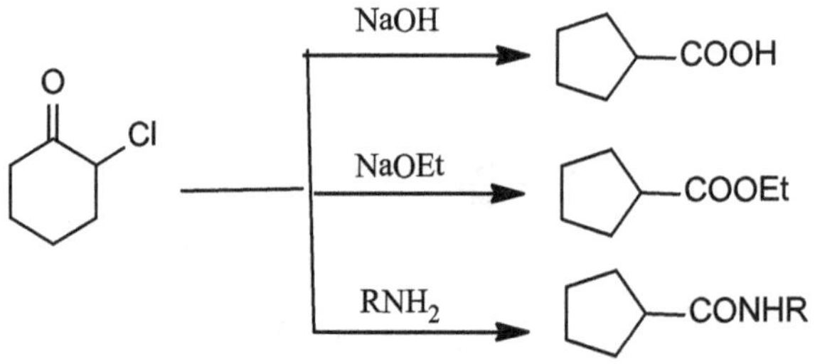

Scheme 179

Though the mechanisms of Favorskii rearrangement has been the subject of much investigation, at least nine different mechanisms have been proposed. Some of these mechanisms involving alkoxy oxiranes, carbine or ketenes were rejected earliar. The accepted mechanism involves a cyclopropnanone intermediate. This cyclopentanone intermediate is formed by the abstraction of α-proton to form a cabanion followed by its nucleophilic attack on the carbon atom carrying halogen atoms and to expel halogen. Opening of this intermediate cyclopropanone yields the product (Scheme 180).

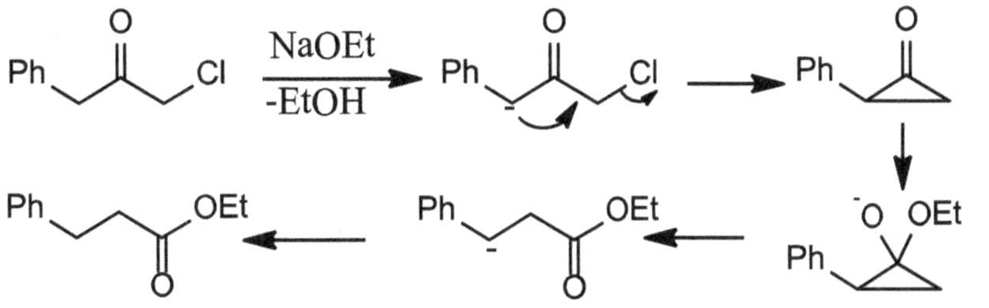

Scheme 180

Evidence for cyclopropanone intermediate mechanism comes from the fact that two isomeric ketones A (PhCH$_2$COCH$_2$Cl) and B (PhCHClCOCH$_3$) gives the same ester while under the conditions of the experiment they are not interconvertible. This can be explained only by assuming cyclopentanone intermediate (Scheme 181).

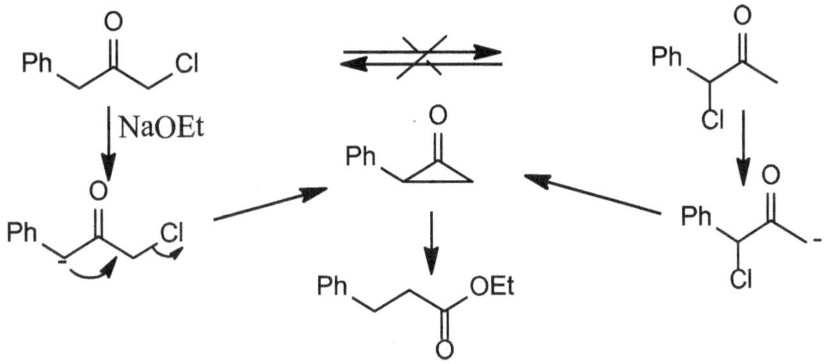

Scheme 181

The most convincing evidence comes from labeled experiments. When Favorskii rearrangement of 2-chloro-cyclohexanone labeled with C^{14} chlorinated carbon ic carried out, it is found that the half labeled carbon is present at Cα in the ester while the other half Cβ indicating the presence of an intermediate in which Cα and Cβ are similarly are identically situated.

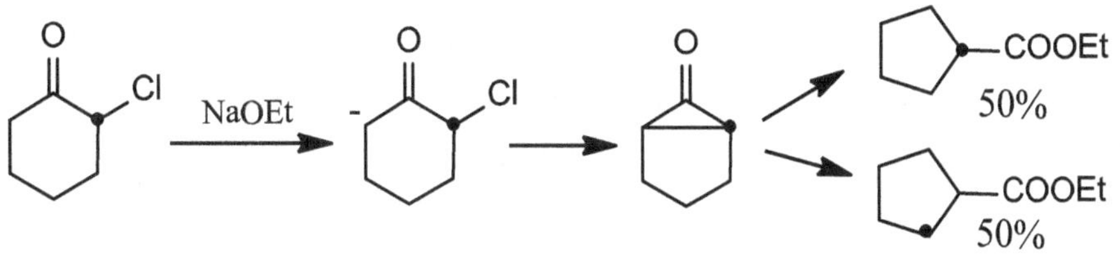

Scheme 182

In the above case, the cyclopentanone formed is symmetrical one, therefore it may open with equal easy either 1,6-bond or 1,2-bond. Still the product is same (except in labeling experiment0. But in an unsymmetrical cyclopropanone intermediate, the ring opening takes place to give more stable carbanion, i.e a phenyl substituent favour breaking of the bond to the more substituted carbon while alkyl groups favour the bond to the less substituted carbon atom.

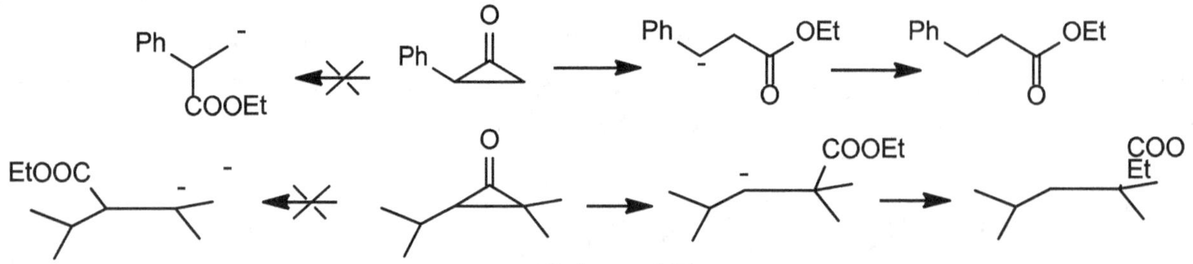

Scheme 183

Although the cycloropanone intermediate mechanism explains most of the steps, it fails to explain some. There are cases in which haloketones which do not posses α-hydrogen atom still undergo this reaction. A cyclopropanone formation is thus precluded since there is no proton to abstract and an alternate scheme as proposed by Jchouban and Sackur.

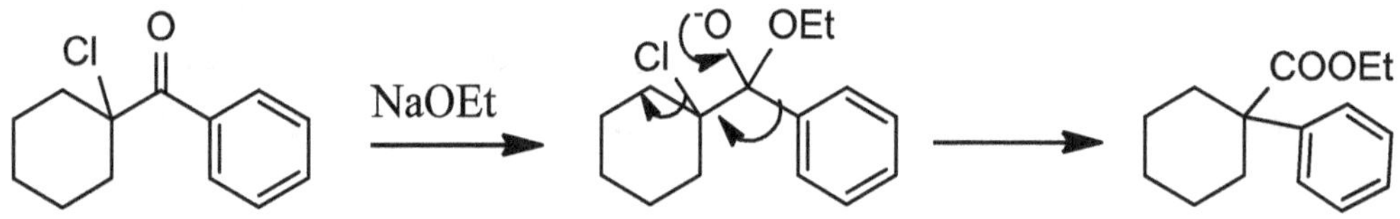

Scheme 184

This sequence is very similar to the benzilic acid rearrangement whereby the anion attacks the carbomyl group pushing the phenyl group to the adjacent carbon atom where it in turn displaces the halide. This mechanism is supported by Stevans and Farkas and is also termed the semi-benzylic mechanism.

4.6   ACYL ANION EQUIVALENTS

In aldehyde, the carbonyl group is electrophilic in nature thereby C-H bond is tightly held. So it is very difficult to break this bond. That means to say that it is very difficult to generate acyl anion. Since acyl anions are in general unattaianable, a great deal of effort has been expended to find masked acyl anion (acyl anion equivalents). The benzoin condensation has been recognized as belonging to the general class of reactions that involve masked acyl anions as intermediates [1991Misc541, 2006Ark119]. The use of masked functional groups such as masked acyl anion equivalents in the formation of C-C bonds has proved to be a powerful strategy in the development of new synthetic methods. For instance, an aldehyde is converted into an addition product RCH(OX)Y, which renders the C-H acidic. Then under basic conditions, a masked acyl anion can be formed and may react with an electrophilic component E^+. Decomposition of the product RCE(OX)Y should generate the carbonyl group with formation of RCOE. Intermediate such as $RC^-(OX)Y$ are used in the conversion of aldehydes into α-hydroxy ketones, α-diketones and 1,4-dicarbonyl compounds proving to be a powerful strategy in the development of new synthetic methods.

The most systematically investigated acyl anion equivalents have been the TMS ethers of aromatic and heteroaromatic aldehyde cyanohydanins, TBDMS protected cyanohydrins, benzoyl protected cyanohydrins, ethoxyethyl protected cyanohydrins, alkoxycarbonyl protected cyanohydrins, THP protected cyanohydrins, α-(dialkylamino)nitriles, cyanophosphates, diethyl 1-(trimethylsiloxy)phenylmethyl phosphates, dithioacetals,

1,3-dithiane derivatives, diselenoacetals, Tosmic acid etc., (Figure 1). Deprotonation of these masked acyl anions under the action of strong base, usually LDA, followed by treatment with a wide variety of electrophiles is a great synthetic value. If the electrophile is another aldehyde, α-hydroxyl ketones or benzoin's are formed.

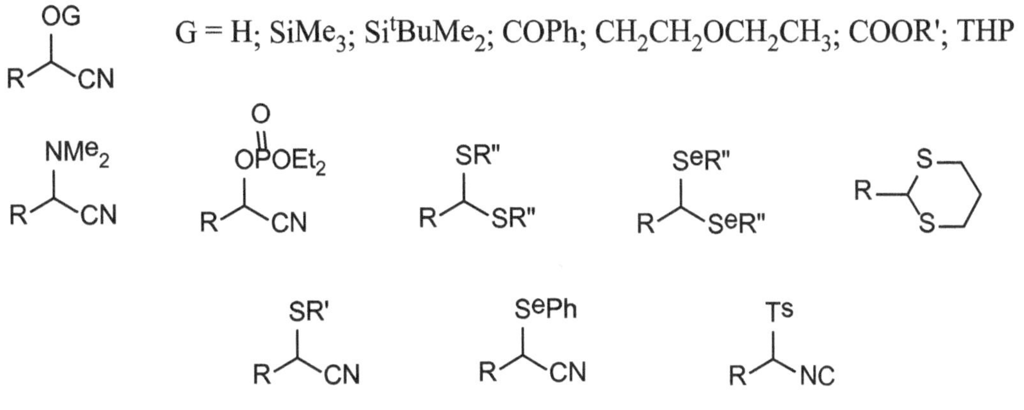

G = H; SiMe$_3$; Si^{t}BuMe$_2$; COPh; CH$_2$CH$_2$OCH$_2$CH$_3$; COOR'; THP

Figure 1

4.6.1 Protected cyanohydrins

Protected cyanohydrins can be easily converted into a variety of valuable synthetic intermediates such as α-hydroxyl ketones, ketones, 1,2-dicarbonyl compounds, 1,4-dicarbonyl compounds, α-ketcarboxylic acid/esters, 4-oxocarboxylic esters, 4-oxonitriles etc. (Scheme 185).

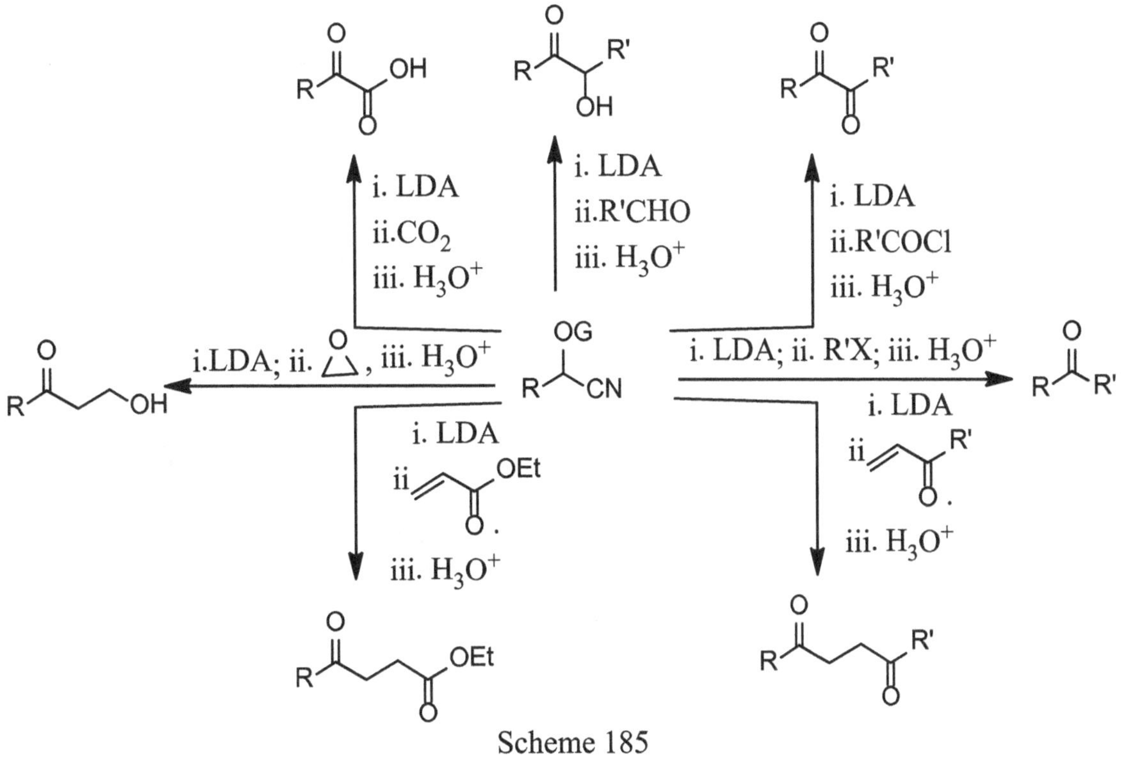

Scheme 185

α-Hydroxyl ketones can be prepared by the addition of carbanions derived from easily accessible α-(dialkylamino)nitriles to carbonyl compounds followed by hydrolysis. A high degree of stereoselectivity is also observed in the condensation of N-benzoyl-2-cyanopiperidine with propanal, a key step in the synthesis of (+)-conhydrine (Scheme 186).

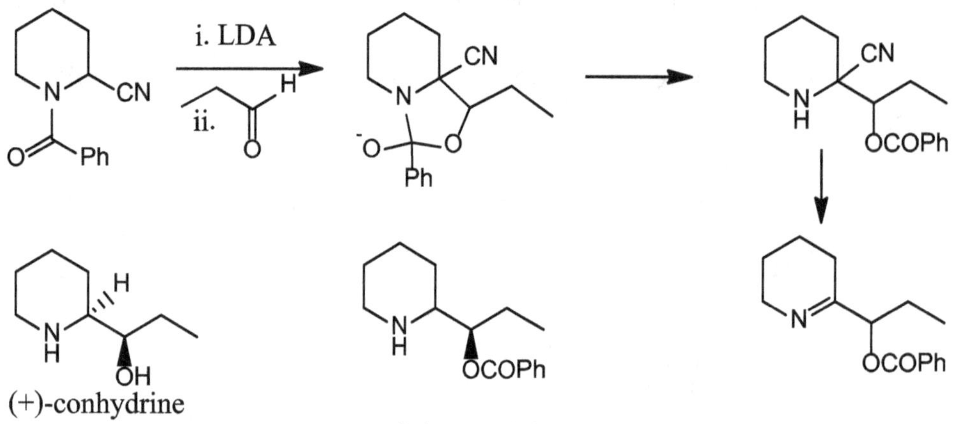

Scheme 186

4.6.2 Azolium stabilized acyl synthons

Studies on vitamin B1 (thiamine) catalysed formation of acyloins from aliphatic aldehydes have established the mechanism for the one catalytic activity of 1,3-thiazolium salts in carbonyl condensation reactions [1976Angew639; 1983T3207]. In the presence of bases, quaternary thiazolium salts are transformed into the ylide structure (**56**), the ylide being able to exert a catalytic effect resembling that of the cyanide ion in the benzoin condensation. Like cyanide ion, the zwitterion (**56**), formed is nucleophilic and reacts at the carbonyl group of aldehydes. The resultant intermediate can undergo base catalysed proton transfer to give a carbanion (**57**), which is stabilized by the thiazoilium ring. This ion behaves as acyl anion equivalent.

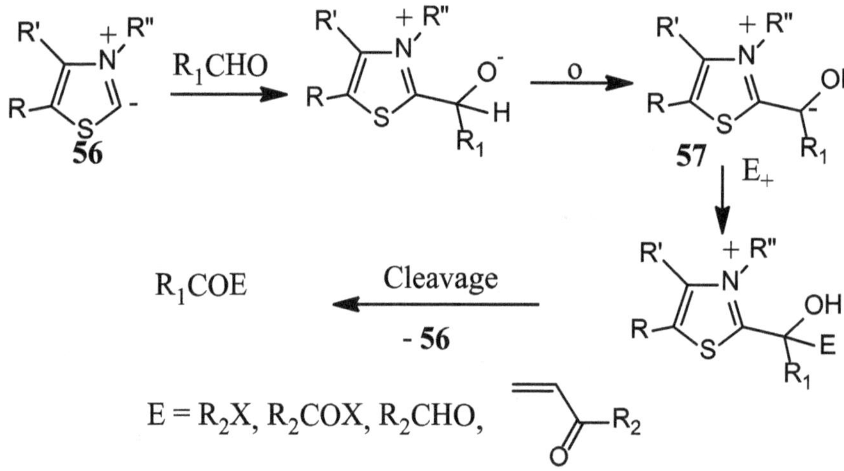

$E = R_2X, R_2COX, R_2CHO,$

Scheme 187

4.6.3 Dithioacetals or 1,3-dithianes

Dithioacetals or 1,3-dithianes are prepared by reaction of the corresponding carbonyl compounds in the presence of acid catalyst with a thiol or 1,3-dithiol. The dithiane derived anion can be generated by the action of n-BuLi in THF at -78°C or with complex bases NaNH$_2$-RONa at room temperature. Dithioacetals or ketals are resistant to acidic or basic hydrolysis. Regeneration of the carbonyl group from the dithioacetal sometimes presents difficulties but can be carried out by hydrolysis in polar solvents in the presence of metallic ions such as Hg^{2+}, Cu^{2+}, Ag$^+$, Ti^{4+} or Tl^{3+}. Alternatively, alkylative hydrolysis uses methyl iodide, methyl triflurosulfonate or oxidative cleavage using chloramine-T leads to carbonyl compounds (Scheme 188).

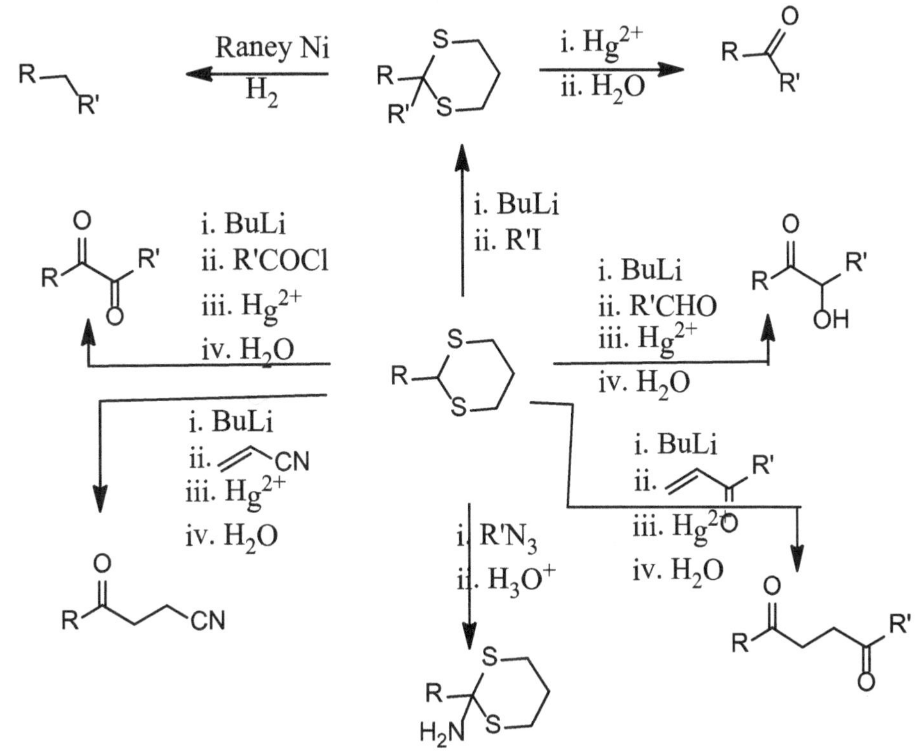

Scheme 188

The acyl anion equivalents formed by the electroreduction of oxazolium salts were found to be useful for the formation of ketones, aldehydes or a α-hydroxyl ketones (Scheme 189). α-Methoxyvinyl lithium also can act as acyl anion equivalent and can be used for the formation of α-hydroxy ketones, a diketones, ketones, γ-diketones and silyl ketones.

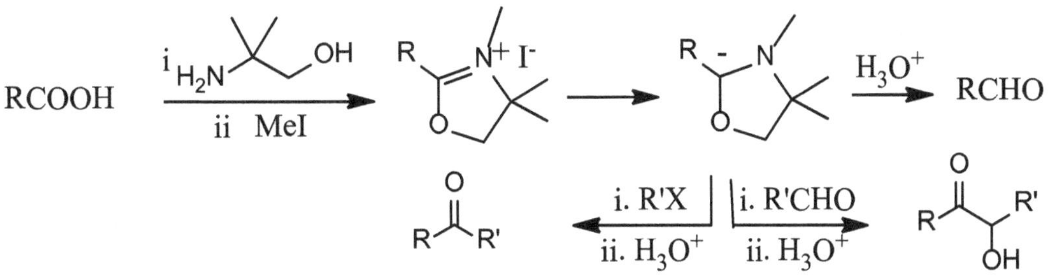

Scheme 189

The addition of lithium and Grignard reagents to isocyanides which do not contain α-hydrogens proceeds an α-addition to produce a metalloaldimine (R-N=CMR', an acyl anion equivalent) (Scheme 190). This metalloaldimines are versatile reagents which can be used as precursors for the generation of aldehydes, ketones, α-hydroxyketones, α-keto acids, α- and β-hydroxy acids, silyl ketones and α-amino acids.

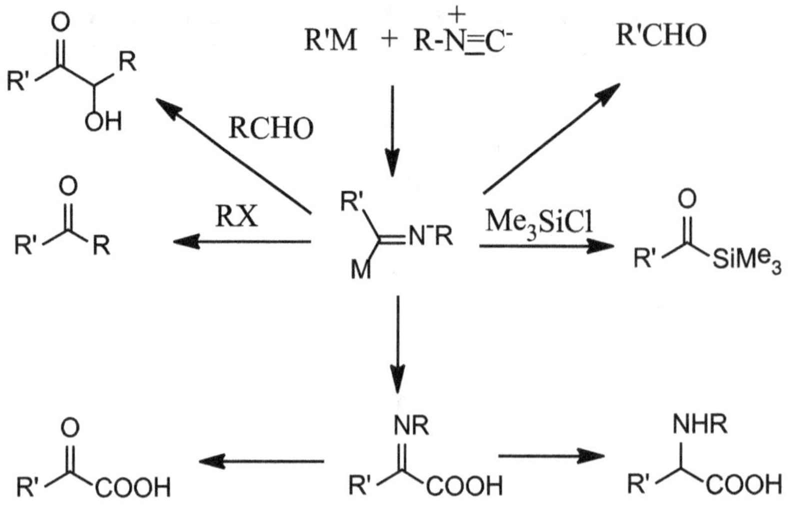

Scheme 190

4.6.4   Vinyl ethers

Alkyl vinyl ethers prepared from aldehydes possessing α-hydrogen have been used as acyl anion equivalents [1980JACS1577]. Deprotonation of this ether with strong base like PhLi or tert-butyllithium followed by treatment with a variety of electrophiles including like alkyl halides, aldehydes, ketones, acid chlorides etc. to yield the intermediates (**58**), which can be hydrolysed to the corresponding ketones by aqueous acid (Scheme 191).

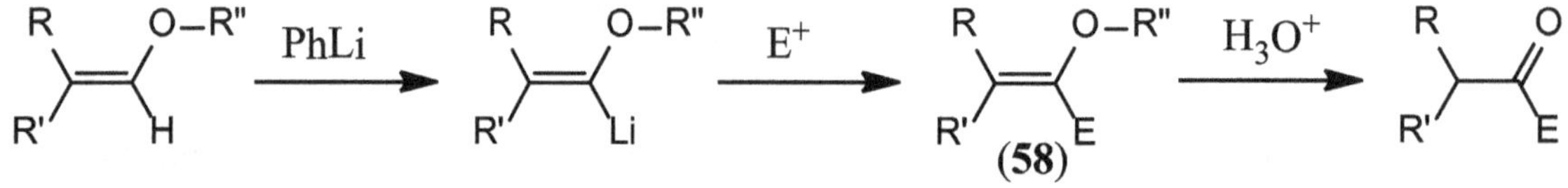

Scheme 191

In a similar manner, 1-(Phenylthio)alkenes (**59**) [1979CL785] or 1-(Phenylseleno)alkenes [1981JACS870] have been used as synthons for the preparation of ketones (Scheme 192 & Scheme 193). In these cases final step, ketones can be obtained by the oxidative hydrolysis.

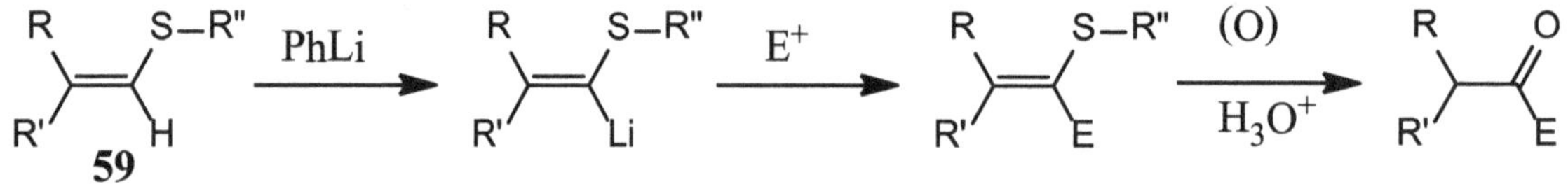

Scheme 192

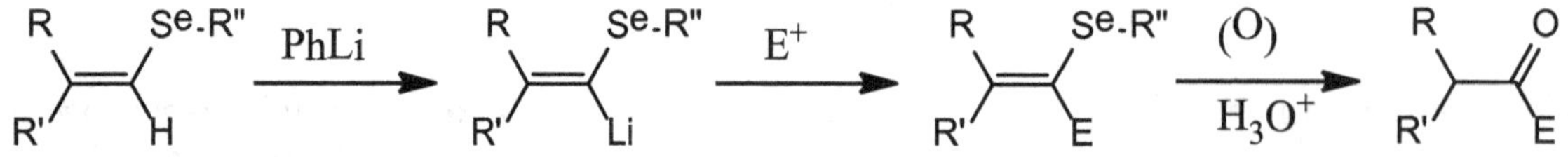

Scheme 193

4.7   α,α-DIOXOKETENE DITHIOACETALS

The α,α-dioxoketene dithioacetals have been recognized as novel 3 carbon electrophile fragments which have been exploited extensively for the construction of five and six membered heterocycles. The desired α,α-dioxoketene dithioacetals were prepared in high yield by reacting the corresponding active ethylene compounds with carbon disulfide in the presence of sodium hydride as a base in benzene followed by alkylation with methyl iodide (Equation 82) [1970ACS1191].

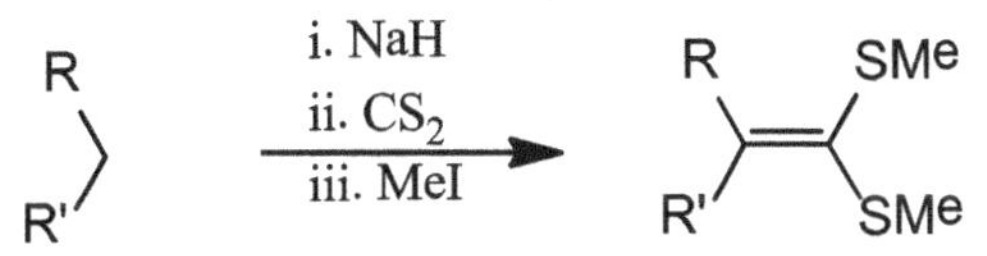

(Equation 82)

a. R = R' = COCH$_3$; b. R = R' = COOEt; c. R = COCH$_3$, R' = COOEt .

When α,α-dioxoketene dithioacetals were treated with hydroxylamine hydrochloride in the presence of pyridine under solvothermal condition afforded the isoxazoles in 71-80% yield [2010SC3569] while with hydrazine hydrate in refluxing ethanol gave therespective pyrazole derivatives in 70-75% yield [2010TL3486] (Scheme 194).

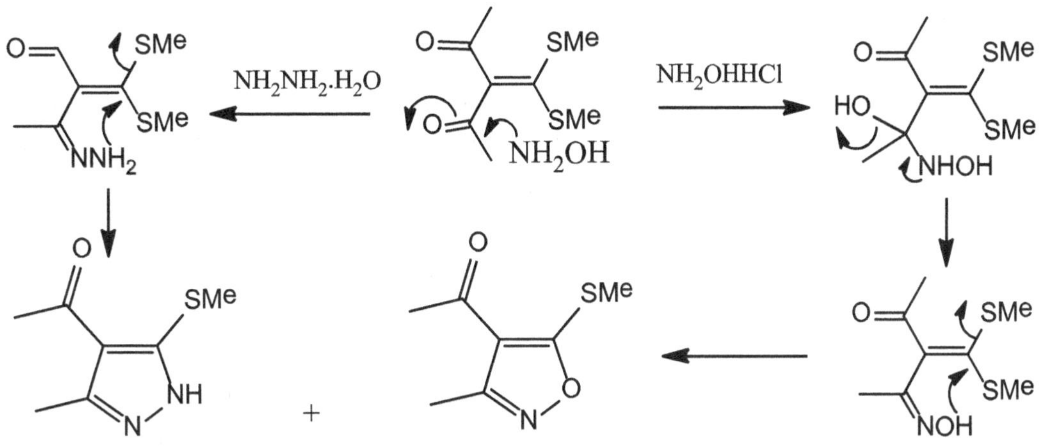

Scheme 194

α,α-Dioxoketene dithioacetals on treatment with urea or thiourea in ethanol as solvent under reflux condition gave pyrimidine derivatives while with aniline gave quinolone derivatives as shown in the scheme 195.

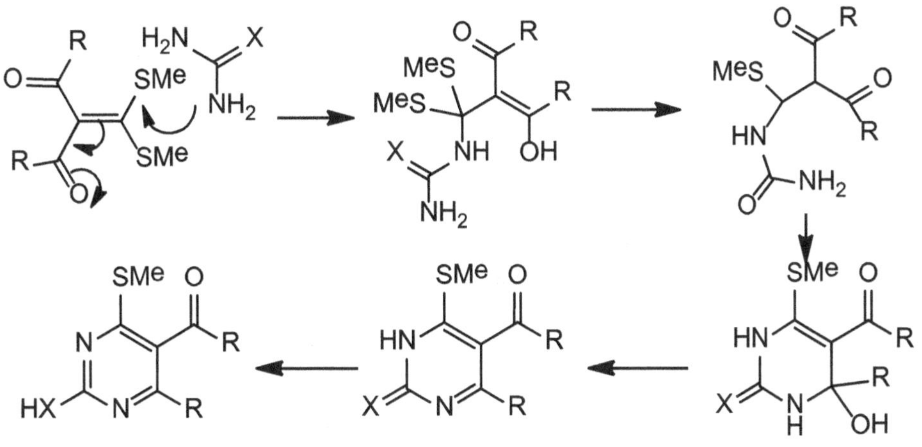

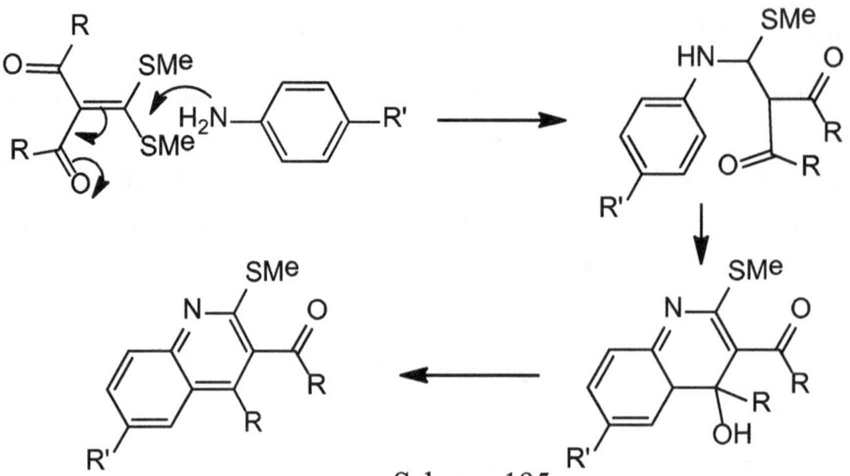

Scheme 195

4.8 α, α-UNSATURATED CARBONYL COMPOUNDS

α,β-Unsaturated carbonyl compounds are an important class of carbonyl compounds with the general structure RCH=CHCOR' for example enones and enals. In general, a compound that contains both a carbon-carbon double bond and a carbon-oxygen double bond has properties that are characteristic of both functional groups. At the carbon-carbon double bond an unsaturated ester or unsaturated ketone undergoes electrophilic addition of acids and halogens, hydrogenation, hydroxylation, and cleavage; at the carbonyl group it undergoes the nucleophilic substitution typical of an ester or the nucleophilic addition typical of a ketone.

The carbonyl group draws electrons away from the alkene, and the alkene group is, therefore, deactivated towards an electrophile. Unlike the case for simple carbonyls, α,β-unsaturated carbonyl compounds are often attacked by nucleophiles at the β carbon. This pattern of reactivity is called vinylogous. Examples of unsaturated carbonyls are acrolein (propenal), mesityl oxide, acrylic acid, and maleic acid. There are several general ways to make compounds of this kind: the aldol condensation, to make unsaturated aldehydes and ketones; dehydrohalogenation of α-halo acids and the Perkin condensation, to make unsaturated acids (Table).

Table: Examples of α,β-Unsaturated Carbonyl Compounds

Name	Structure	m.p	b.p
Acrolein	CH_2=CH-CHO	-88	52
Crotonaldehyde	CH_3CH=CH-CHO	-69	104
Cinnamaldehyde	C_6H_5CH=CH-CHO	-7	254
Mesityl oxide	$(CH_3)_2$C=CH-$COCH_3$	42	131
Benzalacetone	C_6H_5CH=CH-$COCH_3$	42	261
Dibenzalacetone	C_6H_5CH=CH-COCH=CHC_6H_5	113	
Benzalacetophenone (Chalcone)	C_6H_5CH=CH-COC_6H_5	62	348
Acrylic acid	CH_2=CH-COOH	12	142
Crotonic acid	CH_3CH=CH-COOH (trans)	72	189
Isocrotonic acid	CH_3CH=CH-COOH (cis)	16	172

Name	Structure	m.p	b.p
Methacrylic acid	$CH_2=C(CH_3)COOH$	16	162
Maleic acid	HOOCCH=CH-COOH (cis)	130	
Fumeric acid	HOOCCH=CH-COOH (trans)	302	
Cinnamic acid	$C_6H_5CH=CH-COOH$	137	300
Methyl acrylate	$CH_2=CH-COOCH_3$		80
Ethyl cinnamate	$C_6H_5CH=CH-COOC_2H_5$	12	271
Acrylonitrile	$CH_2=CH-CN$	-82	79

The C=O, COOH, COOR, and CN groups are powerful electron withdrawing groups, and therefore would be expected to deactivate a C-C double bond toward electrophilic addition. This is found to be true: α, β-unsaturated ketones, acids, esters, and nitriles are in general less reactive than simple alkenes toward reagents like bromine and the hydrogen halides. But this powerful electron withdrawal, which deactivates a carbon-carbon double bond toward reagents seeking electrons, at the same time activates toward reagents that are electron-rich. As a result, the carbon-carbon double bond of an α,β-unsaturated ketone, acid, ester, or nitrile is susceptible to nucleophilic attack, and undergoes a set of reactions, nucleophilic addition, that is uncommon for the simple alkenes.

4.8.1 Reactions of αβ-unsaturated carbonyl compounds

4.8.1.1 Electrophilic addition:

The presence of the carbonyl group not only lowers the reactivity of the carbon-carbon double bond toward electrophilic addition, but also controls the orientation of the addition. In general, it is observed that addition of an unsymmetrical reagent to an α,β-unsaturated carbonyl compound takes place in such a way that hydrogen becomes attached to the a-carbon and the negative group becomes attached to the β-carbon.

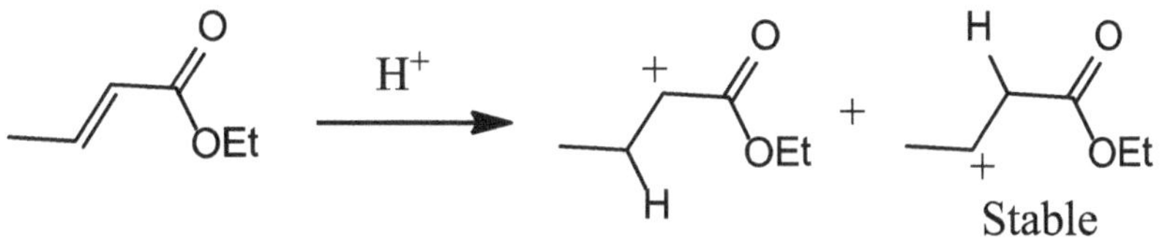

(Eqaution 83)

For instance,

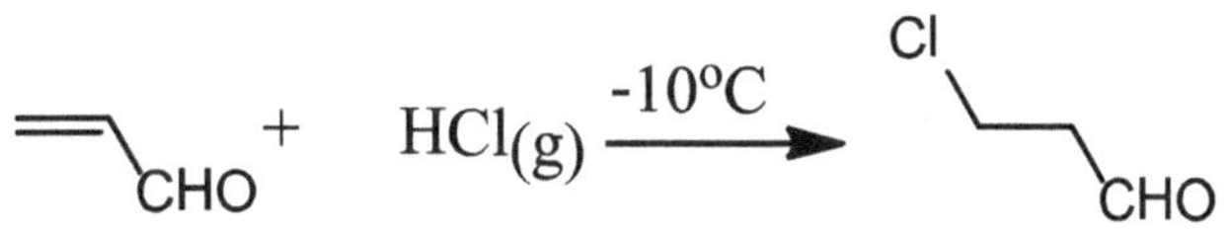

(Equation 84)

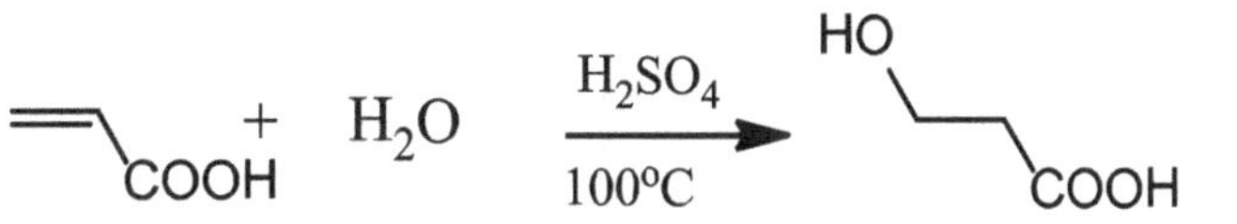

(Equation 85)

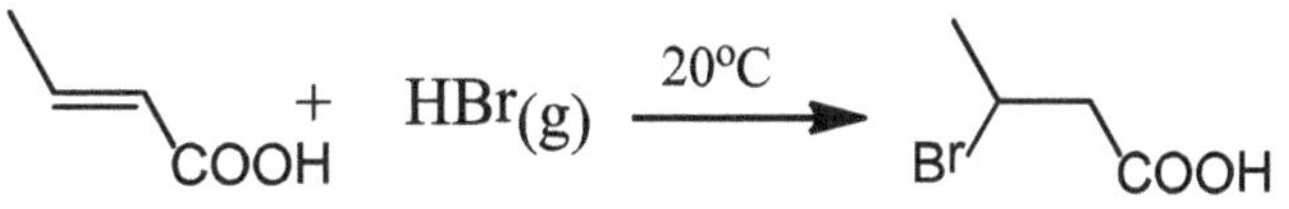

(Equation 86)

4.8.1.2   Nucleophilic addition:

Aqueous sodium cyanide converts α,β-unsaturated carbonyl compounds into β-cyano carbonyl compounds. The reaction amounts to addition of the elements of HCN to the carbon-carbon double bond. For example:

An α,β-unsaturated carbonyl system may be looked upon as a single functional group consisting of two parts namely a carbonyl and a C=C part. When one specific part undergoes the reaction of a single functional group and no other, the reaction is called a regioselective reaction or we may say that the reaction has regioselectivity. For instance, HCN adds to methyl vinyl ketone in two ways (Scheme 196):

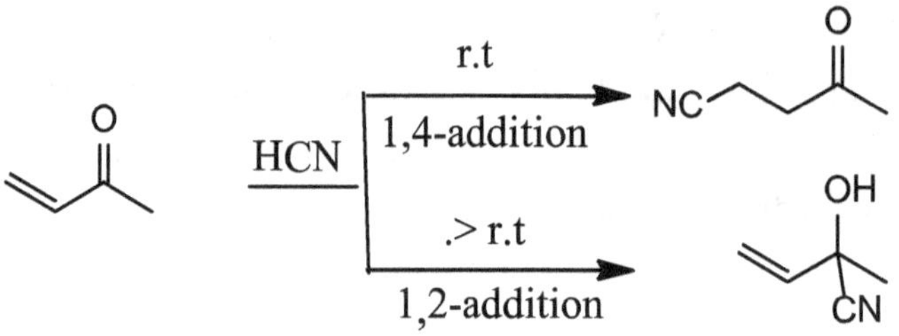

Scheme 196

The elements of HCN appear to have added across the C=C bond rather than across the C=O bond. Yet this is not a normal reaction of the double bond because addition of HCN to simple alkenes does not occur. Addition to the double bond occurs because a resonance-stabilized enolate ion intermediate forms. The enolate ions can protonate on either oxygen or carbon. In either case the carbonyl group is eventually regenerated. The overall result of the reaction is net addition to the double bond (Scheme 197).

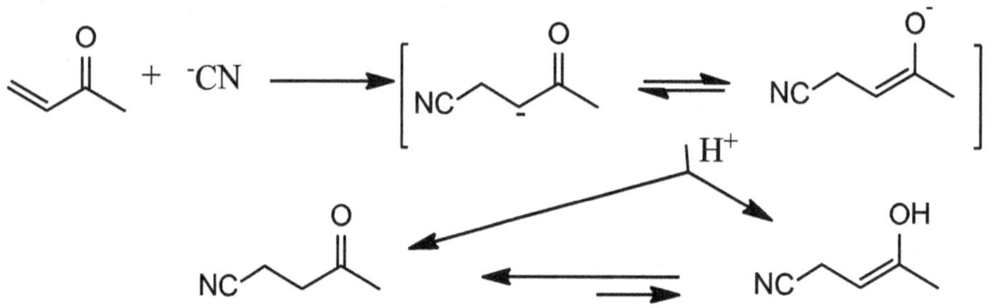

Scheme 197

Here the two addition reactions and their products are regeioselective, once the reaction gives 1,2-product and the other time 1,4-addition product which tautomerises to 3,4-addition product.

Since C=O is more stable than C=C under identical environment the more stable C=O is preserved and the weaker C=C bond is destroyed, 1,4- or conjugate addition product is a thermodynamically controlled product (C-C π bond dissociation energy is 264 kJmol^{-1} but C-O π bond dissociation energy is 390 kJmol^{-1}). Kinetically, C_4 atom is softer and the carbonyl C is a harder one. So, strongly basic or hard nucleophiles tend to undergo 1,2-addition through hard-hard interaction while soft or weakly basic nucleophiles tend to undergo 1,4-addition through soft-soft interaction. The 1,4-addition is more easily reversed than the 1,2-addition. Hence the more stable the nucleophile or the more reversible the reaction, the more conjugate addition is favoured.

4.8.1.3   Metal dissolved reduction

Neither simple alkene nor simple ketones undergo the lithium-ammonia reduction. Conjugated π-systems such as α,β-unsaturated carbonyl compounds, aromatic compounds and dienes are reduced because the radical anion intermediates are resonance stabilized. The double bond of an α,β-unsaturated carbonyl compounds are reduced by a solution of lithium in liquid ammonia. An alcohol is typical added to the reaction mixture as a source of protons. The reaction is very similar to the dissolving-metal reductions of alkynes.

Mechanism: When lithium is dissolved in liquid ammonia, the electrons donated by lithium gets solvated by liquid ammonia. In this reaction, an electron is added to the conjugated π-system of the α,β-unsaturated ketones to give a radical anion, which is stabilized by resonance. Protonation on the oxygen yields a resonance-stabilized radical, which is further reduced to an anion by a second electron. Protonation of the anion by quenching with water give enol which is then isomerized to ketone.

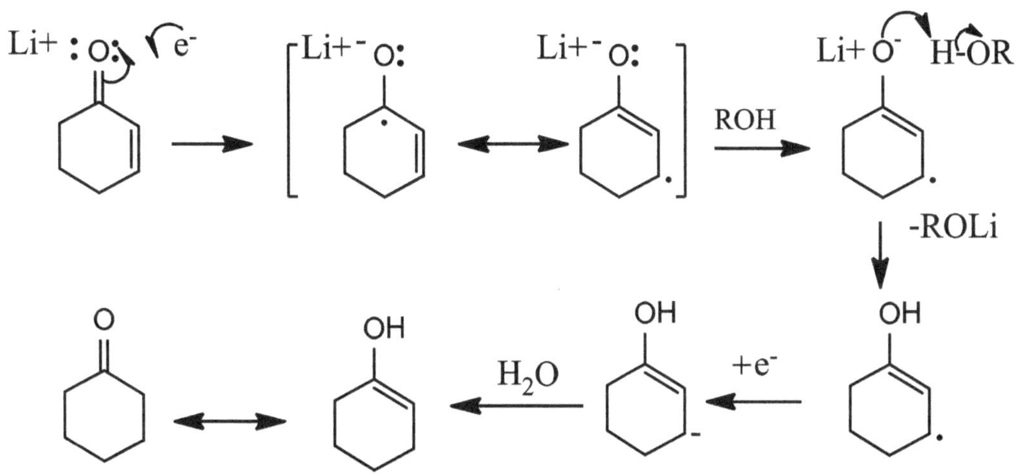

Scheme 198

The Li/NH$_3$ reductions of fused ring alkenes gave the product with trans geometry. For this reason, this reaction is very important to steroid chemistry.

4.8.1.4 Addition of organometallic compounds

In general RLi, $^-$NH$_2$, RO$^-$ and H$^-$ (all are hard nucleophiles) are more likely to add direct; RS$^-$, neutral amines, RMgX and stable carbanions (less hard nulceophiles) are more likely to add Michael like. For the reason noted above when an α,β-unsaturated ester is hydrogenated the weaker C=C gets reduced much readily. On the other hand when an α,β-unsaturated carbonyl compound is reduced with LAH or NaBH$_4$, the harder site C=O gets reduced by the hard H$^-$ ion (Scheme 199).

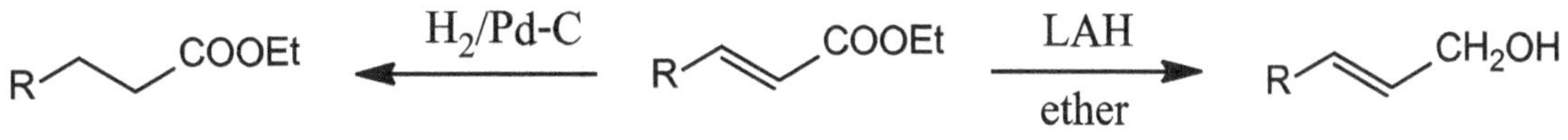

Scheme 199

However, to what extent a given nucleophile undergoes the direct addition and the conjugate addition that depends on the steric and electronic factors. If the carbonyl carbon is more sterically congested and weakly electrophilic, the conjugate addition will occur more readily than the direct addition; on the other hand, if the β-C is more sterically congested, the direct addition will take place predominantly. For instance:

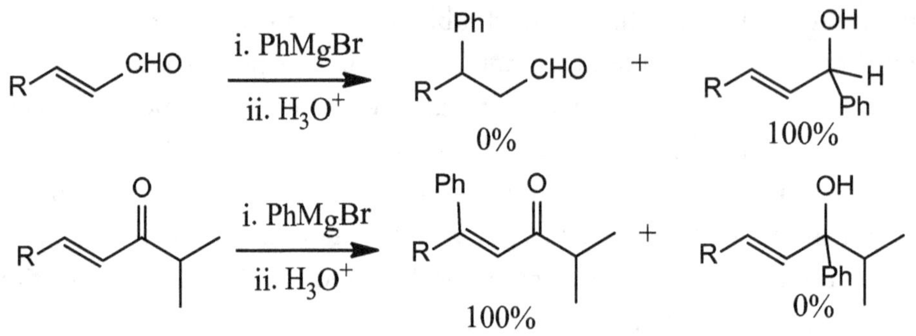

Scheme 200

4.8.1.5   Reaction with organocuprate reagent:

Reaction of lithium organocuprate with α,β-unsaturated carbonyl esters and ketones exclusively yields 1,4-addition products like the addition of Grignard reagents.

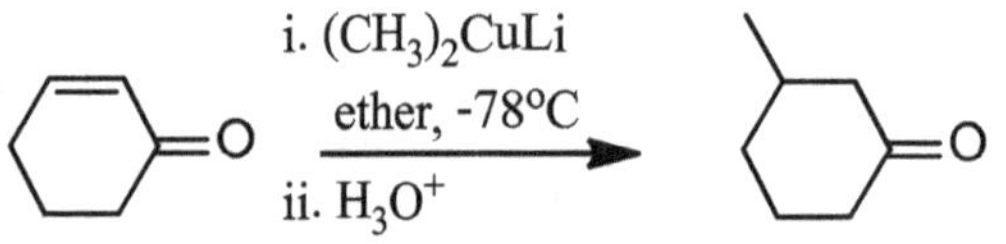

(Equation 87)

The mechanism for the conjugate addition of lithium organocuprtaes is believed to involve radical anion intermediates. Organocopper compounds are excellent sources of reducing electrons. The copper donates a single electron to the α,β-unsaturated carbonyl compound to yield a radical anion. An R.is then transferred to the terminus of the double bond to yield an enolate ion and R-Cu. Addition of water converts the enolate into the saturated ketone.

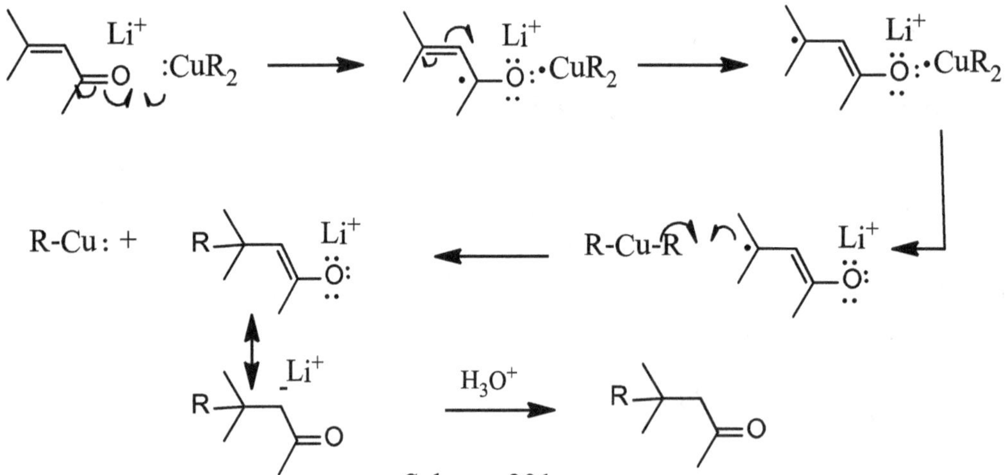

Scheme 201

In principle, the R. could also be transferred to the carbonyl carbon, which shares the unshared electron by resonance. The transfer of R. to the β-carbon is the crucial step that establishes the reaction as a conjugate addition; a transfer to the carbonyl carbon would result in a 1,2-addition.1,4-Addition is preferred because the transfer of R. to the β-carbon gives a resonance stabilized enolate ion, whereas transfer to the carbonyl carbon would give a less stable localized anion. This mechanism bears a close resembles to the dissolving metal reductions.

An addition to the carbon-carbon multiple bond α,β-unsaturated aldehyde, ketone, ester or nitrile is rather general reaction that is observed with many different nucleophiles.

4.8.1.6   Michael addition

An addition to the double bond of an α,β-unsaturated carbonyl compound is called a conjugate addition. The nucleophilic addition of a carbanion or another nucleophile to an α,β-unsaturated carbonyl compound is called Michael addition.

The Robinson annulation is a chemical reaction used in organic chemistry for ring formation. It was discovered by Robert Robinson in 1935 as a method to create a six membered ring by forming three new carbon–carbon bonds *[1935JCS1285, 1971TL4995]*. The first step in the Robinson annulation is a Michael addition to form a 1,5-dicarbonyl compound. In the second step, the Michael adduct undergoes an intramolecular aldol reaction to give cyclic α,β-unsaturated ketones. Basic conditions are required for the reaction. A variety of acceptors and donors can be used in Robinson annulation. This procedure is one of the key methods to form fused ring systems.

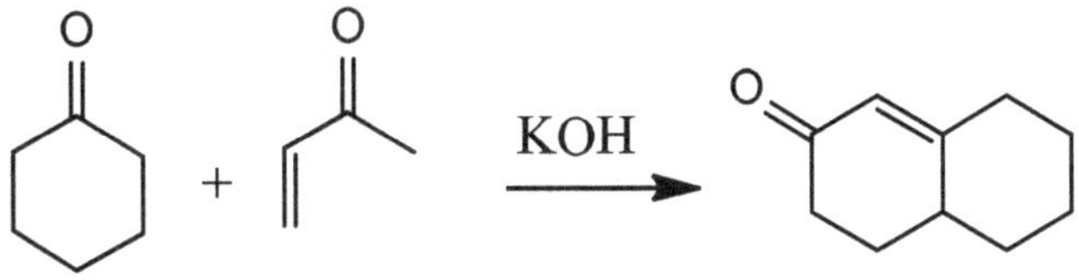

(Equation 88)

Formation of cyclohexenone and derivatives are important in chemistry for their application to the synthesis of many natural products and other interesting organic compounds such as antibiotics and steroids. Specifically, the synthesis of cortisone is completed through the use of the Robinson annulation. It remains one of the key methods for the construction of six membered ring compounds.

In the Michael reaction, the ketone is deprotonated by a base to form an enolate nucleophile which attacks the electron acceptor (vinyl methyl ketone) as we can see in the scheme 202. The resultant enolate undergo rearrangement to give the carbanion, which will undergo intramolecular aldol condensation to form β-keto-cyclic ketone alkoxide ion. Later it undergoes protonation followed by dehydration to lead the product. This sequence is an example of a Robinson annulation.

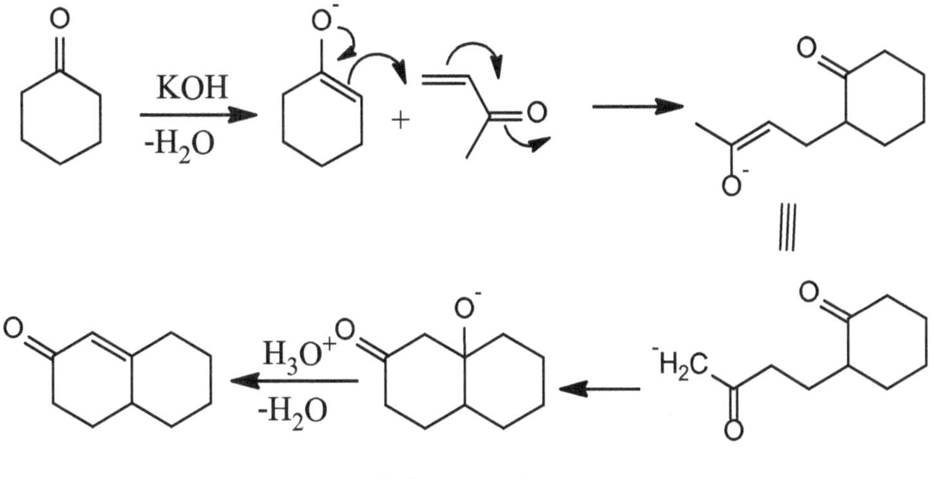

Scheme 202

4.8.1.7   Stetter reaction

The **Stetter reaction** is a reaction used in organic chemistry to form carbon-carbon bonds through a 1,4-addition reaction utilizing a nucleophilic catalyst. The reaction provides synthetically useful 1,4-dicarbonyl compounds and related derivatives from aldehydes and Michael acceptors. Unlike 1,3-dicarbonyls, which are easily accessed through the Claisen condensation, or 1,5-dicarbonyls, which are commonly made using a Michael reaction, 1,4-dicarbonyls are challenging substrates to synthesize, yet are valuable starting materials for several organic transformations, including the Paal–Knorr synthesis of furans and pyrroles. Traditionally utilized catalysts for the Stetter reaction are thiazolium salts and cyanide anion, but more recent work toqward the asymmetric Stetter reaction has found triazolium salts to be effective. The Stetter reaction is an example of umpolung chemistry, as the inherent polarity of the aldehyde is reversed by the addition of the catalyst to the aldehyde, rendering the carbon center nucleophilic rather than electrophilic.

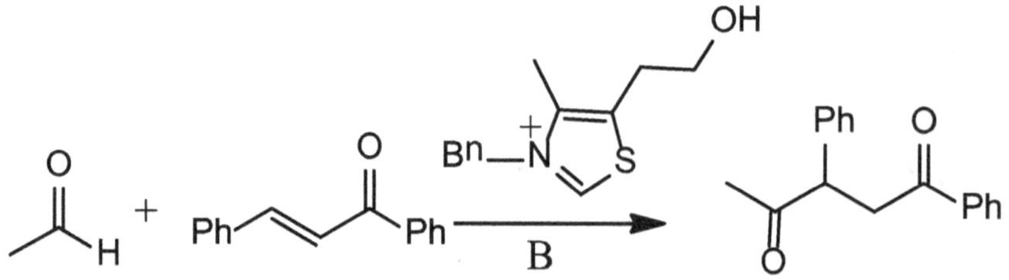

(Equation 89)

The thiazolium ylide formed by the reaction of thiazolium salts with base can add into the aldehyde substrate led to the generation of Breslow intermediate analogous to cyanohydrin anion. This carbanion undergo nucleophlic 1,4-addition to chalocone to form enolate ion, which then rearranges to 1,4-dicarbonyl compound with the release of thiazolium ylide (Scheme 203).

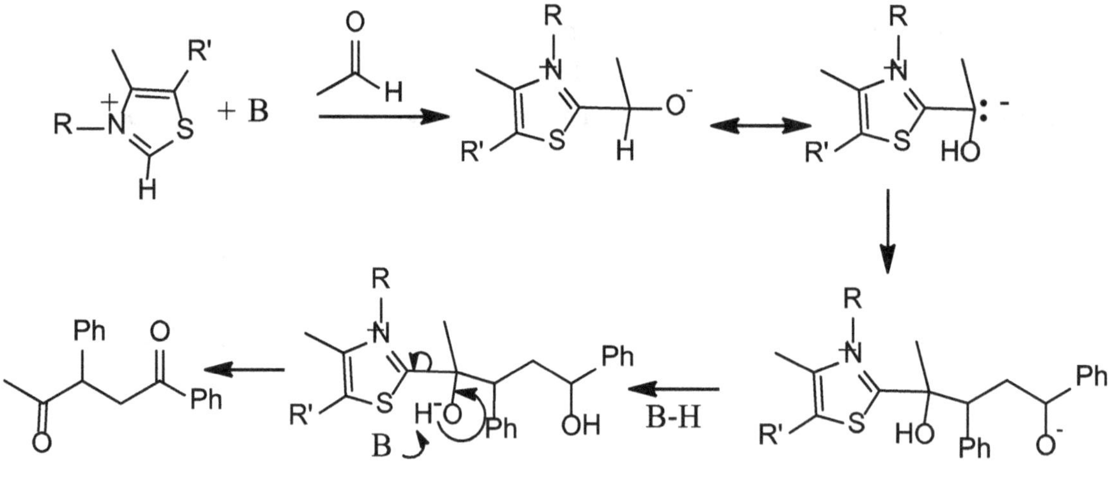

Scheme 203

An example of an asymmetric synthesis by conjugate addition is the synthesis of (R)-3-phenyl-cyclohexanone from cyclohexenone, phenylboronic acid, a rhodium acac catalyst and the chiral ligand BINAP (Equation 90) [2004Misc609].

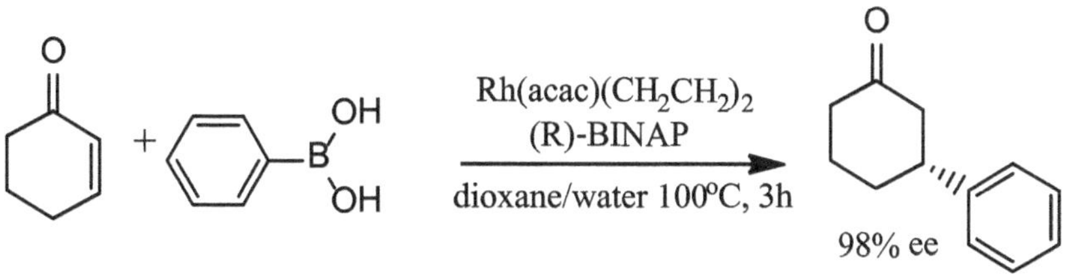

(Eqaution 90)

EXERCISE

1. Outline the synthesis of RCH2R' from RCHO via 2-alkyl 1, 3-dithiane.
2. Alkyl vinyl ethers from aldehydes possessing α-hydrogen have been used as acyl anion equivalents. Justify this with suitable example.
3. Discuss briefly acyl anion equivalents.
4. Rate of iodoform reaction is independent of the concentration of halogen added. Justify this statement.
5. How do you effect the following conversion? Write the mechanism.

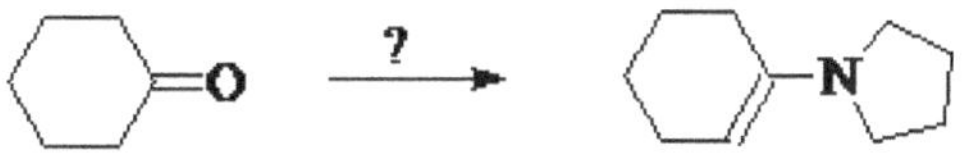

6. The pyrrolidine enamine of α-methyl cycloheanone is failed undergo further alkylation. Why?
7. With suitable example discuss Robinson annulation.
8. Favoroskii rearrangement occurs via cyclopropanone intermediate. Justify.

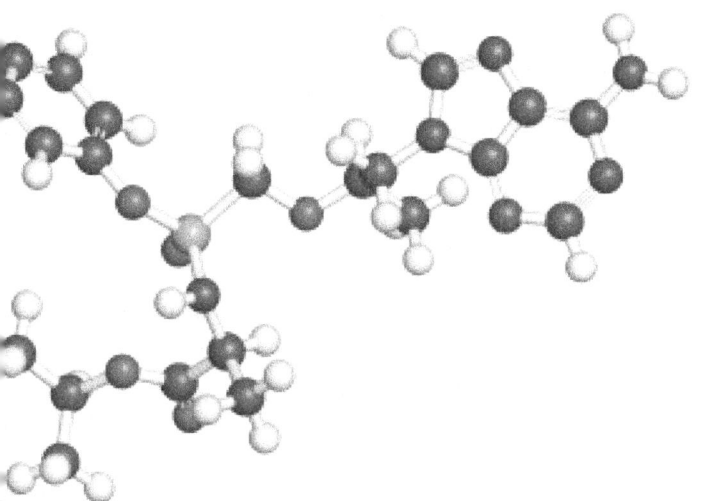

CARBOXYLIC ACID DERIVATIVES

Carboxylic acid derivatives are compounds that can be hydrolyzed under acidic or basic conditions to the corresponding carboxylic acids. Carboxylic acid and their derivatives rank with aldehydes and ketones among the most important organic compounds. Not only do they occur widely in nature, but they also play important role in organic synthesis. Except for nitriles, all carboxylic acid derivatives contain a carbonyl group. Many important reactions of these compounds occur at the carbonyl group. The cyano group of the nitrile also has reactivity very similar to that of a carbonyl group.

As implied by the name, carboxylic acids are among the more acidic organic compounds. Some of their reactions and properties relate directly to this acidity. Carboxylic acids are also carbonyl compounds, however and the reactivity of the carbonyl group plays a very important role in the reactions of carboxylic acid. Finally, the α-position in carboxylic acids and in some acid derivatives) especially esters) is involved in many reactions, as it is in aldehydes and ketones.

5.1   NOMENCLATURE OF CARBOXYLIC ACIDS

Carboxylic acids contain –COOH function. The straight-chain monocarboxylic acids were isolated from natural sources; e.g., from natural fats and waxes by hydrolysis. So these acids in general are known as fatty acids. These were given common names indicating their source. To name a fatty acid, the last syllable from the Latin (L) or Greek (Gr) name of its source is dropped and to that "ic" is added. For instance, formic acid, HCOOH, is an important component in the venom of red ant (Latin, formica = ant); acetic acid is the acidic component of vinegar (Latin, acetus = vinegar) and butyric acid is foul-smelling component of rancid butter (Latin, butyrum = butter).

Common nomenclature is widely used for the simpler carboxylic acids. A carboxylic acid is named by adding the suffix "ic" followed by the word acid to the prefix for the appropriate acyl group give in table. For instance, for CH_3COOH, prefix: acet + ic acid = acetic acid. In the common system, substitution, as usual, is denoted with Greek letter; the position adjacent to the carboxylic acid group is designated as α. For instance, $CH_3CH_2CHBrCOOH$ is α-bromobutryic acid.

In the IUPAC system the principal chain of carboxylic acid is the carbon chain containing the greatest number of carboxyl groups. The name of the carboxylic acid is derived from the hydrocarbon with the same number of carbons as the principal chain by dropping the final –e and adding he suffix –oic followed by the word acid. The final –e is not dropped in the name of dicarboxylic acid.

$$CH_3CH_2COOH \quad \text{propan + oic = propanoic acid}$$

$$HOOCCH_2CH_2CH_2CH_2CH_2CH_2COOH \quad \text{octanedioic acid}$$

An addition to the carbon-carbon multiple bond α,β-unsaturated aldehyde, ketone, ester or nitrile is rather general reaction that is observed with many different nucleophiles.

When the carboxylic acid is derived from a cyclic hydrocarbon, - carboxylic acid is added to the name of hydrocarbon.

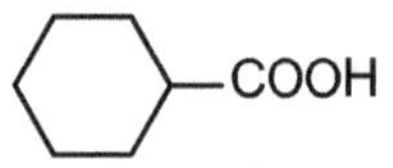

cyclohexanecarboxylic acid 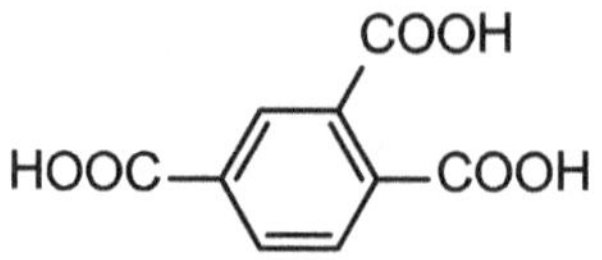   1,2,4-benzenetricarboxylic acid

An addition to the carbon-carbon multiple bond α,β-unsaturated aldehyde, ketone, ester or nitrile is rather general reaction that is observed with many different nucleophiles.

The numbering of the principal chain in substituted carboxylic acid, as in aldehydes, gives the carbonyl carbon the number 1. In carboxylic acids derived from cyclic hydrocarbons, numbering begins at the ring carbon bearing the carboxyl group.

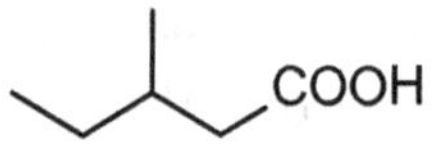

3-methylpentanoic acid 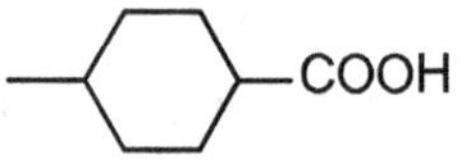   4-methyl cyclohexanecarboxylic acid 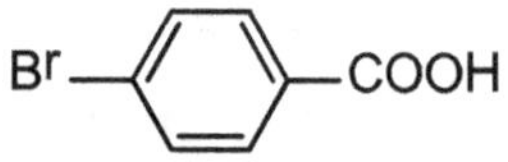   4-bromobenzoic acid

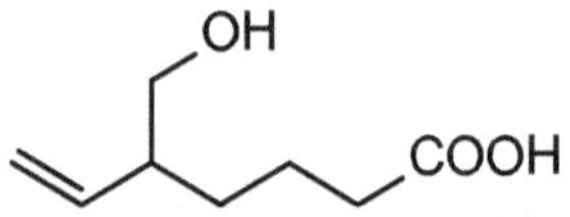

5-vinyl-6-hydroxy   2-(2-hydroxyethyl)-   3-carboxymethyl
hexanoic acid   3-heptanedioic acid   hexanedioic acid

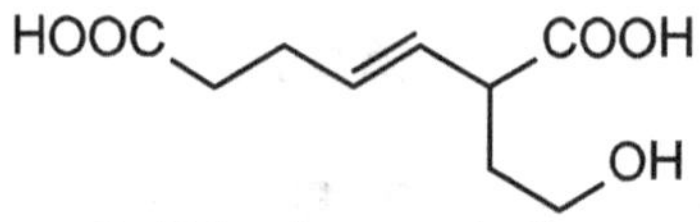

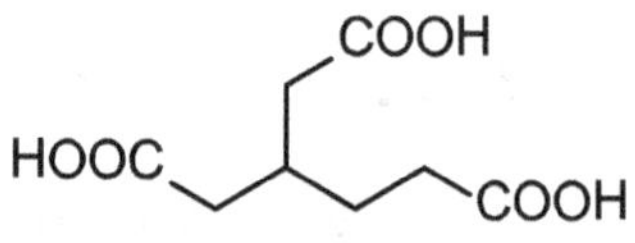

Table: Structure and names of carboxylic acids

Structure	Common name	IUPAC name
HCOOH	Formic acid	Methanoic acid
CH_3COOH	Acetic acid	Ethanoic acid
CH_3CH_2COOH	Propionic acid	Propanoic acid
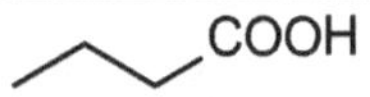 ⌐COOH	Butyric acid	Butanoic acid
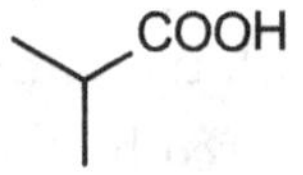 ⌐COOH	Isobutyric acid	2-methylpropanoic acid
 ⌐COOH	Valeric acid	Pentanoic acid

Structure	Common name	IUPAC name
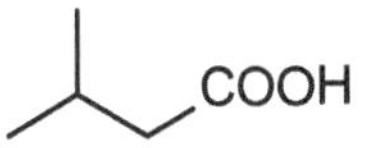 (isovaleric acid structure) COOH	Isovaleric acid	3-methylbutanoic acid
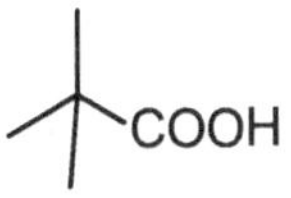 (pivalic acid structure) COOH	Pivalic acid	2,2-dimethylpropanoic acid
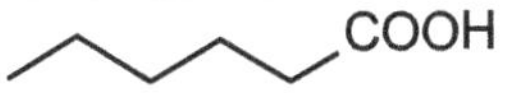 (caproic acid structure) COOH	Caproic acid	Hexanoic acid
$CH_3(CH_2)_5CH_2COOH$	Caprylic acid	Octanoic acid
$CH_3(CH_2)_7CH_2COOH$	Capric acid	Decanoic acid
$CH_3(CH_2)_9CH_2COOH$	Lauric acid	Dodecanoic acid
$CH_3(CH_2)_{11}CH_2COOH$	Myristic acid	Tetradecanoic acid
$CH_3(CH_2)_{13}CH_2COOH$	Palmetic acid	Hexadecanoic acid
$CH_3(CH_2)_{15}CH_2COOH$	Stearic acid	Octadecanic acid
 (oleic acid structure) $(CH)_7COOH$	Oleic acid	Cis-9-octdecaenoic acid
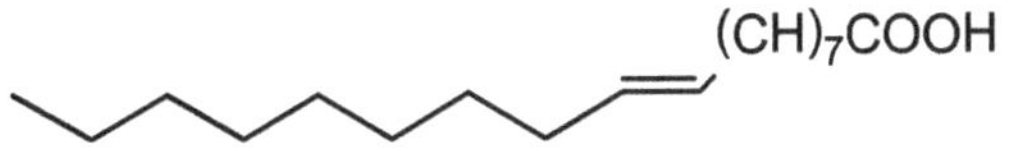 (elaidic acid structure) $(CH)_7COOH$	Elaidic acid	trans-9-octdecaenoic acid
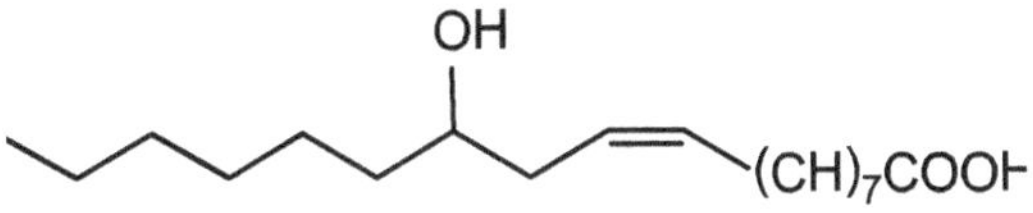 (recinoleic acid structure) OH $(CH)_7COOH$	Recinoleic acid	Cis-12-hydroxy-octadeca-9-enoic acid
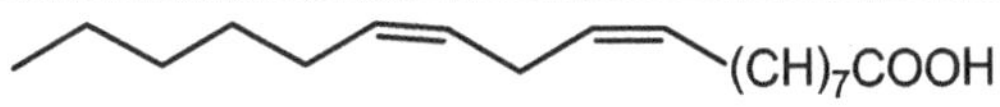 (linoleic acid structure) $(CH)_7COOH$	Linoleic acid	Cis,cis-9,12-octadecadienoic acid
$CH_3(CH_2)_{17}CH_2COOH$		Eicosonoic acid
$CH_3(CH_2)_{19}CH_2COOH$		Doeicosonoic acid
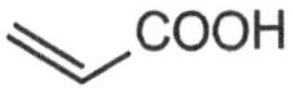 (acrylic acid structure) COOH	Acrylic acid	2-propenic acid
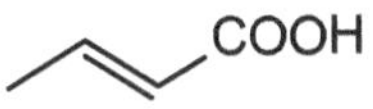 (crotonic acid structure) COOH	Crotonic acid	Trans-2-butanoic acid
HOOCCOOH	Oxalic acid	Ethnaedioic acid
$HOOCCH_2COOH$	Malonic acid	Propanedioic acid
$HOOC(CH_2)_2COOH$	Succinic acid	Butanedioic acid
$HOOC(CH_2)_3COOH$	Glutaric acid	Pentanedioic acid
$HOOC(CH_2)_4COOH$	Adipic acid	Hexanedioic acid
$HOOC(CH_2)_5COOH$	Pimelic acid	Heptanedioic acid
$HOOC(CH_2)_7COOH$	Azeleic acid	Nonanedioic acid
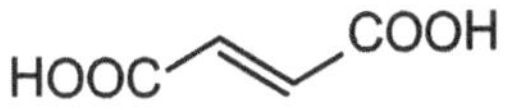 (fumeric acid structure) HOOC—COOH	Fumeric acid	Trans-2-butanedioic acid
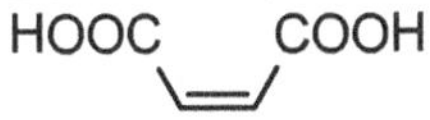 (maleic acid structure) HOOC COOH	Maleic acid	Cis-2-butanedioic acid

5.2 STRUCTURE

The structure of acetic acid illustrates that the carbonyl groups in carboxylic acids have about the same bond lengths as carbonyl groups in aldehydes and ketones. The C-O single bond lengths however are considerably smaller than those of alcohols or ethers (1.36 Å vs 1.43 Å) One reason for this reduction in length is that the C-O bond in an acid is an sp^2-sp^3 single bond, whereas that in an alcohol or ether is an sp^3-sp^3 single bond. Another reason is that carboxylic acids have a resonance structure in which this bond has some double bond character.

The carboxylic acids of lower molecular weight are high-boiling liquids with acrid piercing odors. Many of the aromatic and dicarboxylic acids are solids. Carboxylic acids have significantly boiling points than other types of organic compounds with about the same molecular weight and shape:

| B.P | 117.9° | 82.3° | 56.5° | -6.9° |

The higher boiling point of carboxylic acids can be attributed not only to their polarity but also to the fact that they form very strong hydrogen bonds. In the solid state and under some conditions in both gas phase and solution, carboxylic acids exist as hydrogen-bonded dimers. The equilibrium constants for the formation of such dimers in solution are on the order of 10^6 to $10^7 M^{-1}$

The O-H proton of a carboxylic acid is acidic. The conjugate base of carboxylic acid is a carboxylate ion. Carboxylate salts are named by replacing the -ic in the name of acid with the suffix –ate. An important reason for the acidity of a carboxylic acid is the stability of carboxylate ion, which has two equally important resonance contributors.

Carboxylic acids are much more acidic than aldehydes and ketones because in carboxylates the negative charge resides on a very electronegative oxygen atom instead of a carbon atom. Most carboxylic acids have pKa values in the 4-5 range. However, the acidities of carboxylic acids vary with structure. Electronegative groups near the carboxyl group substantially increase the acidity by an inductive effect. For instance, trifluroacetic acid is almost strong as some mineral acids.

HCOOH (^{Ka} = 3.75)   CH$_3$COOH (4.76)   ClCH$_2$COOH (2.85)   FCH$_2$COOH (2.85)

CH$_3$CH$_2$COOH (2.85)   PhCH$_2$COOH(4.31)   F$_2$CHCOOH (1.24)   F$_3$CCOOH   (0.23)

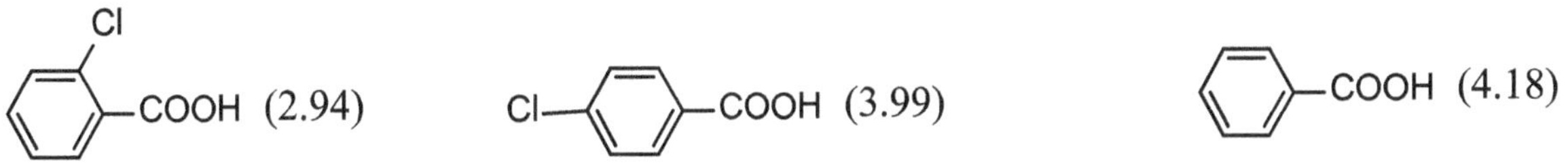

Many carboxylic acids of moderate molecular weight are not soluble in water. Their alkali metal salts however are ionic compounds that in many cases are water soluble. Therefore, many water-insoluble carboxylic acids dissolve in solutions of alkali metal hydroxides (NaOH, KOH) because the insoluble acids are converted completely into their soluble salts (RCOO⁻Na⁺). Even 5% sodium bicarbonate solution is basic enough to dissolve carboxylic acid. A typical carboxylic acid can be separated from mixtures with other water-insoluble, nonacidic substances by extraction with NaOH, Na$_2$CO$_3$ or NaHCO$_3$ solution. The acid selectively dissolves in the aqueous solution and it is readily isolated when the solution is acidified with mineral acid, from which it can be filtered or extracted with organic solvents.

Carboxylic acids, like aldehydes and ketones are weekly basic toward the proton. The basicity of carboxylic acids plays an extremely important role in many of their reactions.

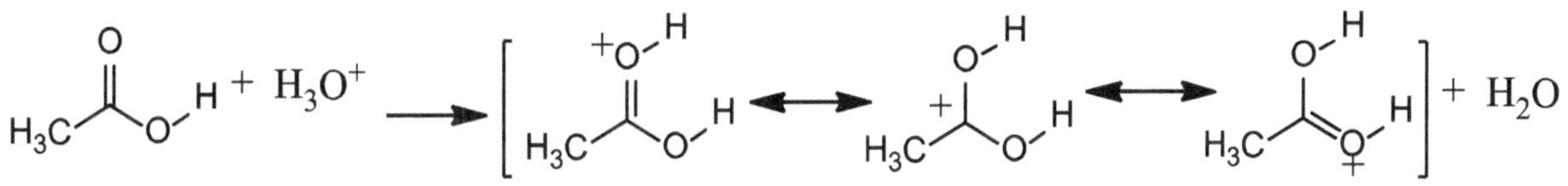

Scheme 204

An addition to the carbon-carbon multiple bond α,β-unsaturated aldehyde, ketone, ester or nitrile is rather general reaction that is observed with many different nucleophiles.

Above scheme shows that protonation on the carbonyl oxygen gives a resonance-stabilized cation. Protonation on the hydroxyl oxygen is much less favorable because it does not give a resonance-stabilized cation.

5.3   SYNTHESIS OF CARBOXYLIC ACID

Synthesis of carboxylic acids can be formed by two ways: i. partial synthesis and ii. total synthesis.

Partial synthesis: partial chemical synthesis is a type of chemical synthesis that uses chemical compounds isolated from natural sources (e.g., microbial cell cultures or plant material) as the starting materials to produce other novel compounds with distinct chemical and medicinal properties. For instance, conversion of unsaturated carboylic acid to saturated carboxylic acid by redcution belongs to one of the method for partial synthesis. The other methods includes-

One carbon degradation:

Barbier–Wieland degradation: The Barbier Weiland degradation is a procedure for shortening the carbon chain of a carboxylic acid by one carbon. It only works when the carbon adjacent to the carboxyl is a simple methylene bridge. The reaction sequence involves conversion of the carboxyl and alpha carbon into an alkene, which is then cleaved by oxidation to convert the former alpha position into a carboxyl itself. [1955OS38]

$$RCH_2COOH \xrightarrow[H^+]{EtOH} RCH_2COOEt \xrightarrow[ii.\ H_3O^+]{i.\ PhMgBr} RCH=CPh_2 \xrightarrow{(O)} RCOOH$$

Scheme 205

The second method for the chain degradation by one carbon involves Hoffmann rearrangement followed by diazotization and oxidation as shown below:

$$RCH_2COOH \xrightarrow[ii.NH_3]{i.PCl_5} RCH_2CONH_2 \xrightarrow[ii.HNO_2]{i.\ NaOH/Br_2} RCH_2OH \xrightarrow{(O)} RCOOH$$

Scheme 206

Two carbon degradation:

In the Gallagher–Hollander degradation [1946JBC549] pyruvic acid is removed from a linear aliphatic carboxylic acid yielding a new acid with two carbon atoms fewer. The reaction scheme is as follows.

$$RCH_2CH_2COOH \xrightarrow[ii.CH_2N_2]{i.PCl_5} RCH_2CH_2COCH=N=N \xrightarrow[\ ]{+\ -\ HCl} RCH_2CH_2COCH_2Cl$$

$$\downarrow {\scriptstyle HAc,\ Zn\ reflux,\ 2h}$$

$$RCOOH + CH_3COCOOH \xleftarrow[ii.\ (O)]{i.\ heat} RCH_2CHBrCOCH_3 \xleftarrow{Br_2,\ HAc} RCH_2CH_2COCH_3$$

Scheme 207

The second method involves the alpha bromination of fatty acids possessing two methylene group adjacent to the carboxylic group followed by oxidation leads the formation of fatty acid with two carbon fewer.

$$RCH_2CH_2COOH \xrightarrow{Br_2} RCH_2CHBrCOOH \longrightarrow RCH=CHCOOH$$

$$RCOOH \xleftarrow{(O)} \rfloor$$

Scheme 208

One carbon elongation;

$$RCOOH \xrightarrow[H^+]{EtOH} RCOOEt \xrightarrow[Ether]{LAH} RCH_2OH \xrightarrow[H^+]{NaBr} RCH_2Br$$

$$RCH_2COOH \xleftarrow[ii.\ CO_2,\ iii\ H^+]{i.Mg/Ether} RCH_2Br \xrightarrow[ii.\ H^+]{i.\ NaCN} RCH_2COOH$$

Scheme 209

Two carbon elongation:

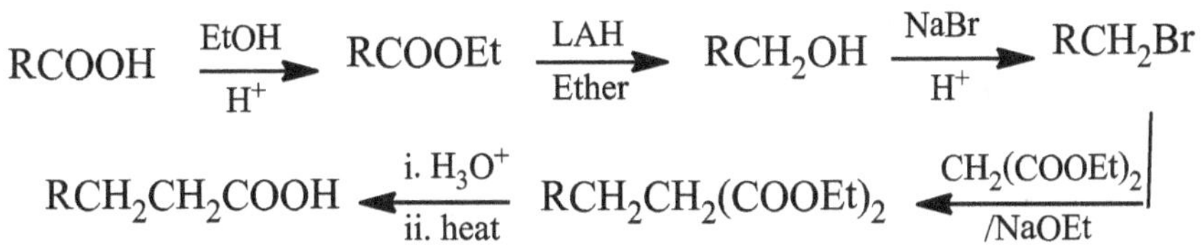

$$RCOOH \xrightarrow[H^+]{EtOH} RCOOEt \xrightarrow[Ether]{LAH} RCH_2OH \xrightarrow[H^+]{NaBr} RCH_2Br$$

$$RCH_2CH_2COOH \xleftarrow[ii.\ heat]{i.\ H_3O^+} RCH_2CH_2(COOEt)_2 \xleftarrow[/NaOEt]{CH_2(COOEt)_2}$$

Scheme 210

Total synthesis

Total synthesis is the complete chemical synthesis of a complex molecule, often a natural product, from simple, commercially available precursors. It usually refers to a process not involving the aid of biological processes, which distinguishes it from semisynthesis. The target molecules can be natural products, medicinally important active ingredients, or organic compounds of theoretical interest. Often the aim is to discover new route of synthesis for a target molecule for which there already exist known routes. Sometimes no route exists and the chemist wishes to find a viable route for the first time. One important purpose of total synthesis is the discovery of new chemical reactions and new chemical reagents.

$$RMgBr + ClCH_2(CH_2)nCN \longrightarrow RCH_2(CH_2)nCN \longrightarrow RCH_2(CH_2)nCOOH$$

Scheme 211

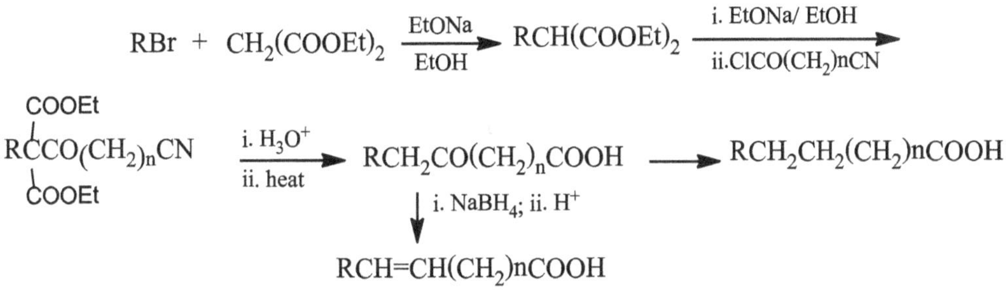

$$RBr + CH_2(COOEt)_2 \xrightarrow[EtOH]{EtONa} RCH(COOEt)_2 \xrightarrow[ii.ClCO(CH_2)nCN]{i.\ EtONa/\ EtOH}$$

$$\underset{\underset{COOEt}{|}}{\overset{\overset{COOEt}{|}}{R}}CCO(CH_2)_nCN \xrightarrow[ii.\ heat]{i.\ H_3O^+} RCH_2CO(CH_2)_nCOOH \longrightarrow RCH_2CH_2(CH_2)nCOOH$$

↓ i. NaBH_4; ii. H^+

$$RCH=CH(CH_2)nCOOH$$

Scheme 212

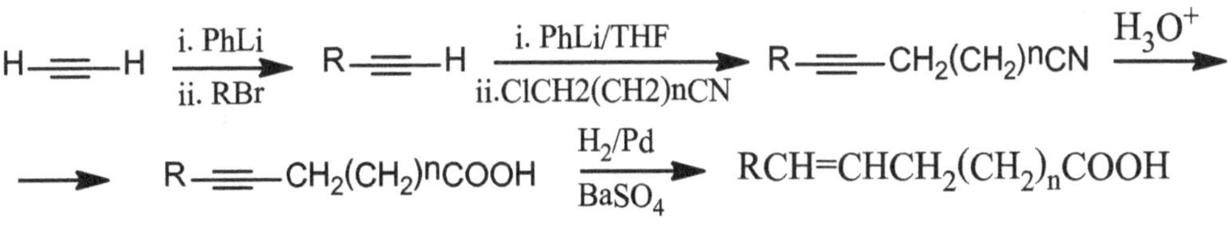

$$H-\!\!\!\equiv\!\!\!-H \xrightarrow[ii.\ RBr]{i.\ PhLi} R-\!\!\!\equiv\!\!\!-H \xrightarrow[ii.ClCH2(CH2)nCN]{i.\ PhLi/THF} R-\!\!\!\equiv\!\!\!-CH_2(CH_2)nCN \xrightarrow{H_3O^+}$$

$$\longrightarrow R-\!\!\!\equiv\!\!\!-CH_2(CH_2)nCOOH \xrightarrow[BaSO_4]{H_2/Pd} RCH=CHCH_2(CH_2)_nCOOH$$

Scheme 213

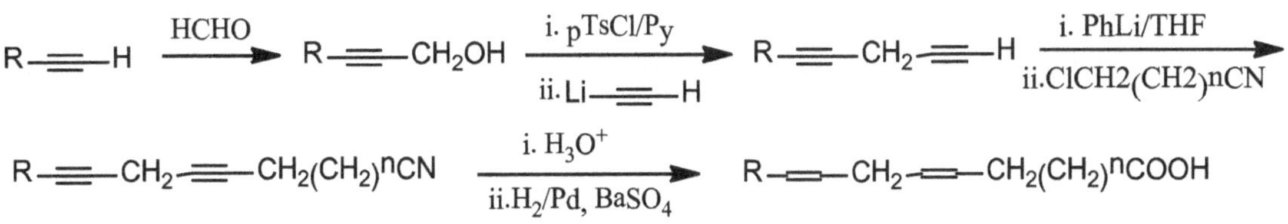

$$R-\!\!\!\equiv\!\!\!-H \xrightarrow{HCHO} R-\!\!\!\equiv\!\!\!-CH_2OH \xrightarrow[ii.Li-\!\!\equiv\!\!-H]{i.\ pTsCl/Py} R-\!\!\!\equiv\!\!\!-CH_2-\!\!\!\equiv\!\!\!-H \xrightarrow[ii.ClCH2(CH2)nCN]{i.\ PhLi/THF}$$

$$R-\!\!\!\equiv\!\!\!-CH_2-\!\!\!\equiv\!\!\!-CH_2(CH_2)nCN \xrightarrow[ii.H_2/Pd,\ BaSO_4]{i.\ H_3O^+} R-\!\!\!\equiv\!\!\!-CH_2-\!\!\!\equiv\!\!\!-CH_2(CH_2)nCOOH$$

Scheme 214

5.4   REACTIONS OF CARBOXYLIC ACIDS

Carboxylic acids undergo many reactions. Important among these are esterification, formation of acid chlorides, amides, condensation and decarboxylation. Although perhaps esterification of acids should be studied first, principle of microscopic reversibility tells us that the mechanism for esterification is just the reverse of that for ester hydrolysis under the same conditions [principle of microscopic reversibility]. In the course of reaction, the nuclei and the electrons assume positions which at each point correspond to the lowest free energy possible. If the reaction is reversible, the position must be the same in the reversible process also. This means that the forward and reverse reaction (run under the same conditions) must proceed by the same mechanism. This is called principle of microscopic reversibility. For example, if in a reaction A→B there is an intermediate C, then C must also be intermediate in the reaction B→A. Accordingly we should deal with ester hydrolysis to which most study has been given.

5.4.1   Hydrolysis of esters

When ester is treated with aqueous acid, it gets hydrolyzed. This is a reversible reaction. The reverse process is known as Fischer esterification. The equilibrium may be displaced on either side depending upon whether water is present in excess (for hydrolysis) or whether it is removed from the reaction in some way (for esterification).

$$RCOOR' + H_2O \rightleftharpoons RCOOH + R'OH \quad \text{(Equation 91)}$$

Ester may be hydrolyzed with base. This process is called saponification (generally used in soap industry).

$$RCOOR' + {}^-OH \rightleftharpoons RCOO^- + R'OH \quad \text{(Equation 92)}$$

Generally this is not a forward reaction, since carboxylic acid forms a salt which displaces the equilibrium towards the forward reaction. Mechanism by which esterification and hydrolysis takes place is not a single one, because:

i. hydrolysis of an ester may be accomplished by acids or by alkaline. And depending upon the reagent, the species undergoing hydrolysis or esterification may be a neutral molecule or a conjugate acid.

neutral   conjugate acid

R' = H, acid
R' = Me, ester

ii. Esterification and hydrolysis may be proceeding either by a unimolecular or by a bimolecular mechanism;

iii. Esterification and hydrolysis of ester may proceed by the cleavage of an ester/acid molecule in two ways.

acyl-oxygen cleavage [Ac]

acyl-alkyl cleavage [AL]

Based on the above factors and also bearings in mind that acid catalyzed reaction is a reversible one, there should be eight possible ($2^3 = 8$) mechanisms for acid catalyzed hydrolysis of ester and ester formation. Out of eight, six have been observed so far. But for alkaline hydrolysis of an ester, there will be only four possible mechanisms because this is not a reversible reaction as the carboxylic acid is converted into resonance stabilized carboxylate anion.

Ingold has suggested a shorthand notations to describe these bewildering number of mechanism in which A = acid medium; B = basic medium; Ac = acyl-oxygen cleavage; A_L = alkyl-oxygen cleavage; 1= unimolecular; 2 = bimolecular. Thus, $B_{Ac}2$ hydrolysis means a bimolecular basic hydrolysis of an ester proceeding through acyl-oxygen cleavage.

Type of mechanism	Hydrolysis	Esterification
$B_{Ac}1$	Not observed	-
$B_{Ac}2$	Very common	-
$B_{AL}1$	special case(hydrolysis of esters of t-alcohols)	-
$B_{AL}2$	Not observed	-
$A_{Ac}1$	Rare – special cases	special cases
$A_{Ac}2$	Very common	Very common
$A_{AL}1$	Very common for t-alcohols)	Very common for t-alcohols
$A_{AL}2$	Not observed	-

Ester saponification - $B_{Ac}2$ – mechanism

$$RCOOR' \ + \ ^-OH \ \rightleftharpoons \ RCOO^- \ + \ R'OH$$

Hydrolysis of esters by hydroxide ions in aqueous solutions is found to be kinetically a second order reaction. i.e., the reactions are bimolecular.

$$\therefore \ Rate = k \ [ester][-OH]$$

The transition states in two reactions may then be assumed to contain one molecule of ester and one molecule of ^-OH ion.

The site at which the ester molecule is broken is established by the experiments of Polanyi et al. They showed that the alkaline hydrolysis of pentyl acetate in water enriched with ^{18}O gave n-pentyl alcohol contain no ^{18}O and the acid obtained contained ^{18}O (Equation 93).

(Scheme 93)

This clearly show that acyl-oxygen bond is cleaved and not alkyl-oxygen bonds. Therefore mechanism is consistent with the observed rate law and bond observed site of bond could be proposed as follows (Scheme 215).

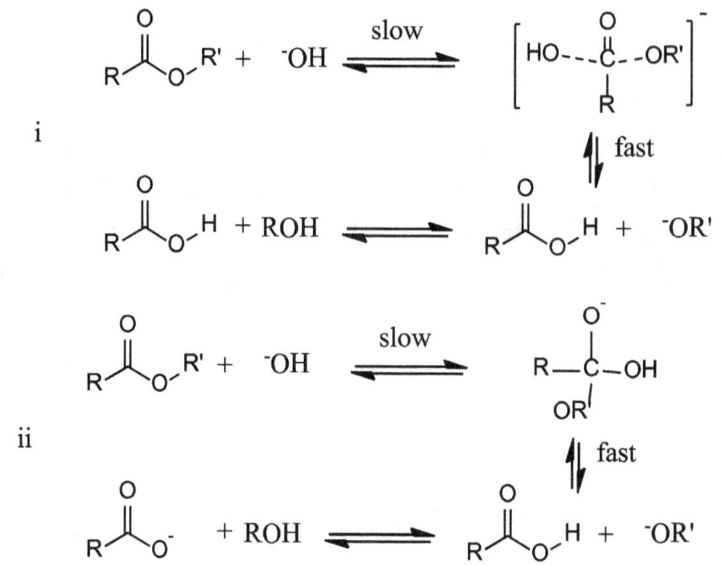

Scheme 215

The overall reaction is irreversible although the first steps are reversible and while the third step is not. Now the anion (i) is it merely an activated complex or actually an intermediate. Which of the above mechanisms is correct?

First one involves a transition state with high content of energy. Second one involves a true intermediate with minimum energy content. Bender solved their problem by carrying a set of experiments carrying labeled oxygen. This time instead of ^{18}O enriched water, the acyl carbonyl oxygen was labeled. He took benzoate (ethyl, isopropyl and t-butyl) labeled at the benzoyl oxygen and subjected to hydrolysis (Equation 94).

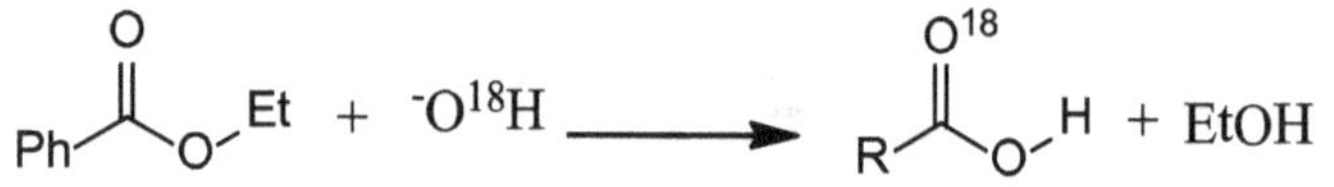

(Equation 94)

An addition to the carbon-carbon multiple bond α,β-unsaturated aldehyde, ketone, ester or nitrile is rather general reaction that is observed with many different nucleophiles.

He did not allow the hydrolysis to go to the completion, he stopped the reaction in the middle and isolated the unreacted ester and measured the ^{18}O content in the ester. The ^{18}O content was found to be less compared to the ^{18}O content in the original ester used for hydrolysis. How could this loss of ^{18}O happen? He explained this assuming oxygen exchange in the following sequence (Scheme 216).

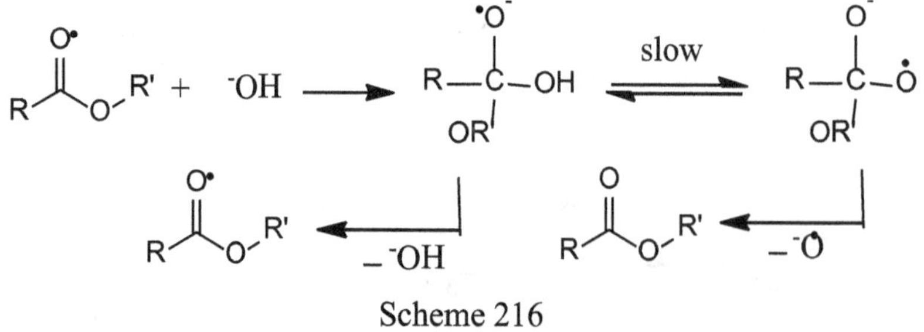

Scheme 216

An addition to the carbon-carbon multiple bond α,β-unsaturated aldehyde, ketone, ester or nitrile is rather general reaction that is observed with many different nucleophiles.

In the exchange reaction, labeled oxygen must depart as -OH departs only after isomerization has taken place. That means the life time of the ion must be sufficient enough to allow this isomerization. It is known that any activated complex not formed and it must be an intermediate (longer life time) that is formed. Thus second mechanism is the most probable one.

If the above mechanism is accepted for saponification two consequences arises: polar effect and steric effect. Presence of electron with drawing groups both on acyl and alkyl sections of the ester molecule should increase the positive charge on the carbonyl carbon atom and facilitate the attack of ⁻OH. Thus such ester should undergo hydrolysis at a faster rate and presence of electron donating group on acyl or alkyl section should increase the negative charge on carbonyl carbon and hence retard the reaction and reduce the rate of hydrolysis. Infact this is observed.

CH_3COOCH_3	1
$CH_3COOCH_2CH_3$	0.6
$ClCH_2COOCH_3$	761
$Cl_2CHCOOCH_3$	16,000
$CH_3OOCCOOCH_3$	170,000

What about CH_3COOCH_3 - faster and $PhCOOCH_3$ – slower. Why?

resonance stabilized

Therefore –OH attack is not facilitated. No such resonance stabilization occurs in CH_3COOCH_3.

In forming intermediate we are increasing the extent of crowding. Anything which increases the crowding should decrease the rate of hydrolysis.

$CH_3COOCH_2CH_3$	1
$CH_3CH_2COOCH_3$	0.47
$(CH_3)_2CHCOOCH_3$	0.1
$(CH_3)_3CCOOCH_3$	0.01
$CH_3COOCH_2CH(CH_3)_2$	0.7
$CH_3COOCH_2C(CH_3)_3$	0.18
$CH_3COOCH_2C(Et)_3$	0.031

In the first series, drop in saponification rate may be attributed mainly due to electron donating inductive effect but in second series, the alkyl groups are far away (four atoms removing) from the carbonyl groups to provide polar effect. Therefore it will be due to steric hindrance only.

Stabilization of the alkoxide ion by electron withdrawing substituents shifts the partitioning in the direction favoring products, leading to an increase in the overall rate. For this reason, exchange of carbonyl oxygen with solvent does not occur in basic hydrolysis when the alkoxy group is a good leaving group (weakly basic). This has been demonstrated in particular for esters of phenols. Because phenols are much stronger acids than simple alcohols, their conjugate bases better leaving group than the simple alkyl groups. Aryl esters are hydrolyzed faster than alkyl esters and without observable exchange of carbonyl oxygen with the solvent.

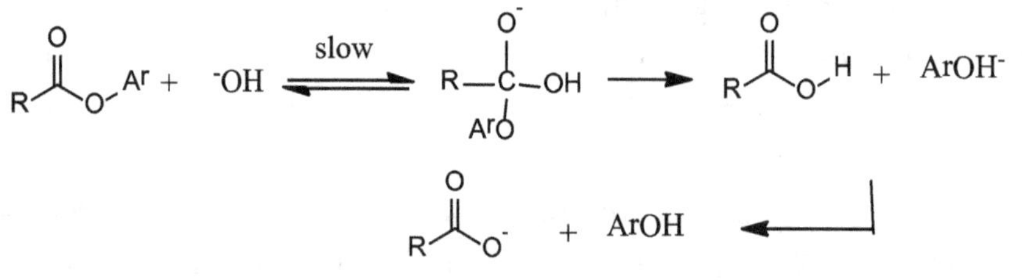

Scheme 217

5.4.1.1   Acid catalyzed hydrolysis and formation of esters $A_{Ac}2$ Mechanism

As mentioned earlier, acid catalyzed hydrolysis of esters differ from that of bases catalyzed hydrolysis in that the former is reversible (experimentally). Its reversal being acid catalyzed esterification, and therefore according to the principle of reversibility, the mechanism of acid catalyzed hydrolysis of esters will automatically establish the mechanism of acid catalyzed esterification but in reverse.

Conjugate acid which takes part in the reaction is proceed by NMR by Franenkal (1961) that, it is the carbonyl oxygen atom that is protonated and not the ester oxygen.

Let us first consider evidences similar to that upon which our mechanism $(B_{AC}2)$ for saponification is based. The rate law for acid catalyzed hydrolysis has been shown to be first order in both ester and hydrogen ion conc, i.e , rate ∞ [ester][H$^+$].Evidence for acyl-oxygen heterolytic cleavage is obtained in several ways.

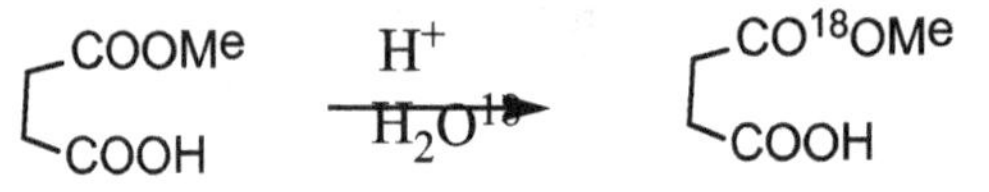

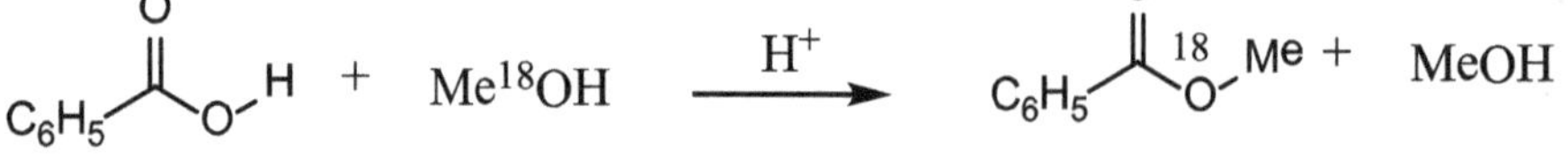

That it proceeds through an intermediate which has enough time to undergo proton shift.

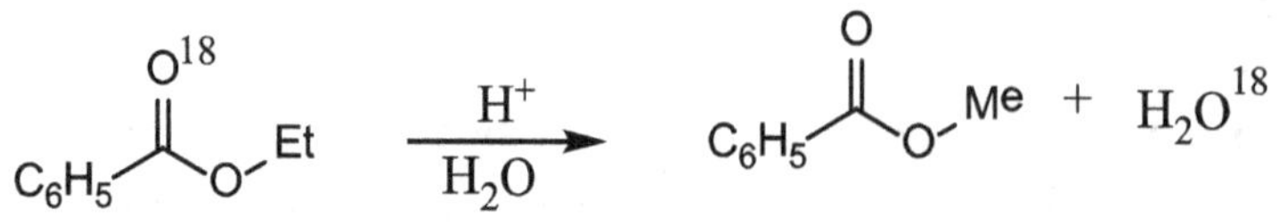

When the reaction was stopped in the middle and checked for oxygen incorporation. Since water is usually in excess, it is difficult to specify the role of water. But when the reaction was done in acetone and known amount of water was added, the rate was found to be proportional to [H_2O]. Further in acid solution of 0.1M concentration, it is ∞ [H_3O^+]. Thus whenever a mechanism is proposed for ester hydrolysis it should account for. All steps should be significantly reversible. Acyl oxygen cleavage should occur. T.S must involve water, H+ and ester. Reaction should pass through an intermediate capable of undergoing proton exchange. Thus the mechanism that fits all this fact is-

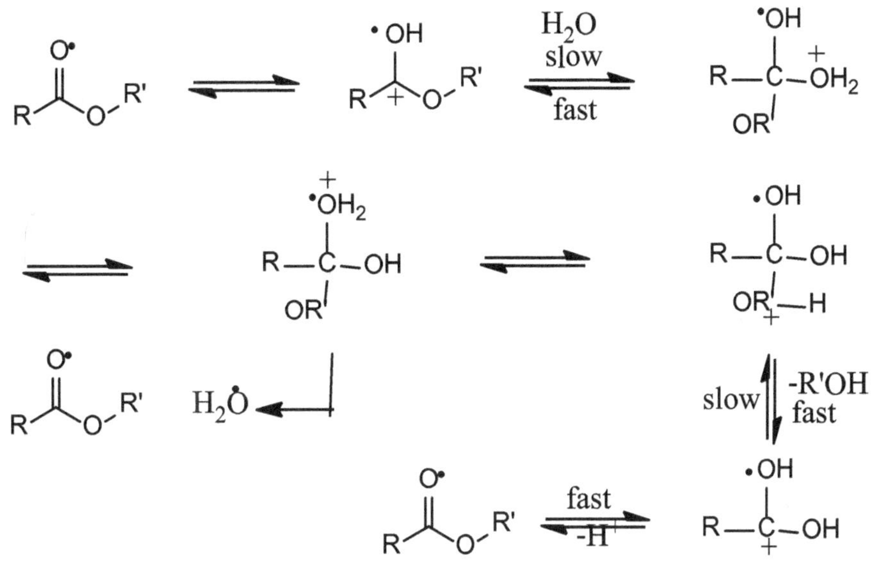

Scheme 218

Therefore rate determining step for ester hydrolysis is addition of water and for esterification is addition of alcohol to acid. Rate determining step involves change of hybridization of the carbon atom of the carbonyl group from sp² to sp³ and consequently steric retardation can be expected to increase as R increase in size. In fact this is observed.

Eg. Relative rate determined (MeOH, 40°C).

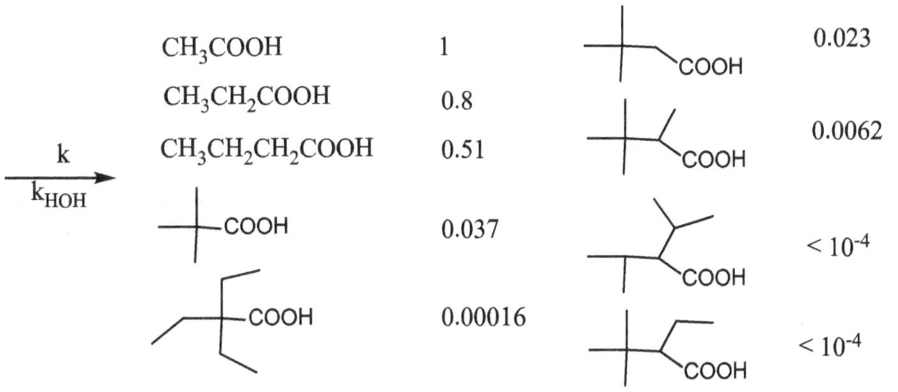

Sharp decrease in rate with the increasing in chain branching β-alkyl substituents are significantly more effective "sterically hindrance" than α-alkyl substituents. For instance, incorporation of 3-methyl groups (α-alkyl) into acetic acid lowers the esterification rate by a factor of 27. Incorporation of methyl group 3β-alkyl into propionic acid lowers the esterification rate by a factor of 36. Newman rule of six: Those atoms which are most effective in providing steric hindrance to addition are separated from the attacking atom in transition state by a chain of 4 atoms.

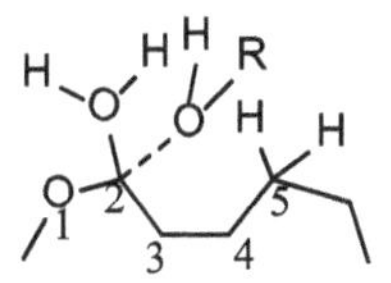

If there are only a few (3 or 6) atoms in sixth position they may be moved out of the way. But if there are (9 or 12) atoms in 6th position it becomes increasingly difficult to twist the acid chain into a permissable conformation.

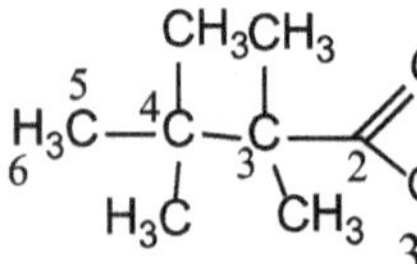

0.00013 3+3+3

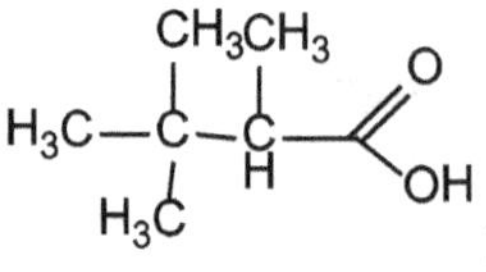

0.00062 3+3+3

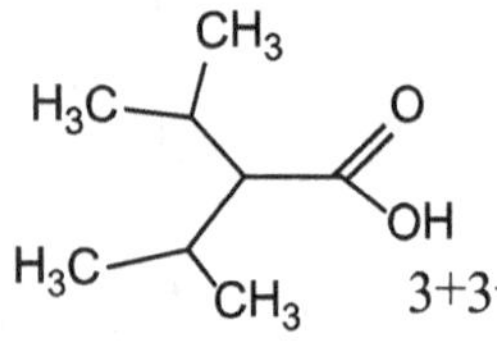

3+3+3+3

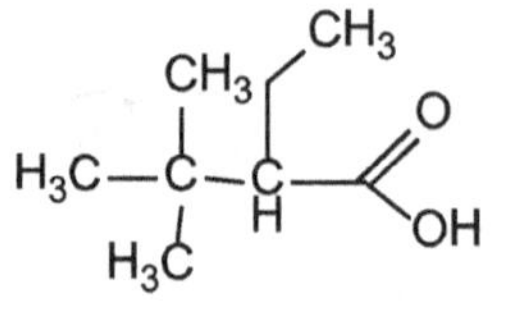

3+3+3+3

In case of aromatic acids, introduction of substituents in ortho position to the –COOH retards the rate of esterification or hydrolysis of ester. This is independent of the polar nature of the substituent (electron attracting or electron repelling).

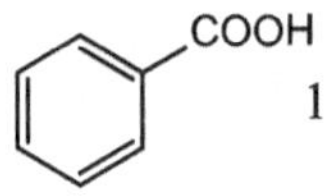

1

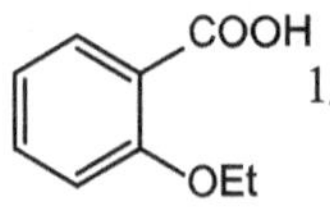

1/5

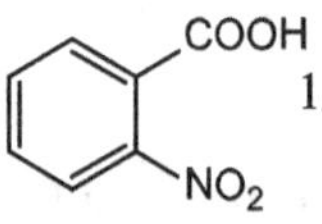

1/32

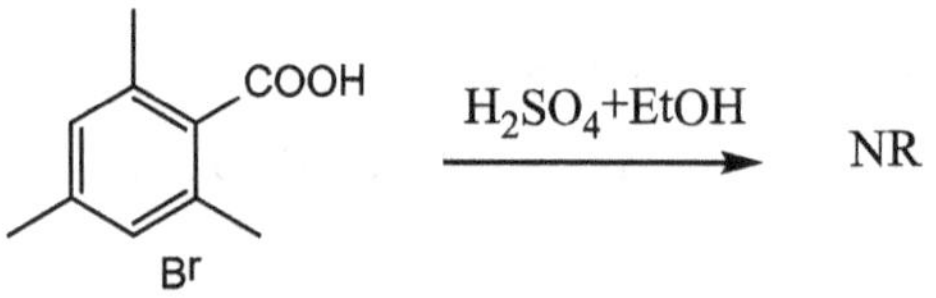

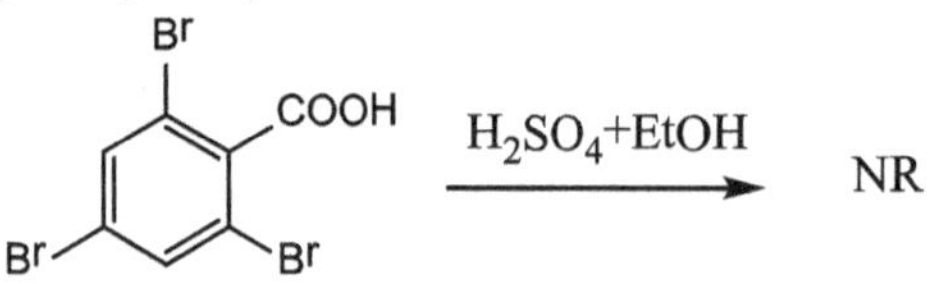

It is possible to shift ester hydrolysis away from the normal $A_{Ac}2$ or $B_{Ac}2$ mechanism by structural changes in the substrate. When the ester is derived from a tertiary alcohol, acid catalyzed hydrolysis often occurs by a mechanism involving alkyl-oxygen fission. The change in the mechanism is due to the stability of the carbonium ions that can be formed by C-O heterolysis and probably also to a decrease in the rate of nucleophlic attack at the carbonyl group because of steric factor. Alkenes as well as alcohols may be produced from the carbonium ion, since water can function either as a nucleophile or a Bronsted bases. This mechanism is referred to $A_{AL}1$ reflecting the fact that the alkyl-oxygen bond is cleaved (Scheme 219).

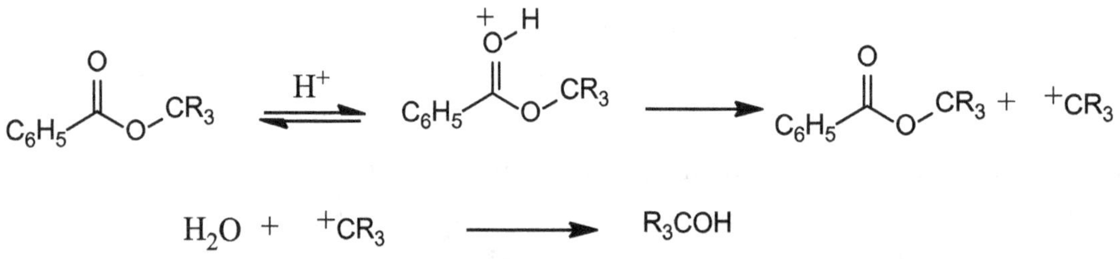

Scheme 219

The activation parameter for the acid-catalyzed hydrolysis of $MeCOOCMe_3$ are found to be $\Delta H = 112$ KJ mol^{-1}; $\Delta S^* = +55$ KJ mol^{-1}. The positive value of ΔS^* suggests that this step is dissociative process, as exemplified in the reaction pathway above the breakdown of the protonated ester into two separate species, the carboxylic acid and carbon cation. This mechanism is generally referred to as $A_{Ac}2$ (Acid-catalyzed, alkyl-oxygen cleavage, unimolecular). It also occurs with ester alkyl groups such as $PhCH_2$, etc. When attempts are made to transesterify $RCOOCMe_3$ with R'OH, the product is not now the expected ester RCOOR', but RCOOH places, the ether $R'OCMe_3$, the latter ester arises from attack of on the carbocationic intermediate.

In practical synthetic terms this change of mechanism can be of value since it allows certain types of esters to be converted to the corresponding acids very selectively. The usual situation involves the use of t-butyl esters which can be cleaved to acids by the action of acids such as tolunesulphonic acids by the action of CF_3COOH under conditions where other types of esters are stable.

In very strongly acidic conditions, a unimolecular mechanism involving acyl-oxygen cleavage of the conjugate acid can operate. The mechanism is the result of decreased availability of nucleophilic water in the strongly acid medium. The products of heterolysis are the alcohol (which is subsequently protonated) and an acylium ion (Scheme 220).

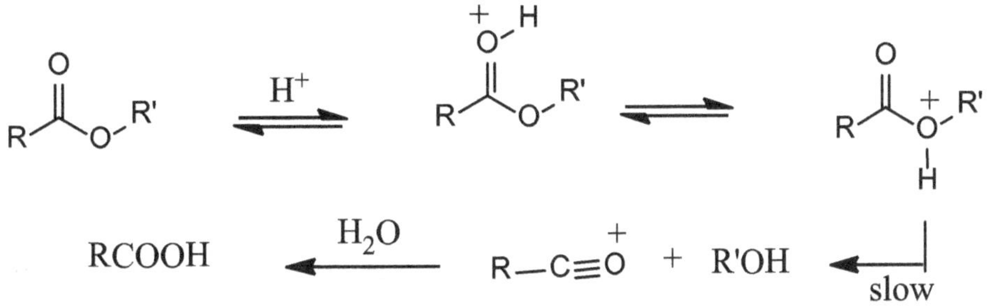

Scheme 220

The mechanistic designation is $A_{Ac}2$. This mechanism is the basis for a useful method of hydrolyzing esters that are very severely sterically hindered, such as ester of 2,4,6- trimethyl benzoic acid. The ester is dissolved in strong sulfuric acid and then quenched into water. Hydrolysis occurs via an acylium ion which is formed in the strongly acid solution (Scheme 221].

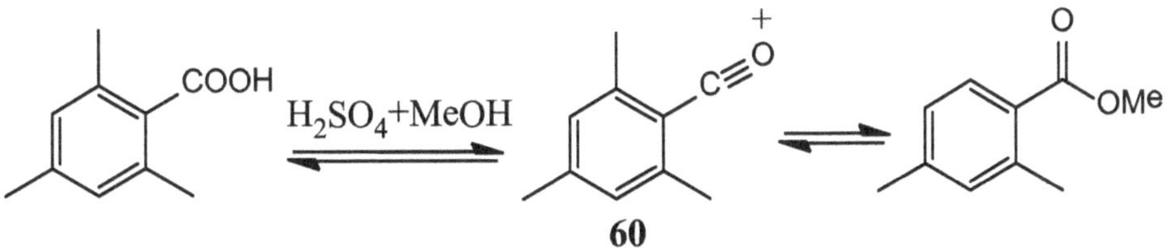

Scheme 221

Evidence for the formation of **60** is provided by the observation that which dissolution of unhindered benzoic acid itself in conc. H_2SO_4 results in the expected two-fold freezing point depression, while dissolution of the hindered acid results in four-fold depression [Scheme 222].

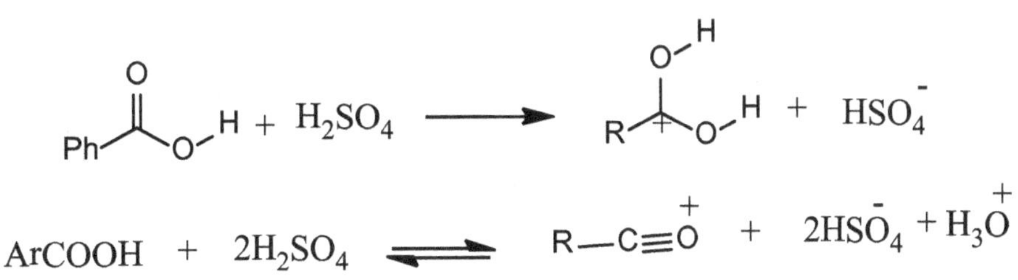

$$ArCOOH + 2H_2SO_4 \rightleftharpoons R-C\overset{+}{\equiv}O + 2H\bar{S}O_4 + H_3\overset{+}{O}$$

Scheme 222

Furthmore, if the 2,4,6-triphenyl ester is dissolved in conc. H_2SO_4, the brilliant colour of the diphenyl fluroenone at once observed –obtained via ring closure (intramolecular Friedel-Crafts acylation) of the aryl cation (Scheme 223).

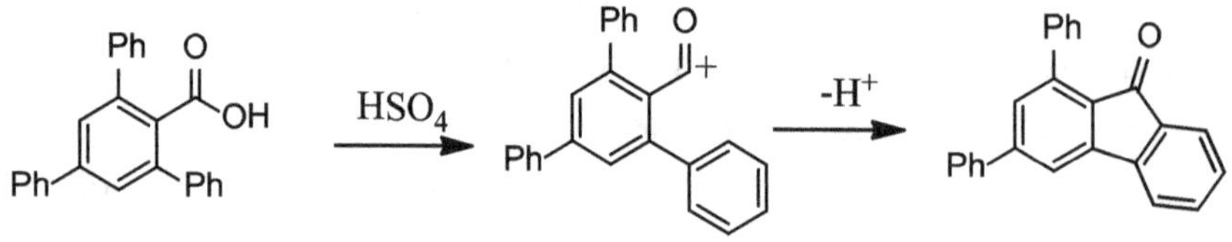

Scheme 223

If the trisubstituted acid was protonated in the normal position (on the carbonyl oxygen atom), the two bulky ortho methyl groups would force the two adjacent OH group into a plane virtually at right angles to the plane of the ring.

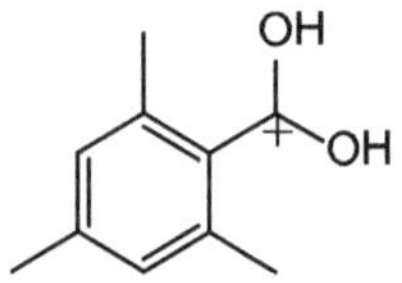

Nucleophilic attack on the catoinic carbon atom by for example, MeOH is thereby prevented from taking place from all directions. By contrast, abnormal protonation on the hydroxyl oxygen atom in $Me3C_6H_2COOH$ allows formation of the planar aryl cation atom by MeOH can now take place from either of two directions at right angles to the plane of the ring. That two different pathways $A_{Ac}2$ and $A_{Ac}1$ are indeed operating in acid-catalyzed hydrolysis of simple ester of benzoic acid and 2,4,6-trimethyl benzoic acid respectively.

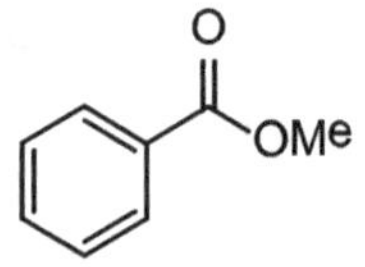

	$\triangle H^*$ (KJ mol^{-1})	$\triangle S^*$ (KJ mol^{-1})
	-79	-110

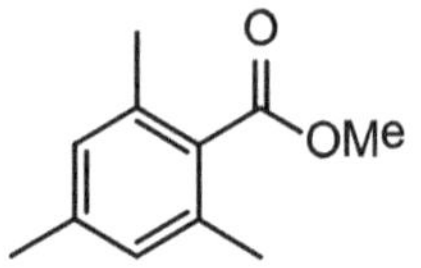	$\triangle H^*$ (KJ mol^{-1})	$\triangle S^*$ (KJ mol^{-1})
	115	+57

That the major factor responsible for this shift in reaction pathway is indeed a steric one is demonstrated by the e observation that the acids **61** and **62** and their simple esters undergoes ready esterification /hydrolysis by the normal $A_{Ac}2$ mode.

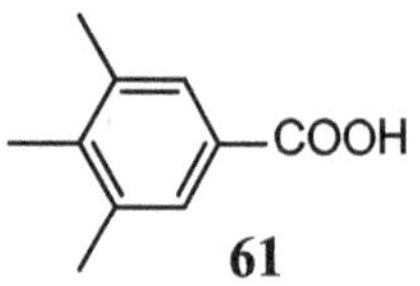

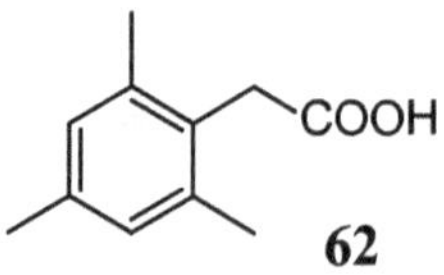

The breakage of the acyl-oxygen bond to form an acylium ion is favored by the presence of electron attracting substituents in R' and electron repelling substituents in R, whereas the reverse is true if the alkyl-oxygen bond is to be broken, forming a carbonium ion.

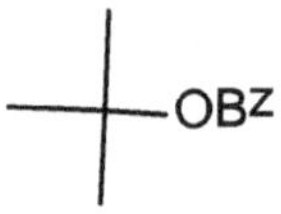

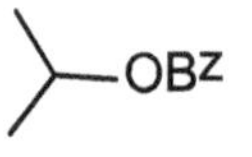

Cyanide anion is an effective nucleophile that opens lactone ring to form new carbon-carbon bond. When a lactone is treated with sodium cyanide, the product is a cyano acid [1963JOC1933]. The reaction is best explained in terms of the cyanide ion at the carbonyl group and an irreversible reaction probably much slower, nucleophilic attack of cyanide on the alkyl carbon of the internal ester. Here the first reversible reaction reaction regenerates the starting materials and hence the only product possible is that derived from the slower irreversible S_N2 type nucleophilic attack of cyanide ion on the methylene group of the lactone to yield more stabilized carboxylate anion (Scheme 224).

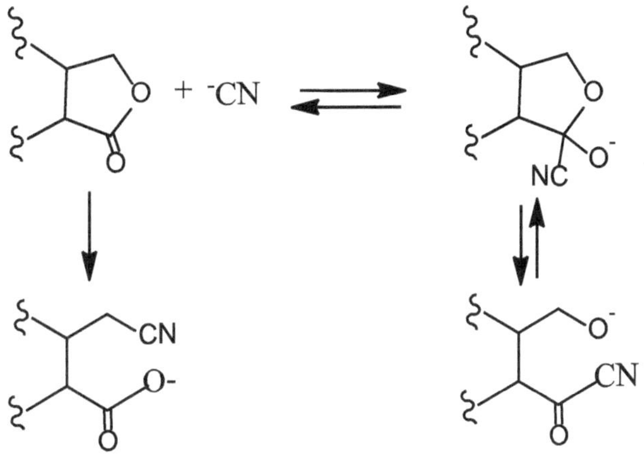

Scheme 224

This reaction prompted, Rai et al [1985IJCB502] to synthesize six membered lactone from the existing five membered lactone ring in podophyllotoxin. The sequence of reaction involves the following steps. First, nucleophilic attack of cyanide anion on podophyllotoxin furnishes the cyano acid, which on hydrolysis afforded dicarboxylic acid. Dehydration of this acid followed by reduction with 2% sodium amalgam yields picropodophyllin homolactone (Scheme 207)

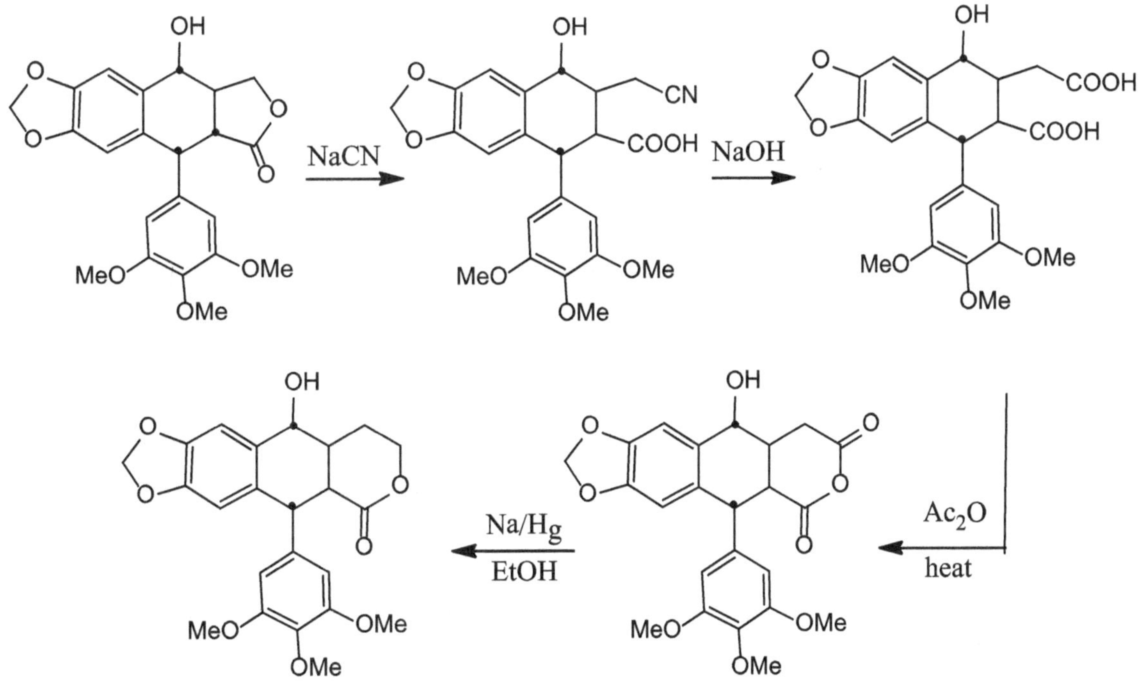

Scheme 225

5.4.2   Claisen Condensation

The self condensation reaction of esters containing an α-hydrogen in the presence of a strong base to give a β-keto esters is known as the Claisen Condensation which may be termed as Claisen ester condensation. It is named after Rainer Ludwig Claisen, who first published his work on the reaction in 1887 [1887Ber651; 1972Misc266]. Claisen condensation between a ketone and an ester gives rise to a 1,3-diketone.

Mechanism:

In the first step of the mechanism, an α-proton is removed by a strong base (usually an alkoxide, LDA or NaH), resulting in the formation of an enolate anion, which is made relatively stable by the delocalization of electrons. The reversible formation of the carbanion is experimentally proved by deuterium exchange reaction using a little amount of EtOD in the reaction mechanism.

Next, the enolate attacks the carbonyl group of the other ester to form a tetrahedral intermediate, which breaks down in the third step by ejecting a leaving group. The alkoxide removes the newly formed doubly α-proton to form a new highly resonance-stabilized enolate anion, which is much less reactive than the ester enolate generated in the first step. Aqueous acid (e.g. sulfuric acid or phosphoric acid) is added in the final step to neutralize the enolate and any base still present. The newly formed β-keto ester or β-diketone is then isolated. That is, Claisen condensation does not work with substrates having only one α-hydrogen because of the driving force effect of deprotonation of the β-keto ester in the last step (Scheme 226).

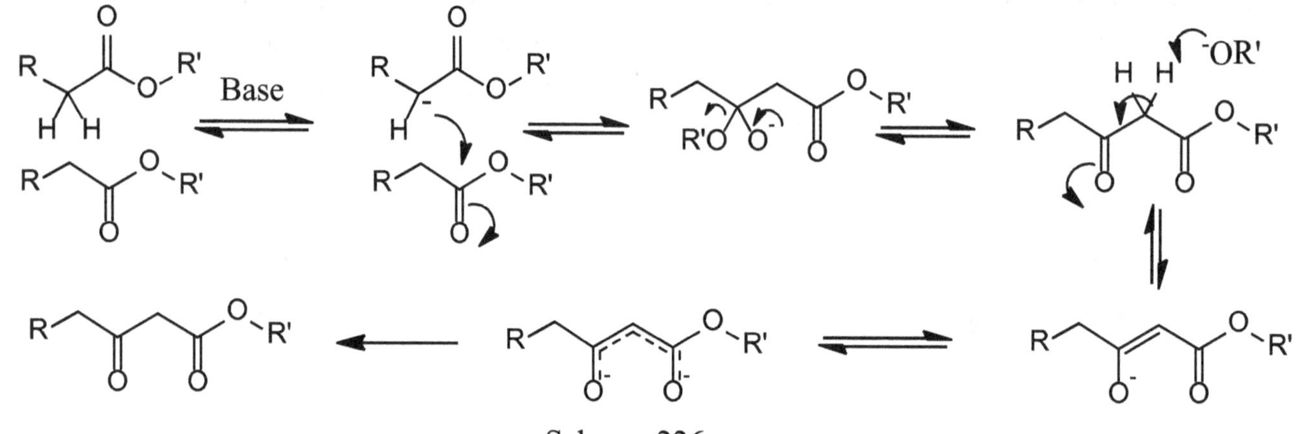

Scheme 226

5.4.2.1   Dickmann cyclization

The intramolecular or internal Claisen Condensation reactions involving esters of dibasic acids containing α-H atoms and 5 to 6 C atoms give rise to cyclic compounds and the reaction is known as the Dickmann cyclization [1894Ber102; 1991Misc795].

Each step of the Dickmann condensation is completely reversible. The driving force of the reaction is the generation of the resonance-stabilized enolate of the product β-keto ester. Te rate determining step, however is te ring formation in which the ester enolate attacks the carbonyl group of the second ester fucnitonal group. The resulting tetrahedral intermediate then rapidly breaks down to the enolate of the β-keto ester. Protonation of the enolate affords the final product.

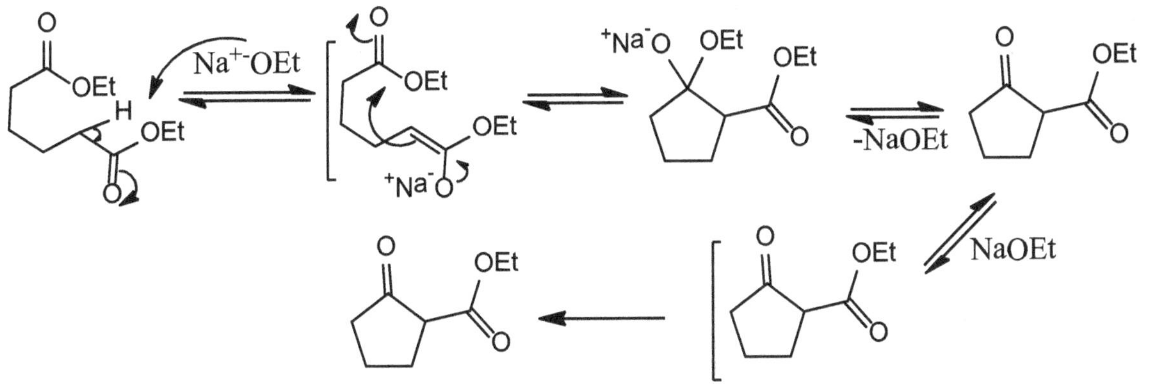

Scheme 227

Due to the steric stability of five- and six-membered rings, these structures will preferentially be formed. 1,6 diesters will form five-membered cyclic β-keto esters, while 1,7 diesters will form six-membered β-keto esters.

5.4.2.2 Biginelli reaction

In 1893, **Biginelli** was yhe first to synthesize functionalized 3,4=dihydropyrimidin-2(1H)-ones b*y the one pot three-component cndensation reaction of an aromatic aldehyde, urea and ethyl acetoacetate in the prsence of catalytic HCl in refluxing ethanol. This reaction is called the Biginelli reaction and the products are referred to as Biginelli compounds [1891Ber2962; 2004OrgRe01].

The reaction mechanism of the Biginelli reaction is a series of bimolecular reactions leading to the desired dihydropyrimidinone. The first step in the mechanism of Biginelli reaction is the acid catalyzed condensation of urea with aldehyde affording the N-acyliminium ion intermediate. Subsequently, the enol form of the β-keto ester attacks the N-acyliminium ion to generate an open chain ureide, which readily cyclize to a hexahydropyrimidine derivatives (Scheme 228).

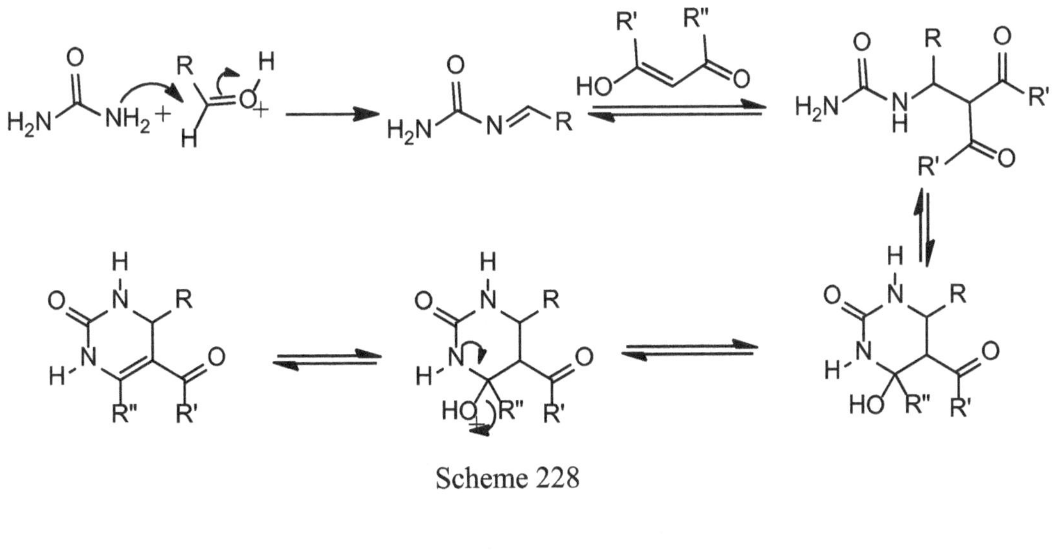

Scheme 228

5.5 DECARBOXYLATION

The term "decarboxylation" literally means removal of the COOH (carboxyl group) and its replacement with hydrogen. The term relates the state of the reactant and product. Decarboxylation is one of the oldest organic reactions, since it often entails simple pyrolysis, and volatile products distilled from the reactor. Heating is required because the reaction is less favorable at low temperatures. Yields are highly sensitive to conditions.

Alkanoic acids and their salts do not always undergo decarboxylation readily. Exceptions are the decarboxylation of beta-keto acids, α,β-unsaturated acids, and α-phenyl, α-nitro, and α-cyanoacids. Such reactions are accelerated due to the formation of a zwitterionic tautomer in which the carbonyl is protonated and the carboxyl group is deprotonated. Typically fatty acids do not decarboxylate readily. Reactivity of an acid towards decarboxylation depends upon stability of carbanion intermediate formed in above mechanism.

The decarboxylation of an acid should occur more readily if within group R-, there is strongly electron-attracting substituent such as $-NO_2$, $-CCl_3$, $-CN$ or $-CO-$. This is to be expected for a decarboxylation is ordinarily a heterolysis in which group R- departs with an elctron pair. One of the first decarboxylation to be studied kinetically was that of acetoacetic acid.

(Equation 95)

This reaction as well as the closely related decomposition of α,α-dimethylacetoacetic acid into methyl isopropyl ketone and CO_2, follows a rate law of the following type.

Rate = k(keto acid) + k'(keto-acid anion)

This suggests that there are two different modes of decarboxylation, neither of which involves the enol form of the acid or its salt (since α,α-dimethylacetoacetic acid cannot exist in an ordinary enol form). It seems likely that the two terms in the rate low correspond simply to unimolecular decompositions of the keto acid (equation 96) and its anion (Scheme 229) respectively. There is one minor difficulty with both keto acids, k for the keto acid is much greater than k' for the anion, whereas one would expect the proton on the $-COOH$ group to inhibit rather than to accelerate decarboxylation. It is likely therefore that there is partial transfer of the carboxyl proton, through intramolecular bonding to the keto group wher it should aid decarboxylation.

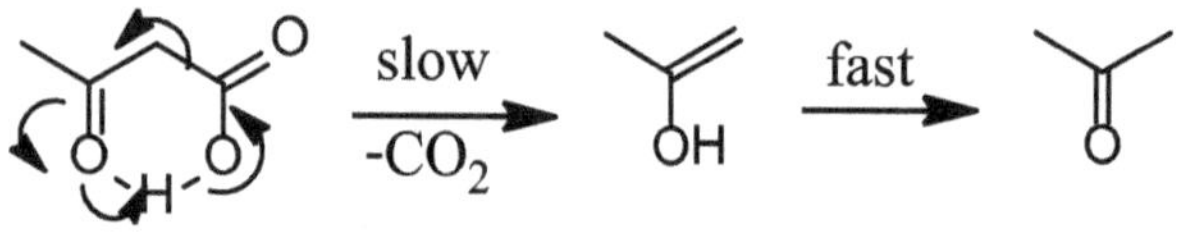

(Equation 96)

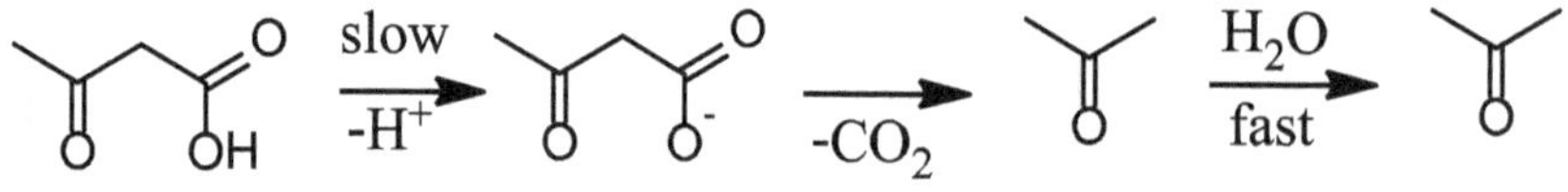

Scheme 229

Furthermore, the intermediate or the anion may be trapped by carrying out the decarboxylation in the presence of bromine or iodine. Here, the reaction product is monobromo or monoiodoketone, but the rate at which the keto acid disappears is unchanged. The halogenation does not occur before the decarboxylation step nor it occur after the final formation of the ketones. For it may be shown that these are not halogenated under decarboxylation conditions.

On this basis, we can understand why bicyclic β-keto acids in which the $-COOH$ group is bound to a bridgehead carbon may be decarboxylated only with extreme difficulty. The hypothetical enol that would presumably result from the decarboxylation of acid may be drawn on paper, but a scale model of this enol having the correct bond angles, cannot be constructed even with pushing and twisting. The four atoms attached to the pair of double bonded atoms must lie very nearly in a common plane – a requirement clearly incompatible with this bicyclic system since one of these four atoms in "**63**" is the carbon atom of the methylene bridge that must be pulled far out of the plane of the remaining three (thic is covered by Bredt's rule). Since this acid cannot, except with rearrangement, yield the necessary enol or an enolate ion intermediate, it rejects decarboxylation.

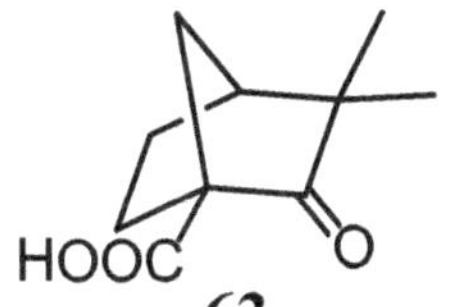

63

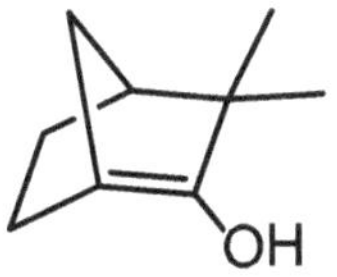

The rates of decarboxylation of a number of additional acids are proportional to the concentrations of the respective carboxylate aions, which undergo heterolysis to acabanion and CO_2 in the same manner as the anions of acetoacetica acid. Among these acids are α-nitroacetic acid, α-nitroisobutyric acid, dibromomalonic acid, PhC=CCOOH, trihaloacetic acid and 2,4,6-trinitrobenzoic acid. With none of these is the rate of decarboxylation of the acid itself appreciable, but malonic acid for which a hydrogen bonded structures analogous to "**64**" may be drawn, decomposes almost ten times as rapidly as its monovalent anion.

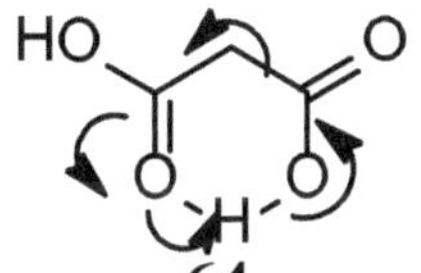

64

Furthermore, a number of nitrogen containing acids exist that undergo first order decarboxylation themselves but anions do not form. However, the dipolar ion is found to undergo decarboxylation readily, it seems very likely that for the other acids, it is the zwitterion that is being decarboxylated (Scheme 230). Here again, the anion like intermediate may be trapped by the addition of aldehyde or ketone.

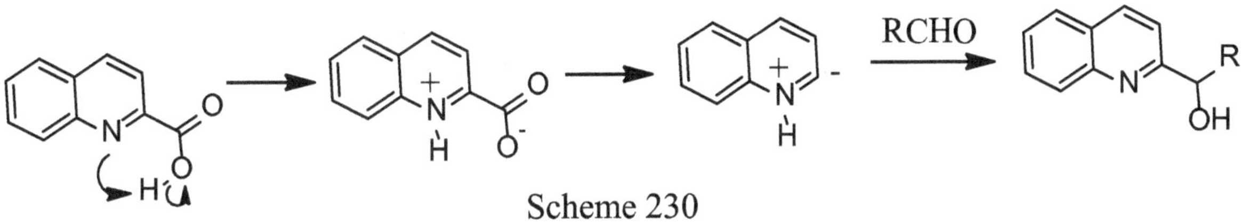

Scheme 230

Decarboxylation should take place even more readily if the –C=N group is situated beta to the –COOH group; in this case, decarboxylation of zwitterion results directly in neutralization of charge (Scheme 231).

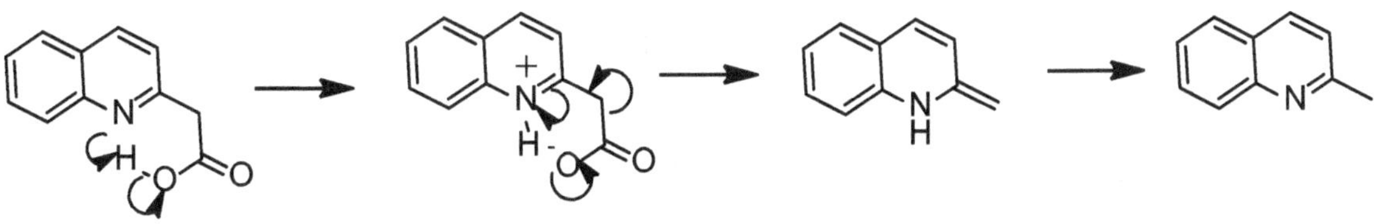

Scheme 231

The keto group is converted by the amine to an imine linkage and the resulting β-imino acids undergoes decarboxylation more rapidly than the original keto acid (Scheme 232).

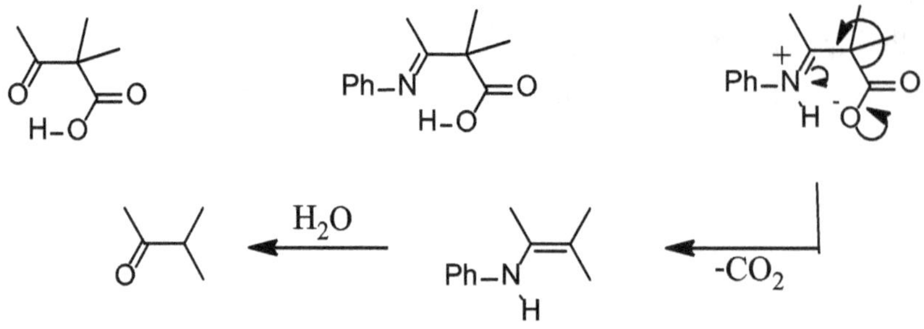

Scheme 232

The decarboxylation of β,γ-unsaturated acids are formally quite similar to those of β-keto acids and β-amino acids (Equation 97).

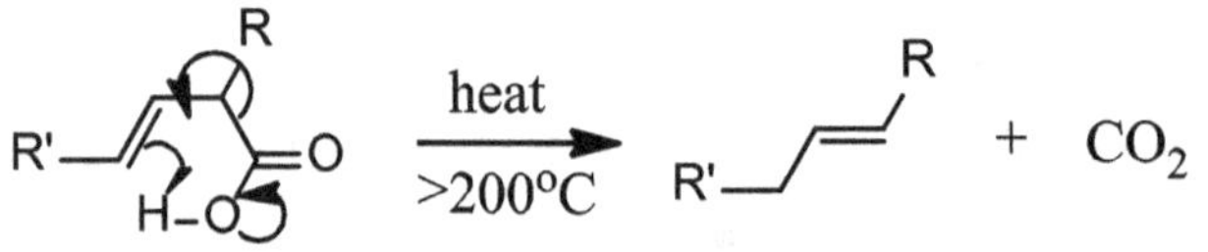

(Equation 97)

It is also likely that the decarboxylations of α,β-unsaturated acids take the same path - that is, that these acids first rearrange to β,γ-unsaturated acids (Scheme 233). It has been shown that number of a α,β-unsaturated acids are in mobile equilibrium with the corresponding β,γ-unsaturated acids at temperature necessary for the decarboxylation, provided that interconversion between the two require transfer of only a proton.

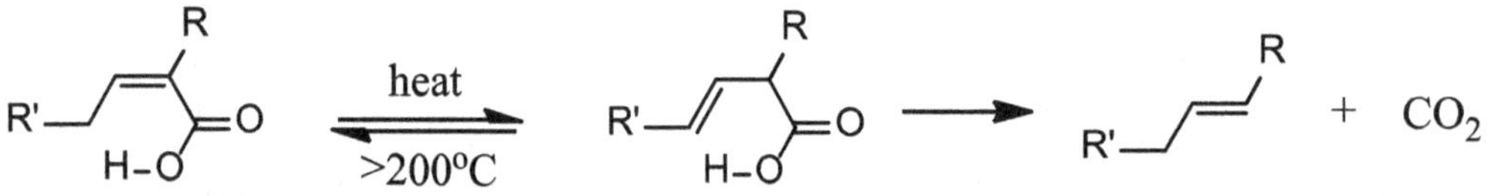

Scheme 233

Like nucleophilic substitutions, decarboxylations display a duality of reaction mechanism. The arge majority of decarboxylations are unimolecular, but a number of bimolecular decarboxylations are now known. These are, in effect, displacements of the carboxyl group by a proton – that is, S_E2 reactions and almost always take place in strongly acid solutions. Such reaction generally occurs at unsaturated carbon atoms; for in such cases, the new CH bond may form without necessity for simultaneous breakage of the old C-C bond. We should expect them to be first order both in carboxylc acid and in H+, and they should favoured by electron donating substituents and aromatic rings bound to the β-carbon, since such groups should stabilize the intermediate carbocation (Scheme 234).

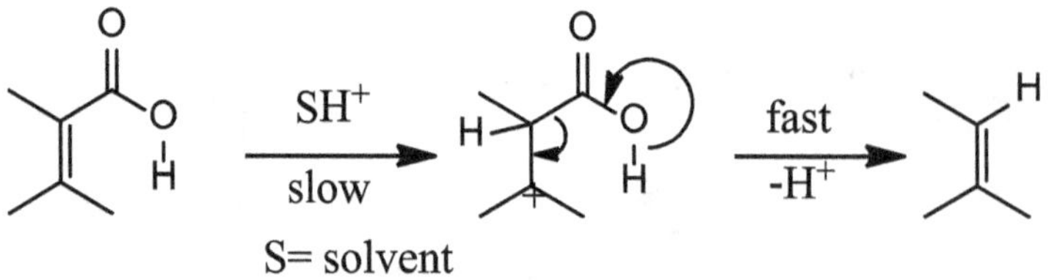

S= solvent

Scheme 234

5.5.1 Decarboxylation of aromatic acids

The reaction pathways for decarboxylation of aromatic carboxylic acids are surprisingly complex and dependent on the reaction condition [1972MIsc303]. The two major pathways for aromatic decarboxylation reactions are ionic and free radical. In aqueous solution, ionic decarboxylations can be catalyzed by acid or base. Acid-catalyzed decarboxylation reactions are the most common, and the reaction pathways are dependent on acid concentration, ionic strength, and substituents on the aromatic ring. In dilute acid, ipso-protonation of the aromatic ring is the rate-determining step (Scheme 235), while in highly acidic solutions, the rate determining step is decarboxylation of the aromatic cation. Electron donating substituents accelerate the acidcatalyzed decarboxylation reaction. In the absence of an acid-catalyst, decarboxylation of carboxylate salts or carboxylic acids with strongly electron withdrawing substituents, such as 2,4,6-trinitrohenzoic acid, occur by rate-determining unimolecular elimination of carbon dioxide from the anion (Scheme 236).

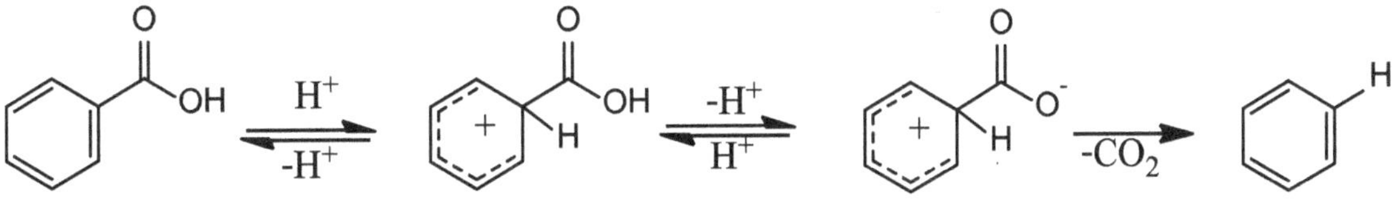

Scheme 235

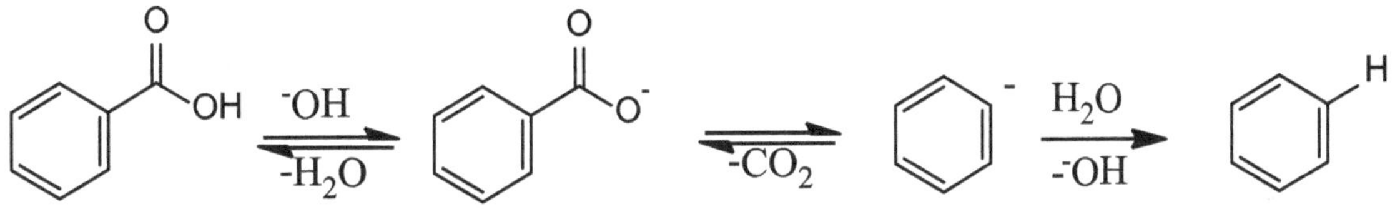

Scheme 236

Decarboxylation of aromatic carboxylic acids can also occur by a free-radical pathway. Since free radicals are known to be formed as reactive intermediates in the thermolysis of coal, the free-radical decarboxylation pathway has been viewed as a possible route to cross-linking. Hydrogen abstraction or electron transfer to an acceptor [1996EF776] can form the benzoyloxyl radical (PhCOy) which will rapidly decarboxylate (log k (s-5 = 12.6 - 8.6 kcal mo1-'/2.303RT)" to form an aryl radical (Scheme 237). This highly reactive intermediate can abstract hydrogen or competitively add to an aromatic ring to form biaryls [1989JA1418]. This aryl-aryl linkage is thermally stable at T 400°C and would constitute a low-temperature cross-link.

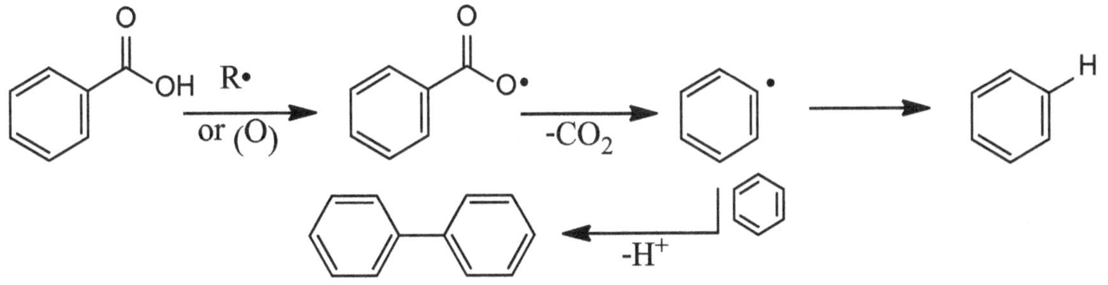

Scheme 237

5.5.2 Kolbe reaction

The Kolbe electrolysis is an organic reaction named after Hermann Kolbe. The Kolbe reaction is formally a decarboxylative dimerisation of two carboxylic acids (or carboxylate ions) The overall general reaction is shown in scheme 238.

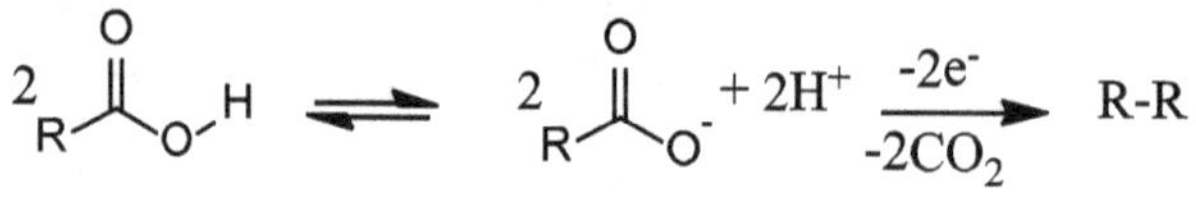

Scheme 238

If a mixture of two different carboxylates is used, all combinations of them are generally seen as the organic product structures:

$$3\ R_1COO^- + 3\ R_2COO^- \rightarrow R_1\text{–}R_1 + R_1\text{–}R_2 + R_2\text{–}R_2 + 6\ CO_2 + 6\ e^-$$

Another example is the synthesis of 2,7-dimethyl-2,7-dinitrooctane from 4-methyl-4-nitrovaleric acid:

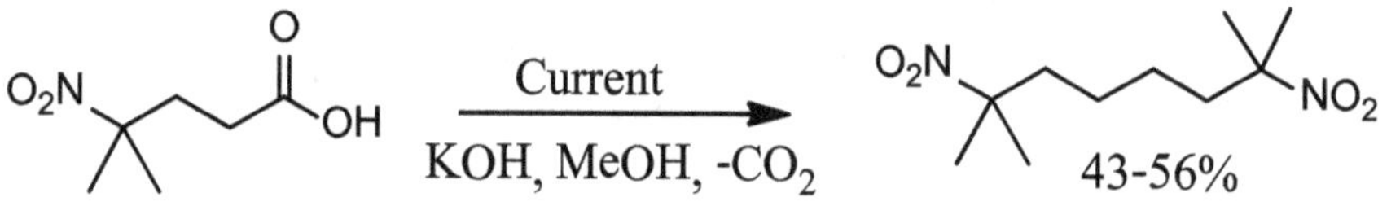

(Equation 98)

5.5.3 Hunsedecker reaction

Treatment of the silver slat of a carboxylic acid with bromine results in the formation of alkyl bromide followed by the loss of CO_2 leads to the formation of alkyl bromide (Scheme 221). This reactin is called the Hunsedecker reaction. The reaction has the characteristic of a radical process. The initial step is the formation of an acyl hypochlorite which may be detected in solution and which is decomposed by a chain mechanism. The reaction is often accomplished by at least partial racemization of an asymmetric group R.

$$RCOOAg\ +\ Br\text{-}Br\ \longrightarrow\ RCOOBr\ +\ AgBr$$
$$RCOOBr\ \longrightarrow\ RCOO\cdot\ +\ Br\cdot$$
$$RCOO\cdot\ \longrightarrow\ R\cdot\ +\ CO_2$$
$$RCOOBr\ +\ R\cdot\ \longrightarrow\ RBr\ +\ RCOO\cdot$$

Scheme 239

5.5.4 Barton decarboxylation reaction

Thiocarbonyl containing compounds are good accepter for radical additions. Radical reactions have been reported on a broad range of compounds such as thioamides, thiocarbanmates, thioureas, thiocarbonates, thioesters and thioketones. A thiophilic radical such as tin radical and silyl radical can add to the thiocarbonyl's sulphur atom to generate a stabilized radical intermediate, providing a rich radical chemistry of thiocarbonyl compounds.

A unique radical chemistry of thiocarbonyl containing compounds has been recognized for a long time, featured by Barton's well known thiohydroxamate ester chemistry. The radical decarboxylation of a Barton ester proceeds to the corresponding alkane after treatment with tributyltin hydride or t-butylmercaptan (Scheme

240). The initiation of the Barton Decarboxylation is effected with a radical initiator, and as with the Barton-McCombie Deoxygenation, the driving force for the reaction itself is the formation of the stable S-Sn bond.

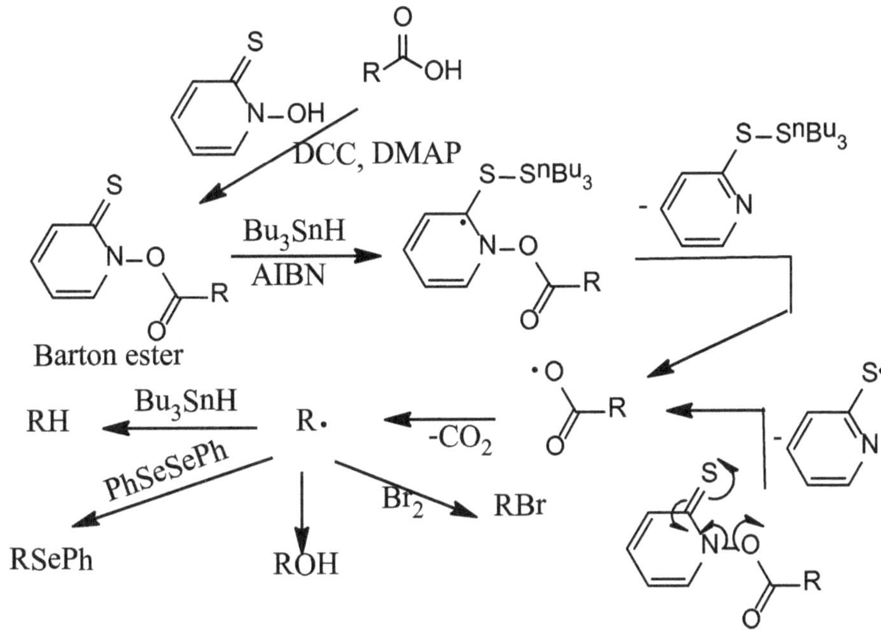

Scheme 240

5.5.5 Biological decarboxylation

Thiamine pyrophosphate (TPP) serves as a coenzyme in enzymatic reactions transferring an activated aldehyde unit. For instance, TPP used as excellant catalyst for the non oxidative and oxidative decarboxylation of α-keto acids.

i. Non-oxidative decarboxylation:

Nucleophilic addition of thiamine pyrophosphate (**65**, TPP) carbonyl group of α-keto acids forms an intermediate, which then decarboxylate to give the product. This rearranges to a zwitterion, which then looses the catalyst to form the respective aldehyde (Scheme 241).

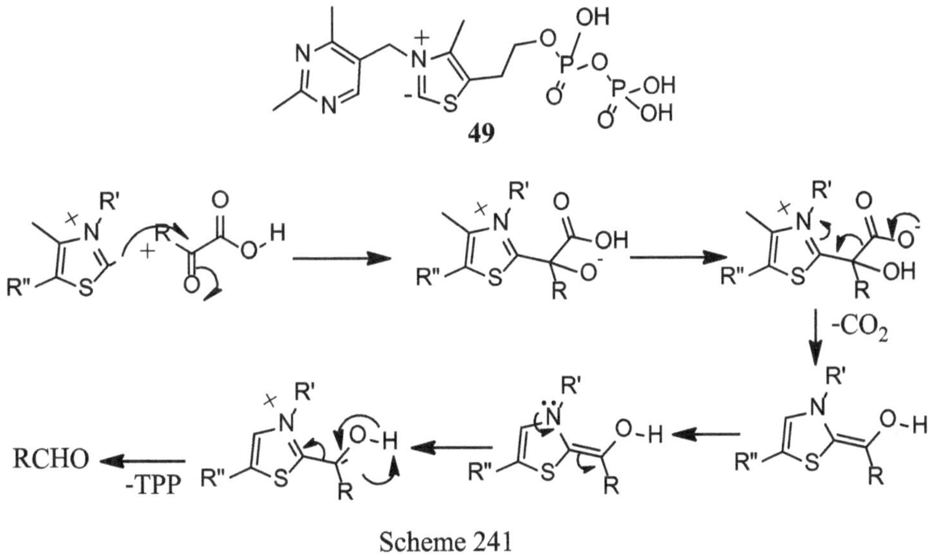

Scheme 241

ii. Oxidative decarboxylation

The role of thiamine pyrophosphate in the oxidative decarboxylation is similar to that described for the non-oxidative process (Scheme 242). The former reaction are quite generally quite complicated involving several coenzymes, lipoic acid (R" = -(CH$_2$)$_4$COOH) usually serving the actual intermediate oxidizing agent. Such reactions are quite common and occur with a variety of α-keto acids. Again, the thiamine diphosphate provides a stable carbanion to react with the α-carbon of α-ketoglutarate.

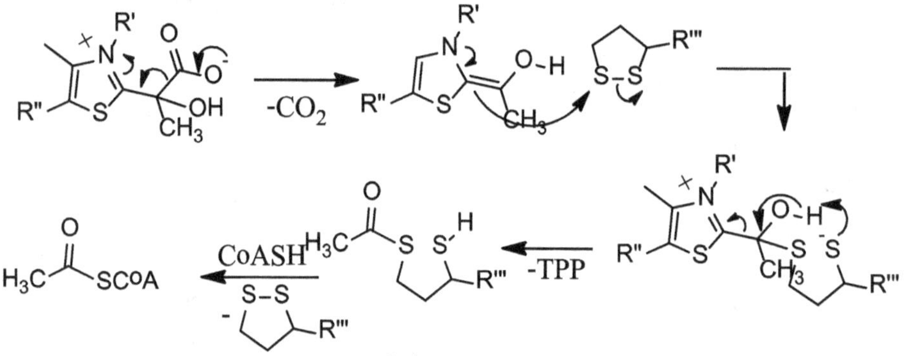

Scheme 242

Pyridoxyl pyrophosphate dependent enzymes catalyze a striking variety of distinct reactions including decarboxylation of amino acids. In all pyridoxyl pyrophosphate dependent reactions of amino acids, the initial step are formation of an enzyme bound Schiff's base intermediate (Scheme 243). This intermediate is stabilized by interaction with a cationic region of the active site, which can be rearranged in a ways that include release of carbon dioxide with the formation of enzyme bound pyridoxyl pyrophosphate (66). Hydrolysis of 66 results in the formation of amine derivative.

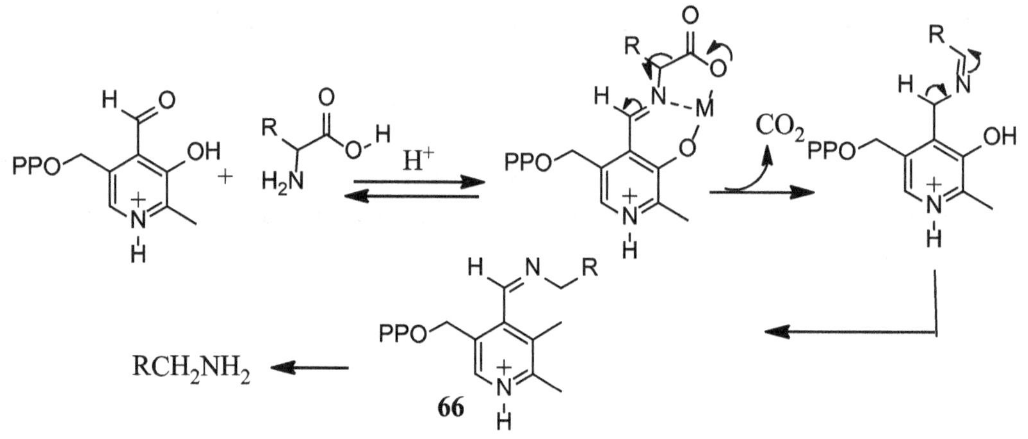

Scheme 243

5.6 AMIDES

5.6.1 Aminolysis of esters

Esters react with ammonia and amins to give amides. Aminolysis of ester often reveals general bases catalysis and in particular a contribution to the reaction from terms that are second order in the amine. The base is believed to function by deprotonating the zwitterionic tetrahedral intermediate. Deprotonation facilitates breakdown of the tetrahedral intermediate, since the increased electron density at nitrogen favours expulsion of an anion.

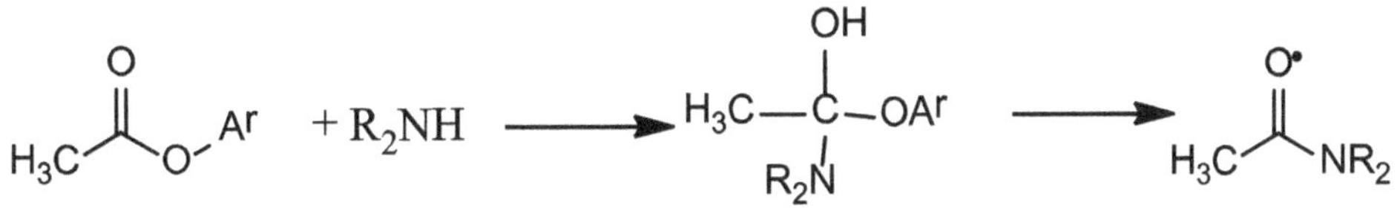

Scheme 244

The underlying mechanism is similar to hydroxide-ion catalyzed ester hydrolysis and involves nucleophilic attack of the amines at the carbonyl group followed by expulsion of alkoxide ion from the tetrahedral intermediate. The identity of the rate the determining step depends primarily on the leaving ability of the alkoxy group. With releatively good leaving groups such as phenolates or acidic alcohols such as trifluroethanol, the slow step is expulsion of the oxygen leaving group from a ziwitterionic tetrahedral intermediate **68** with poorer leaving groups, breakdown of the tetrahedral intermediate occurs only after formation of the aminoic species **69**. For such systems deprotonation step is rate determining.

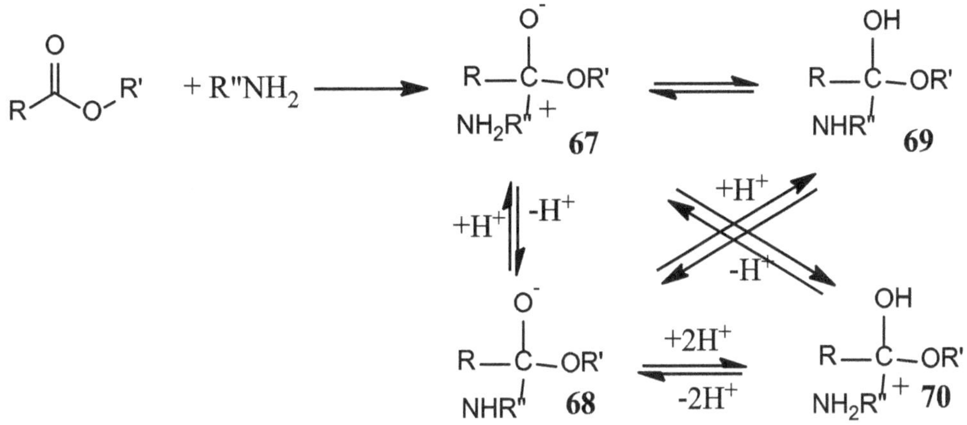

Scheme 244

In **67** & **70**, the best leaving group at the tetrahedral carbon is the neutral amine, whereas in **68** & **69**, the group ⁻OR' would be expected to be better leaving group than R'NH. Furthermore in **68** & **69**, the low pair on nitrogen can assist in elimination. In **67**, the negatively charged oxygen also has the capacity to assist by pushing with reformation of a carbonyl group. Precisely how the intermediate proceeds to product depends upon P^H and the identity of the groups RNH_2 and ⁻OR'. When ⁻OR' is poor leaving group, as would be the case for alkyl esters reaction occur through **68** & **69**.

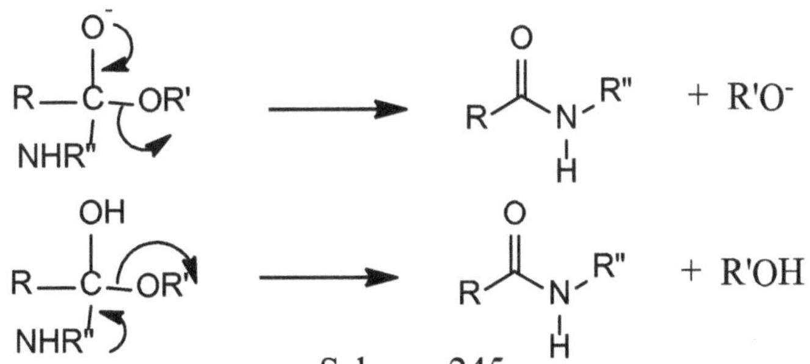

Scheme 245

When the leaving group is better, breakdown can occur directly from **67**. This is when OR is a phenolate anion. It is clear that the presence of acids and bases will be able to affect the mechanism since the overall rate of the reaction will depend upon the position of the equilibrium between the four possible tetrahedral intermediates and the rates of the proton transfer processes.

Insight into the factors which govern breakdown of tetrahedral intermediates has also been gained by studying the hyldrolysis of amide acetals. If the amino group is expelled an ester is formed, whereas elimination of an alcohol group gives an amide (Scheme 246).

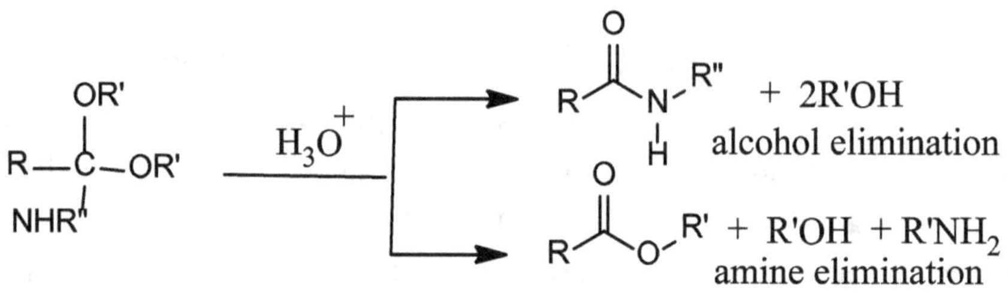

Scheme 246

The P^H of the solution is of overwhelming importance in determining the course of hydrolysis. In basic solution, oxygen elimination is dominant. This is because an unprotonated nitrogen substituent is a poor leaving group and is also more effective, at stabilizing resulting from alkoxide elimination (Scheme 247).

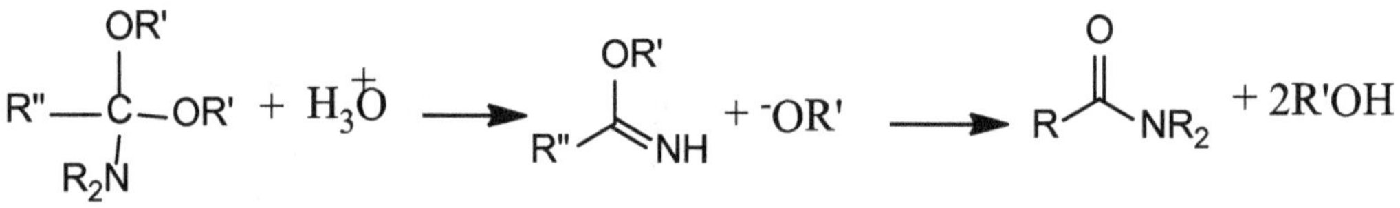

Scheme 247

In acidic solution nitrogen is protonated and becomes a better leaving group and also loses its ability to stabilize the resulting intermediate. In these circumstances, oxygen elimination is favored.

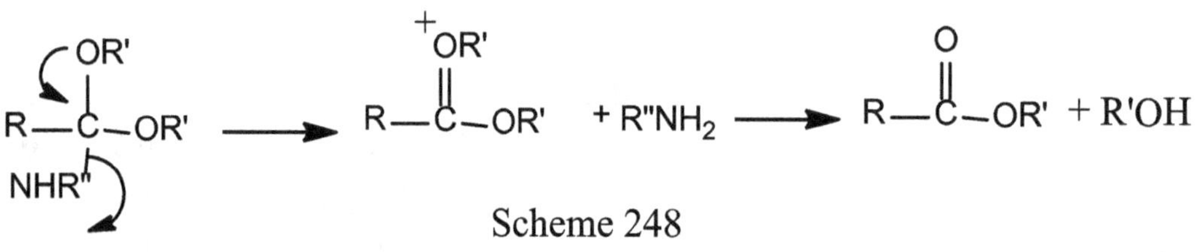

Scheme 248

If a mixture of two different carboxylates is used, all combinations of them are generally seen as the organic product structures:

5.6.2 Amide hydrolysis

The hydrolysis of amides to carboxylic acids and amines require considerably more vigorous condition than ester hydrolysis. The reasons are that the electron releasing nitrogen substituent imports very significant ground state stabilization to the carbonyl group, which is lost in the T.S leading to the tetrahedral intermediate.

In basic solution, a mechanism similar to the $B_{Ac}2$ mechanism for ester hydrolysis is believed to operate.

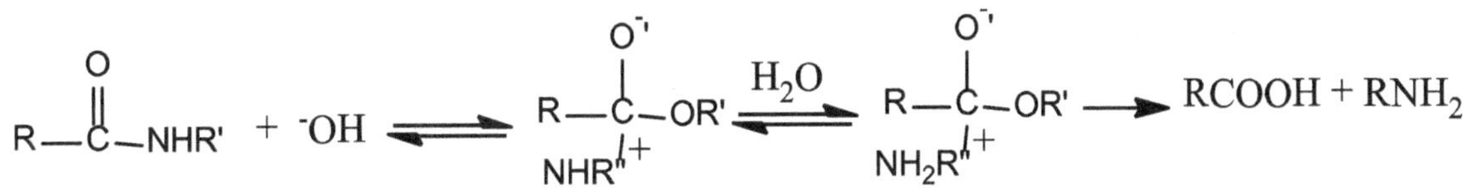

Scheme 249

The principal difference lies in the poor ability of amides ions to act as leaving groups, compared to alkoxide. As a result, protonation of nitrogen is required for breakdown of the tetrahedral intermediate. Also exchange between the carbonyl oxygen and water.

Rai et al developed a simple, rapid, high-yielding and environmentally benign method for the conversion of esters and amides to the corresponding thio compounds using thiourea as thionating agent under solvothermal reaction [2006JMS1391]. The probable mechanism for the formation of thioamide is as follows (Scheme 250).

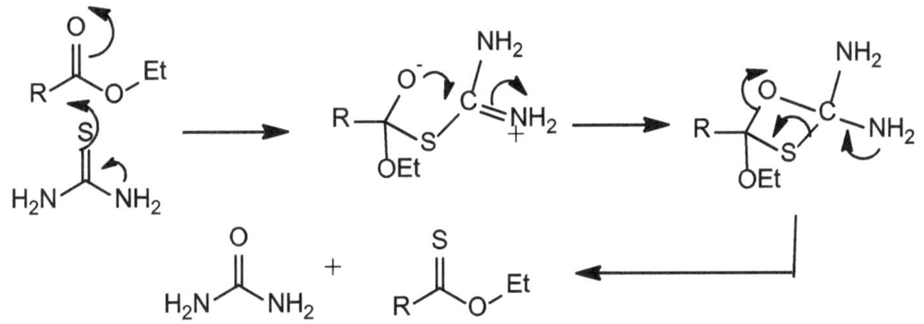

Scheme 250

If a mixture of two different carboxylates is used, all combinations of them are generally seen as the organic product structures:

EXERCISE

1. Predict the products for the following reactions. Justify your answer.

$$RCOOH \xrightarrow[\substack{AgNO_3/Br_2 \\ heat}]{NH_4OH/} \quad ?$$

2. Write the energy profile diagram for the dissociation of acetic acid
3. With suitable mechanism explain Curtius rearrangement, identify the intermediate and how it is trapped?
4. ΔS for acid hydrolysis of methyl acetate is negative while that of methyl benzoate is positive. Justify.
5. Predict the products in the following hydrolysis reactions. Justify your answer with suitable mechanism.
    $PhCOOCH_3 + H_2O^{18} + H^+ \longrightarrow ?$
6. Acid hydrolysis of t-butyl acetate undergo via $A_{AL}{}^1$ mechanism. Justify.
7. Predict the product for the following:

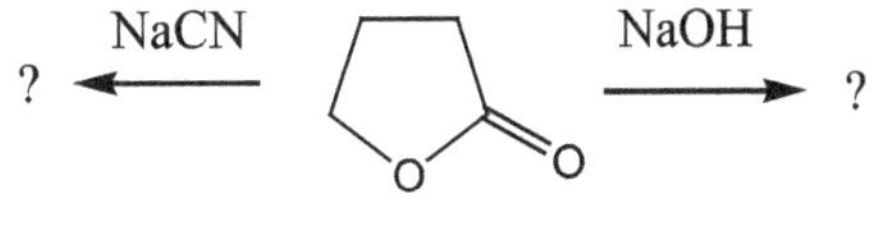

8. How Isotope labeling experiment will help to deduce the possible mechanism for ester hydrolysis.

9. Hydrolysis of ethyl acetate undergoes via acyl oxygen bond cleavage while that of tertiarybutyl acetate undergoes via alkyl oxygen bond cleavage. Justify.

10. How do you say that base catalyzed hydrolysis of ethyl acetate undergo via acyl oxygen bond cleavage?

11. Discuss the mechanism for the anhydride formation using DCC as coupling reagent.

12. Arrange the following in order of increasing acidity.

    CH_3COOH, O_2NCH_2COOH, $ClCH_2COOH$ & $(CH_3)_3CCOOH$

13. Between RCOOH and $RCOCH_2COOH$, which one undergo decarboxylation fast and why?

14. With suitable example, explain hydrolysis of an ester via BAc2 mechanism.

15. Give one example each for hydrolysis of ester by AAc^1 and AAl^1 mechanism.

16. Discuss AAc^2 mechanism of ester formation and its hydrolysis.

17. Acid hydrolysis of t-butyl acetate undergo via $A_{AL}{}^1$ mechanism while ethyl acetate undergo via AAc2 mechanism. Justify

18. Explain Barton decarboxylation reaction.

19. Outline the synthesis of RCOOH using Grignard reagent.

PHOTOCHEMISTRY OF CARBONYL COMPOUNDS

6.1 REACTIVITY OF ELECTRICALLY EXCITED KETONES

Ketones have two readily accessible electronic transitions namely n$\longrightarrow\pi^*$ and $\pi\longrightarrow\pi^*$. In general lowest energy transition is the n$\longrightarrow\pi^*$ transition. This means that the S_1 state of most simple ketones has the n, π^* configuration. In solution, excitation to S_2 will be followed by rapid internal conversion and vibrational equilibrium to S_1. The low lying triplet state T_1 may have either the n, π^* or π,π^* configuration. This comes from the fact that the difference in energy between $^1(\pi,\pi^*)$ state and the corresponding $^3(\pi,\pi^*)$ state is much longer than the energy between $^1(n, \pi^*)$ state and the corresponding $^3(n, \pi^*)$ state. If the energy between $^1(n, \pi^*)$ and the $^1(\pi,\pi^*)$ state is small, it is likely that the T_1 state will be $^3(\pi,\pi^*)$ state. If the energy between $^1(n, \pi^*)$ and the $^1(\pi,\pi^*)$ state is large, it is likely that the T_1 state will be $^3(n,\pi^*)$ state

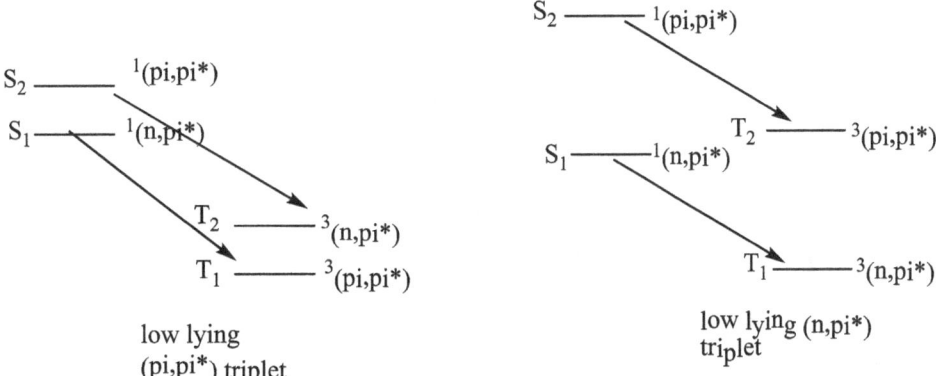

The reactivity of an excited ketone depends upon the multiplicity of the excited state and upon the electronic configuration of the excited state. Electronic excited states with the (n, π^*) configuration show reactivity, that is primarily a result of the singly occupied n-orbital. The vacancy in the orbital means that n, π^* excited state will undergo reaction that places an electron in the orbital. Excited states with the π,π^* configuration are less reactive and longer lived than n, π^* states.

6.2 REPRESENTATION OF EXCITED STATES OF KETONES

Valence bond representation of excited states of ketones is less satisfactory than valence bond representation of ketone ground states. We shall focus attention on the representation of n, π^* states, since these are the states responsible for much of the interesting photochemistry of ketones. In an n, π^* state, the presence of an electron in the antibonding (π^*) orbital reduces the double bond character of the carbon-oxygen bond, while the singly occupied n-orbital conveys radical like reactivity to the oxygen atom. These two ideas have led to representation (**71**), which was popular in the early photochemical literature. This representation suffers from the implication that the double bond is now a single bond and that free rotation about the carbon-oxygen bond is implied. Furthermore, the geometry of the singly occupied orbital on oxygen is not specified, we shall use representation **72** for ketone n, π^* states but with no conviction that this is the best representation.

$^1(n, \pi^*)\ ^3(n, \pi^*)$

The advantage of this representation is as follows: i. the vacancy in the n(2px) orbital of the oxygen atom is apparent, ii. The partial double bond is clearly shown, iii. The spin multiplicity of the state can be shown, iv. The electron in the π^* orbital is shown between carbon and oxygen, which is appropriate for a molecular orbital.

There are three general primary photochemical processes which are commonly encountered-

During excitation of carbonyl compounds, cleavage of bond α to carbonyl function occurs. This reaction is called α-cleavage or Norish-I type reaction.

i.   Abstraction of hydrogen by oxygen atom of n, π^* state from suitable donors
ii. Addition of oxygen atom of n, π^* state to unsaturated linkage.

These are all preliminary processes, each of which the new radical formed may undergo variety of possible reactions like addition, elimination or coupling reactions.

6.3   NORRISH-I TYPE REACTIONS (α-CLEAVAGE)

One consequences of the vacancy in the n-orbital of an n, π^* excited state of a ketone is the tendency to homolytic cleavage of the α-carbon bond. This vacant nonbonding orbital of the n, π^* excited state overlaps the σ-bond orbital between the carbonyl group and the σ-carbon. This overlap facilitates cleavage of the σ-bond. The reaction is named after Ronald George Wreyford Norrish [1937Nat195]. A typical reaction of this type is shown by acetone, which produces carbon monoxide efficiently.

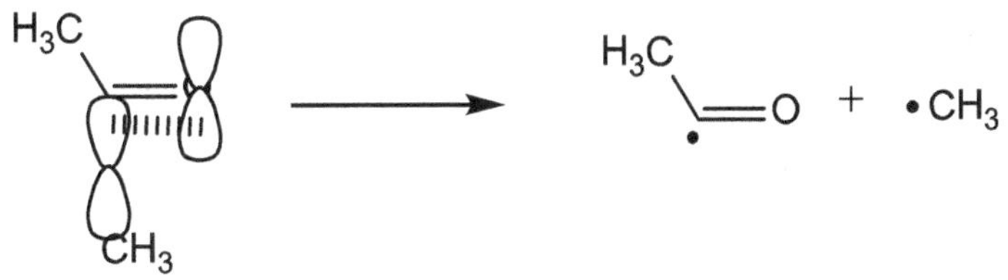

The carbonyl group accepts a photon and is excited to a photochemical singlet state. Through intersystem crossing the triplet state can be obtained. On cleavage of the α-carbon bond from either state, two radical fragments are obtained. The size and nature of these fragments depends upon the stability of the generated radicals; for instance, the cleavage of 2-butanone largely yields ethyl radicals in favor of less stable methyl radicals. Once the cleavage takes place, a number of secondary process occur.

i. The fragments can simply recombine to the original carbonyl compound (path A).

ii. By extrusion of carbon monoxide in path B, two organic residues can recombine with formation of a new C-C bond

iii. When the carbon fragment has an α-proton available it gets abstracted forming a ketene and a saturated hydrocarbon in path C. The presence of ketene can be demonstrated by spectroscopic methods or in the presence of nucleophilic species as water or methanol; it is converted into carboxylic acid or ester derivatives which gets trapped.

iv. When the alkyl fragment contains a β-proton it gets abstracted with formation of an aldehyde and an alkene.

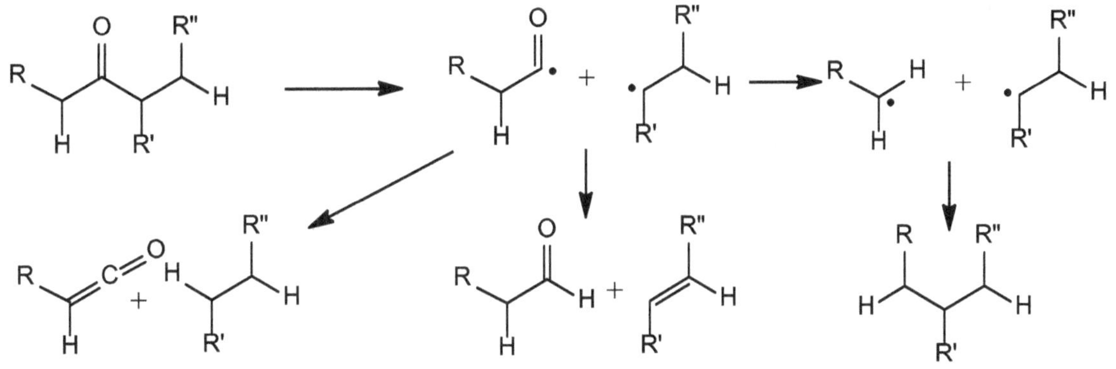

Scheme 251

Photolysis of 2,2,4,4-tetramethyl pentane-2-ene, results in a high yield (~90%) of CO from both singlet and triplet excited state. The life time of singlet excited state 4.5 to 5.6×10^{-9}S as compared with 0.11×10^{-9}S for triplet excited state. The Norrish-type-I cleavage occurs 100 times faster from triplet than the singlet excited state.

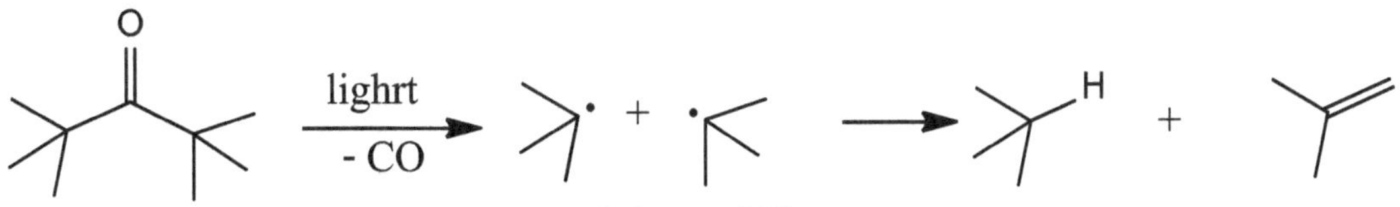

Scheme 252

As the excitation wavelength is reduced (energy increased), the selectivity of carbon bond cleavage decreases. For example, butan-2-ene is little selective of bond cleavage at 254nm, but as the energy of the radiation is reduced, a greater preference for cleavage of the weaker bond is observed at 313nm.

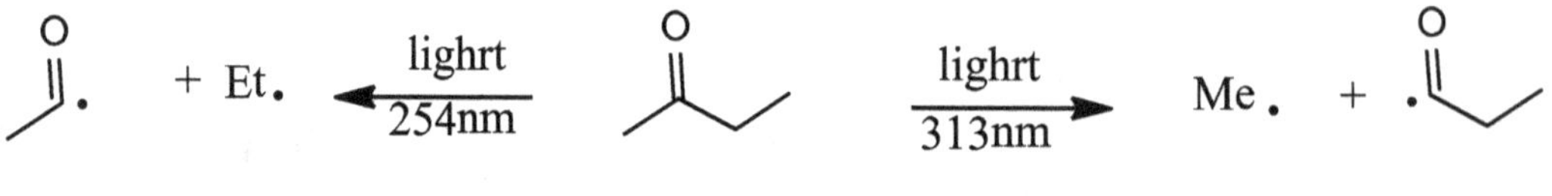

Scheme 253

In molecule with differing degree of substitution on the α-carbon, α-cleavage will occur in such a way that the most stable alkyl radical is formed.

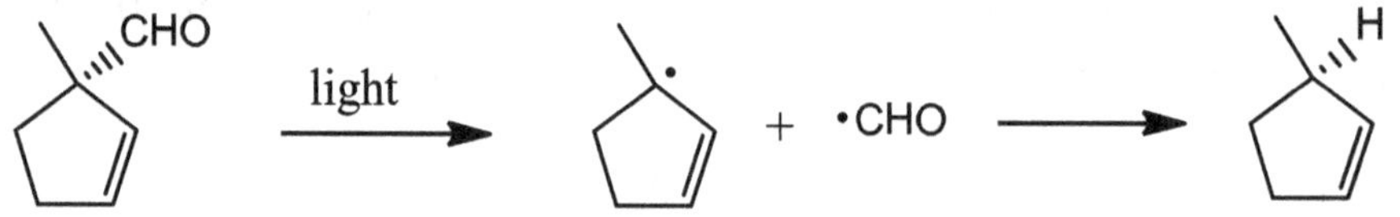

Scheme 254

6.4   PHOTOCHEMISTRY OF CYCLIC KETONES

In saturated cyclic carbonyl compounds, the hydrogen if present on γ-carbon atom will not be in position to transfer to oxygen of carbonyl group. That is why, photochemistry of saturated cyclic compounds is dominated by Norrish type-I process, involving the initial cleavage of a carbon- carbonyl bond. The reaction proceeds from ^{3}(n-π*) excited state because those ketones in which the n-π* triplet state is of lower energy either cleave slowly or not at all. While the n-π* singlet state is generally less reactive to α-cleavage reactions proceeds from both excited states.

The acyclic diradical formed after photolysis undergoes following subsequent processes:

i.   Intramolecular hydrogen abstraction: The terminal alkyl radicals undergo intramolecular hydrogen abstraction to produce a ketene.

ii. Intramoleular hydrogen abstraction by the carbonyl carbon atom to give an unsaturated aldehyde.

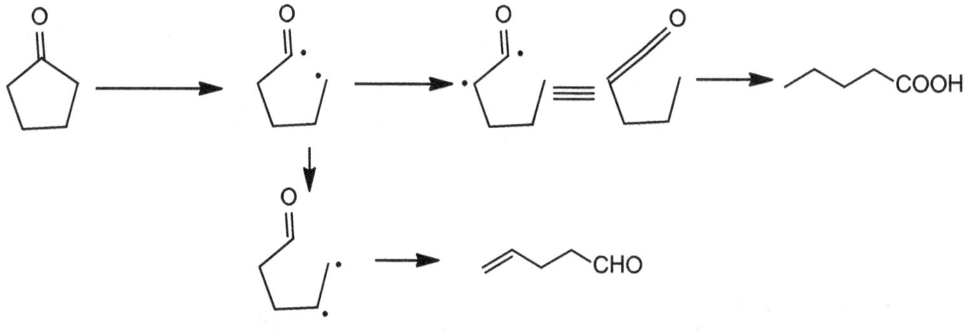

Scheme 255

Photodecarbonylation: Acyclic carbonyl diradical on decarbonylation to give carbon monoxide, an alkene or a cyclic alkane or both.

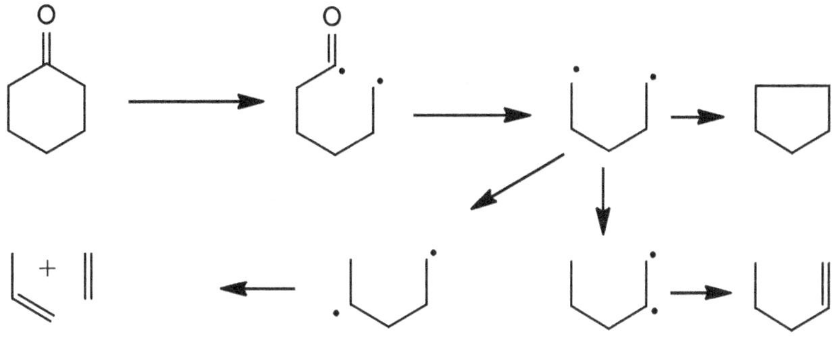

Scheme 256

Alkyl substituents on the α-carbon atom enhance α-cleavage to give more stable alkyl radical. For instance, 2,2-dimethyl cyclohexanone cleaves to give tertiary radical rather than the primary alkyl radical. The ultimate products are formed by intramolecular hydrogen transfer reaction.

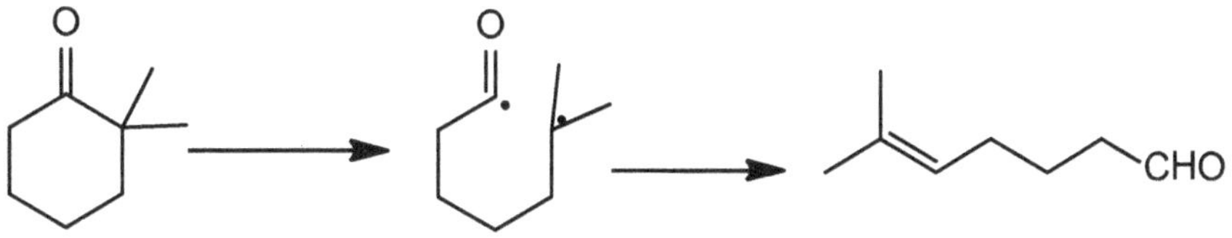

Scheme 257

Suitably substituted double bonds or cyclopropane rings greatly facilitate α-cleavage.

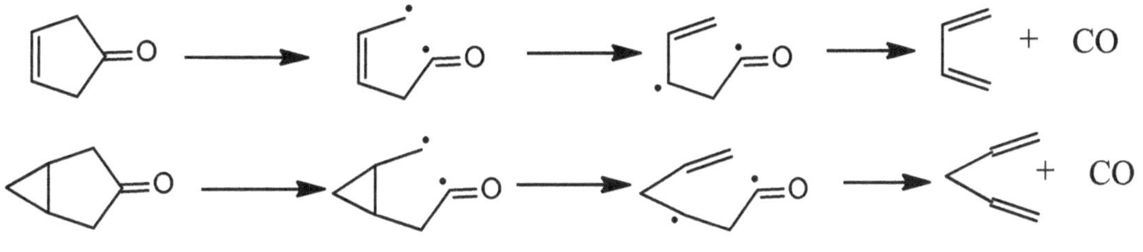

Scheme 258

The photochemistry of 3,5-cycloheptadienone provides a interesting example. Direct irradiation give rise to S1, which has the $^1(n,\pi^*)$ configuration and undergoes α-cleavage leading ultimately decarbonylation. The T1 state produced by energy transfer has the $^3(\pi,\pi^*)$ configuration and leads to isomeriztion of the diene. In the low lying singlet excited state energy is localized primarily on the carbonyl group, whereas in the low lying triplet state energy is localized mainly in the diene system.

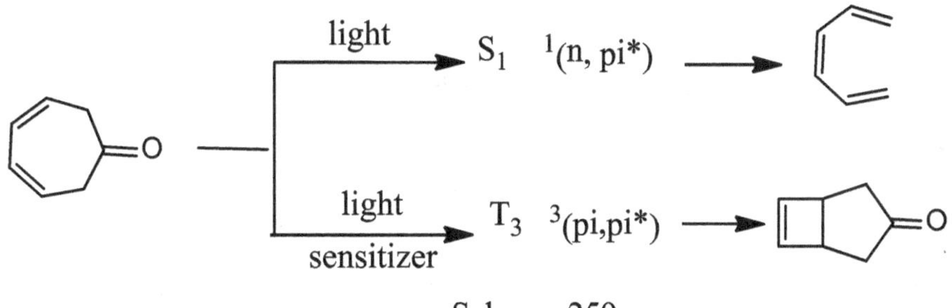

Scheme 259

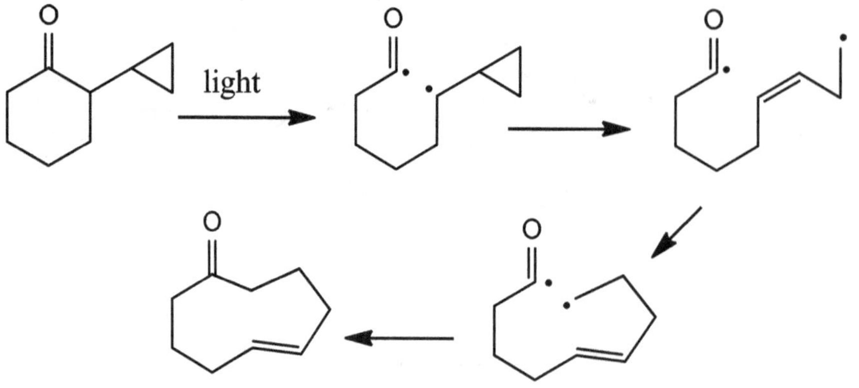

Scheme 260

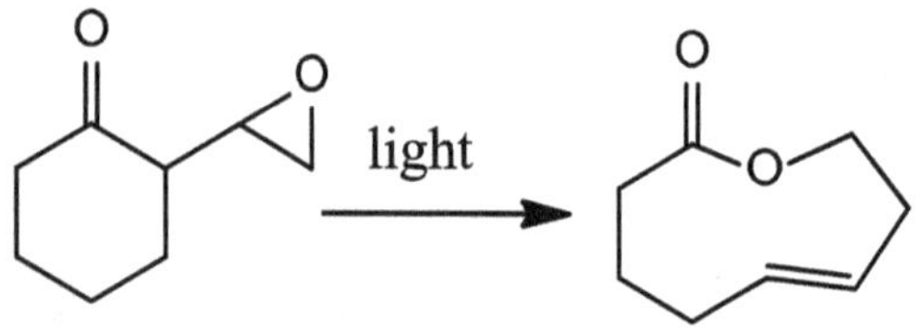

(Equation 99)

Influence of ring size is also important, with smaller ring systems, biradical formed but hydrogen abstraction is inhibited as six or five membered transition state is inhibited.

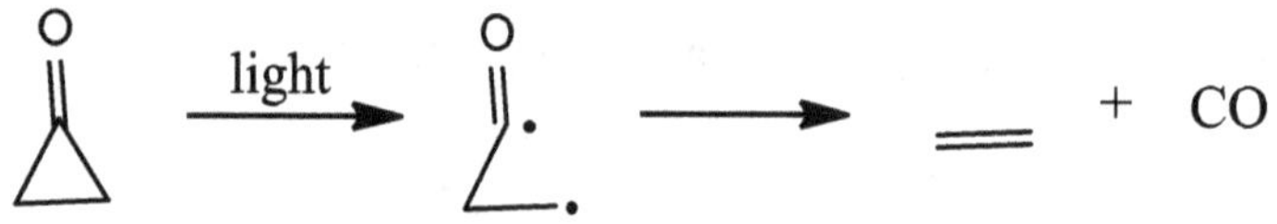

Scheme 261

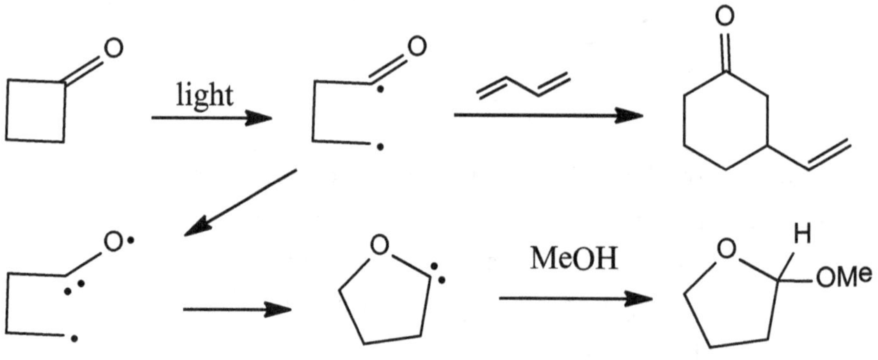

Scheme 262

When α-hydrogen extraction is hindered, then decarbonylation occur.

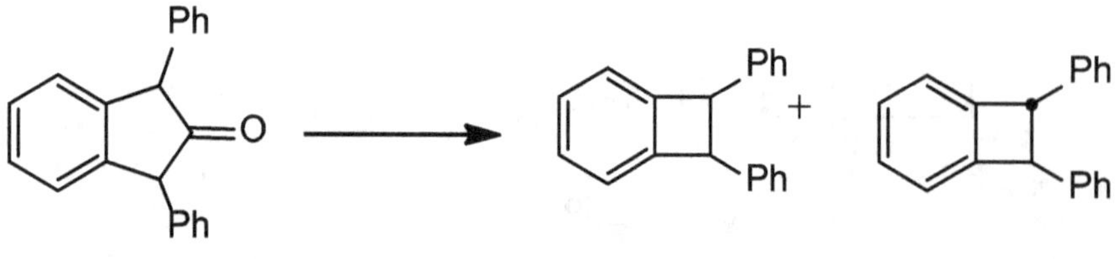

(Equation 100)

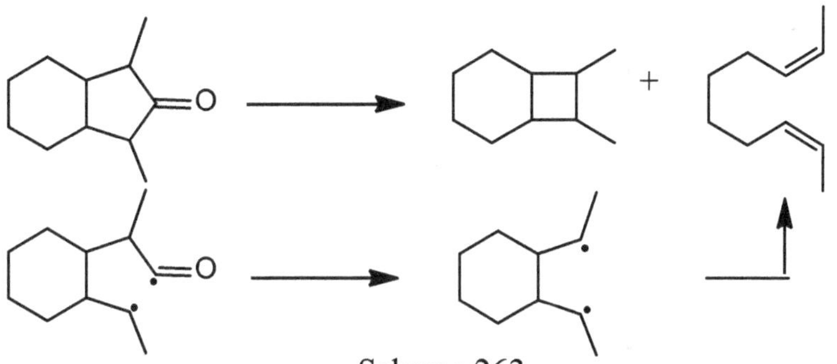

Scheme 263

If a mixture of two different carboxylates is used, all combinations of them are generally seen as the organic product structures:

6.5   Norrish II Type Reactions

A **Norrish type II reaction** is another common reaction in which excited state of the carbonyl group undergo intramolecular abstraction of a γ-hydrogen (which is a hydrogen atom three carbon positions removed from the carbonyl group) to produce a 1,4-biradical as a primary photoproduct (IUPAC definition).

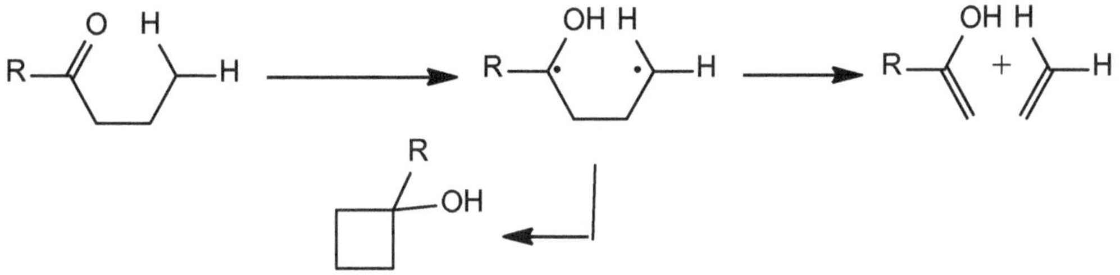

Scheme 264

Secondary reactions that occur are fragmentation to form an enol and an alkene, or intramolecular recombination of the two radicals to a substituted cyclobutane (the Norrish–Yang reaction).

Consequence of hydrogen abstraction by the photo excited carbonyl group from the side chain, is the formation of 1,4-diradical. This 1,4-diradical can revert to starting ketone or close to substituted cyclobuatanol or can be cleaved into olefin and an enol of a ketone. The latter process is known as Norrish II type cleavage. In this reaction, both n$\longrightarrow\pi^*$ singlet and triplet states are involved. Because when triplet in the reaction is quenched using 1,3-pentadiene as a quencher, quenching was observed to be partially effective.

Cleavage between Cα and Cβ is often referred as type II photo elimination to distinguish the carbonyl carbon Cα. Type II photo elimination is observed for both S_1 and T_1 are involved for aliphatic ketones but one of the carbonyl substituent is aryl, intersystem crossing is very fast and T_1 is the reactive state. Usually, cleavage is the dominant reaction with the cyclobuatanol yields being below 20%, but there are exceptions. Solvents that can hydrogen bond to the hydroxyl group of the 1,4-diradical stabilize it and retard the reverse reaction.

Formation of 1,4-radical was confirmed by treating the reaction mixture with D_2O, which replaces γ-hydrogen by deuterium yielding the 1,4-diradical with an O-D bond. The energy required to break an O-D bond is larger than the energy required to break an –H bond and consequently the reverse reaction is slower and an increase in the efficiency of the overall reaction is observed. Though the reaction proceeds through singlet and triplet state, there are certain differences.

Irradiation of S(+)-5-methyl-2-heptanone gives initially the $^1(n,\pi^*)$ state, which can intersystem cross to the $^3(n,\pi^*)$ state return to optically active starting material or react to give products. The products probably are formed from a singlet biradical. The $^3(n,\pi^*)$ state give rise to a triplet biradical, which can decay to racemic starting material or go on to products. Racemization of the starting material occurs only by the triplet reaction.

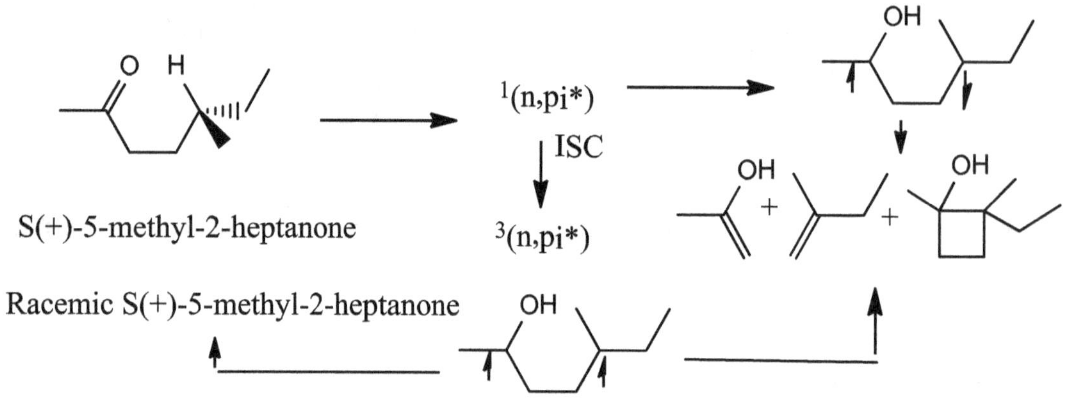

Scheme 265

The Norrish reaction has been studied in relation to environmental chemistry with respect to the photolysis of the aldehyde heptanal, a prominent compound in Earth's atmosphere. Photolysis of heptanal in conditions resembling atmospheric conditions results in the formation of 1-pentene and acetaldehyde in 62% chemical yield together with cyclic alcohols (cyclobutanols and cyclopentanols) both from a Norrish type II channel and around 10% yield of hexanal from a Norrish type I channel (the initially formed n-hexyl radical attacked by oxygen).

In one study the photolysis of an acyloin derivative in water in presence of hydrogen tetrachloroaurate (HAuCl$_4$) generated Nano gold particles with 10 nanometer diameter. The species believed to responsible for reducing Au^{3+} to Au^0 is the Norrish generated ketyl radical [2006JACS15980].

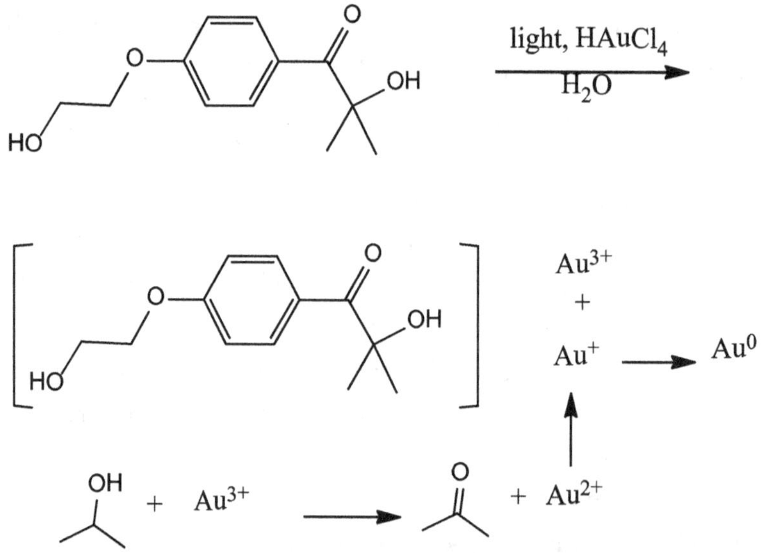

Scheme 266

An example of a synthetically useful Norrish type II reaction can be found early in the total synthesis of the biologically active cardenolide ouabagenin by Baran and coworkers [2013Sci 59].

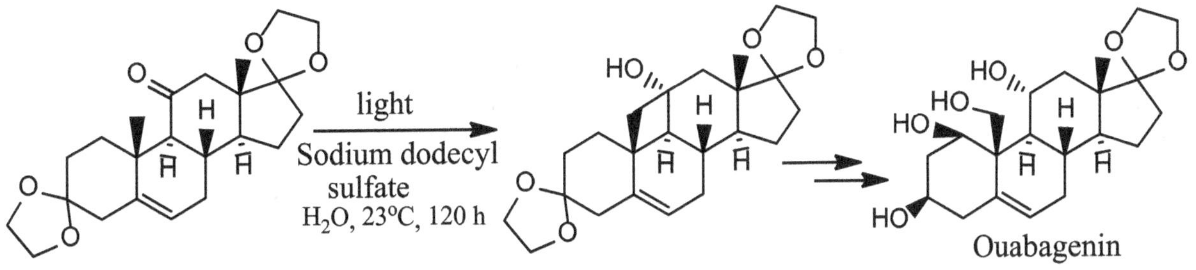

Scheme 267

6.6   PHOTOREDUCTION

In Norrish type II reaction, reduction of a ketone takes place to alcoholic group by extracting hydrogen from within the molecule i.e. intramolecular one. Intermolecular hydrogen abstraction is also possible from a suitable donor. It is because the oxygen of a carbonyl group in (n,π^*) excited state is electrophilic in nature.

The well known example is the exposure of a mixture of benzophenone and benzhydrol to sunlight when benzpinacol is formed. The first step is the photochemical reduction is excitation of the benzophenone to the $^1(n,\pi^*)$ state, which inter system crosses to the $^3(n,\pi^*)$ state. Hydrogen abstraction from benzhydrol gives two diphenylhydroxymethyl radicals, which combines to form benzphinacol.

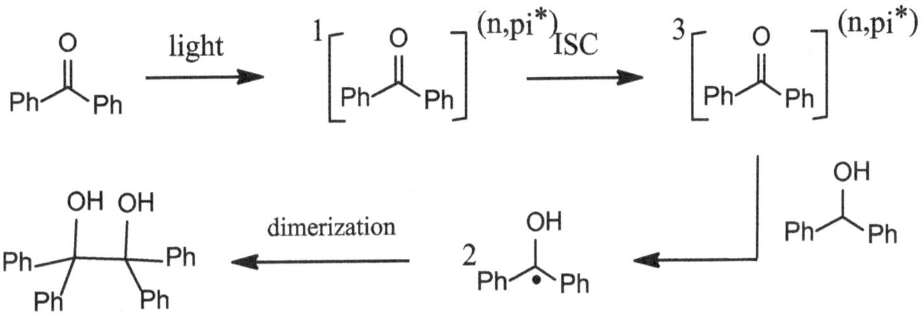

Scheme 267

The photolysis of propan-2-ol results in hydrogen atom transfer to the ketone and two identical ketyl radicals are formed. The formed ketyl radical undergo disproportion and dimerization reaction to get stabilized.

Ketones undergo photoreduction in the presence of variety of hydrogen atom donor other than secondary alcohols. Apart from secondary alcohols other hydrogen donor solvents could be used in this reaction, e.g. amines, thiols etc. The key reaction in this case is electron transfer from the amine to the ketones producing the ketyl radical and the cation radical from the amine.

Carbobnyl compounds like ketones, aldehydes and quinines can also add photochemically to activated methylene groups by hydrogen atom abstraction and subsequent radical recombination.

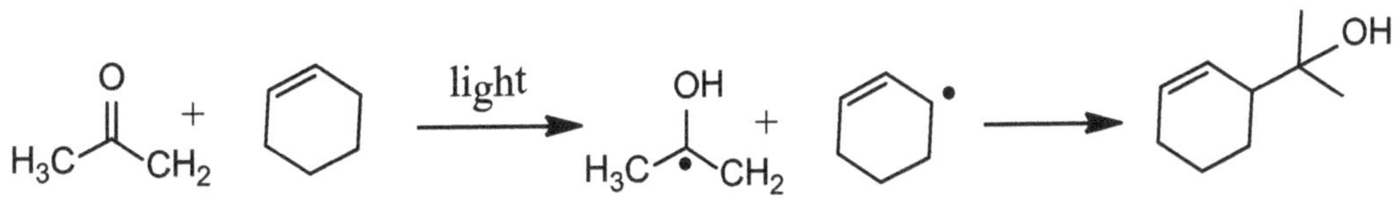

Scheme 268

Photoreduction generally proceeds from $^3(n,\pi^*)$ state. However, with aromatic ketones both $^3(n,\pi^*)$ and $^3(\pi,\pi^*)$ states can be populated and $^3(\pi,\pi^*)$ state is more reactive and sufficient in photoreduction.

The reactivity of these two triplet states is different. The $^3(n,\pi^*)$ state behave like alkoxy radical with one unpaired electron on the oxygen atom. While on $^3(\pi,\pi^*)$ state, the unpaired electrons are delocalized over both the carbon and the oxygen atom. For example, photoreduction of acetophenone proceeds from $^3(n,\pi^*)$ state while of 2-acetylnaphthalene not.

All ketones are not reduced. Substituents on the aryl ring play very important rule.

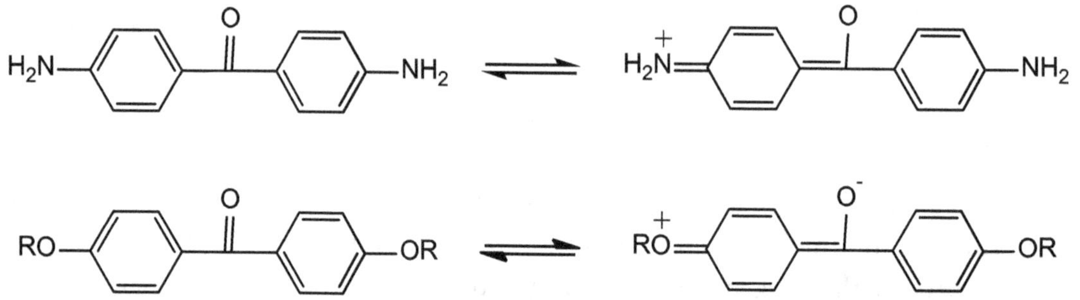

Scheme 269

We know that photoreduction takes place from (n,π^*) transition where oxygen is electrophilic in nature due to half vacant n-orbital and (n,π^*) is of the lower energy, But due to resonance negative charge on oxygen makes it nucleophilic and also (π,π^*) transition as the lowest energy state. Therefore reduction is forbidden. Since hydrogen atom abstraction is much more efficient for (n,π^*) state of ketones than for (π,π^*) states.

In the case of ortho alkyl substituted ketones, photoenolization takes place but not reduction. This is proved by using deuterated solvent R'OD.

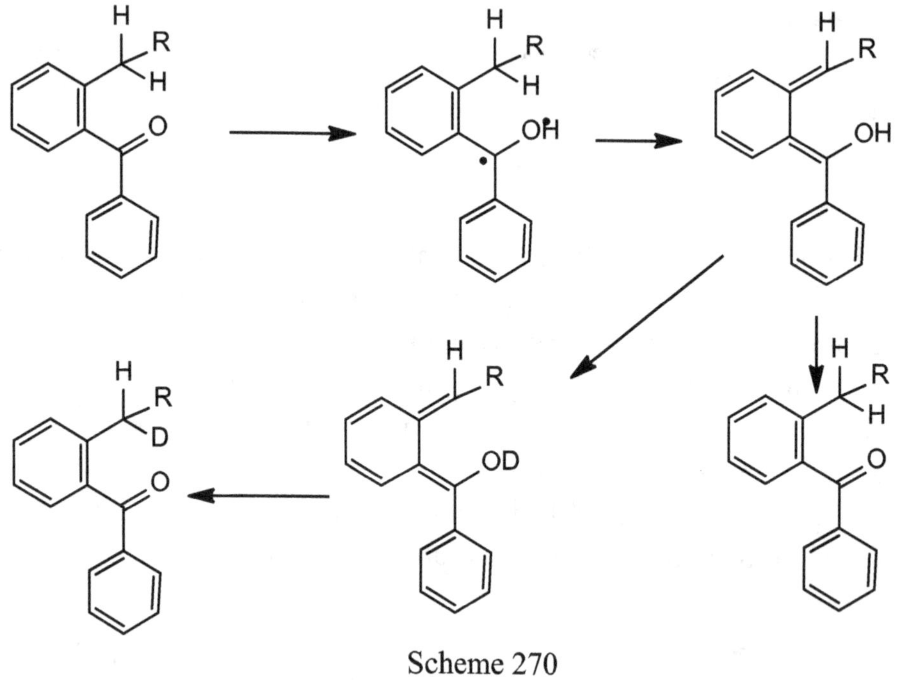

Scheme 270

If a mixture of two different carboxylates is used, all combinations of them are generally seen as the organic product structures:

6.7   PATTERNO-BUCHI REACTION

The [2 + 2] photocycloaddition between an electronically excited carbonyl compound and an alkene leading to oxetanes is one of the most investigated organic photochemical reactions. This reaction is known as Patterno-Buchi reaction.

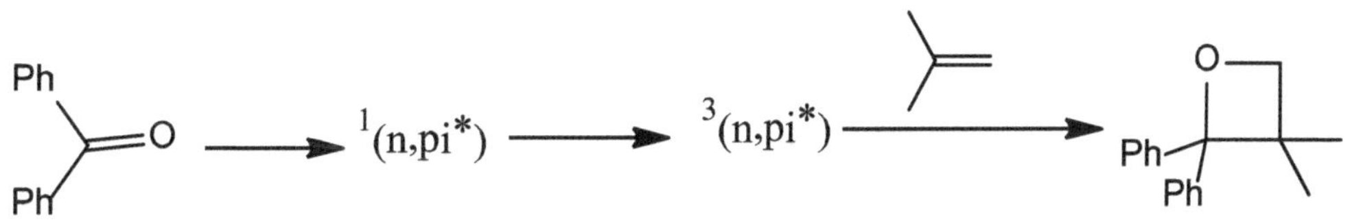

Scheme 271

Only ketones with low lying (n,π*) state are reactive in this process. Most of the Patterno-Buchi reaction reported to involve ^{3}(n,π*) state ketones. The major mode of addition can be predicted by assuming that the radical like oxygen atom of the ^{3}(n,π*) ketone adds to the olefin to give preferentially the most stable biradical intermediate. Thus the reaction is stereoselective, favouring the more stable adduct. For instance, in the addition of ^{3}(n,π*) state benzophenone to trimethylene, the element of choice lie between a secondary radical and a tertiary radical, since both radicals are the same in other respects. These tertiary radical is more stable and the mode of addition is preferred.

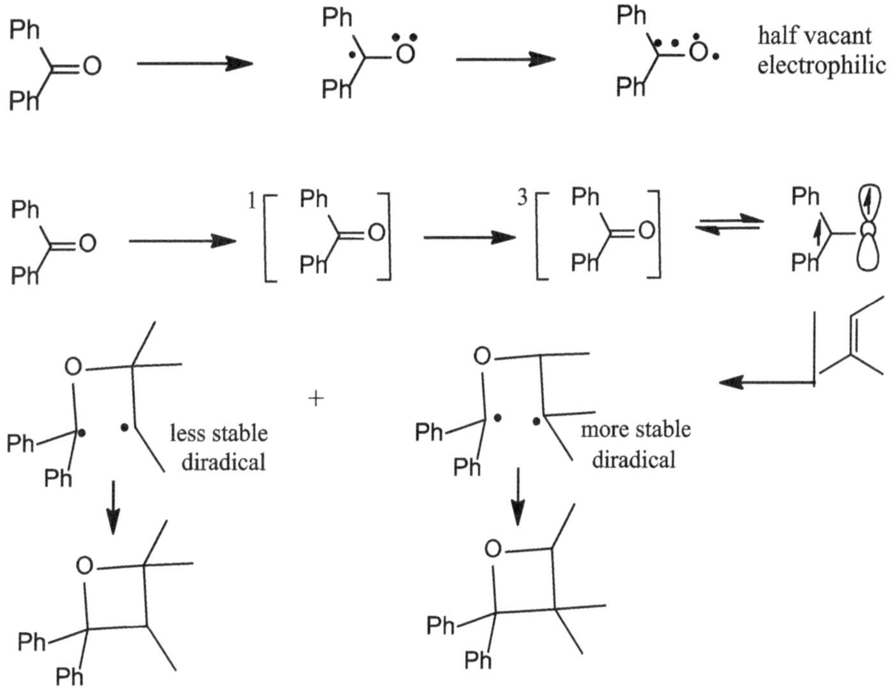

Scheme 272

The biradical hypothesis is useful in predicting the major product in a Patterno-Buchi reaction, but it is not adequate as a mechanism. The rate constant for the reaction of excited ketone with ground state olefin is very high (-10^{-9}mol^{-1}sec^{-1}) in few cases that have been studied. This value is several orders of magnitude greater than rate constants for the addition of any radicals to olefins. It is probable that the reaction involves an exciplex (complex between excited ketone and olefin) that collapses to the biradical. The biradical is an appealing intermediate because addition of

$$\text{Ketone} \xrightarrow{\text{light}} {}^{1}(\text{Ketone})$$

$${}^{1}(\text{Ketone}) \xrightarrow{\text{ISC}} {}^{3}(\text{Ketone})$$

$${}^{3}(\text{Ketone}) + \text{olefin} \longrightarrow \text{exciplex}$$

$$\text{exciplex} \longrightarrow \text{biradical}$$

$$\text{biradical} \longrightarrow \text{oxetane}$$

$$\text{biradical} \longrightarrow \text{ketone} + \text{olefin}$$

${}^{3}(n,\pi^{*})$ state benzophenone to cis or trans 2-butene gives the same mixture of addition in each case.

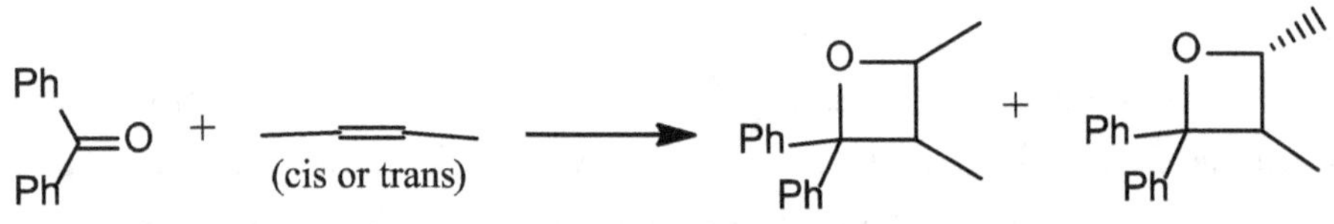

(Equation 101)

Two important rules for successful formation of oxetanes: i. only compounds with low lying (n,π^{*}) state will form oxetanes; ii. The energy of the carbonyl exciplex must be lower than that of olefin to prevent easy transfer from ${}^{3}(n,\pi^{*})$ state ketone to olefin.

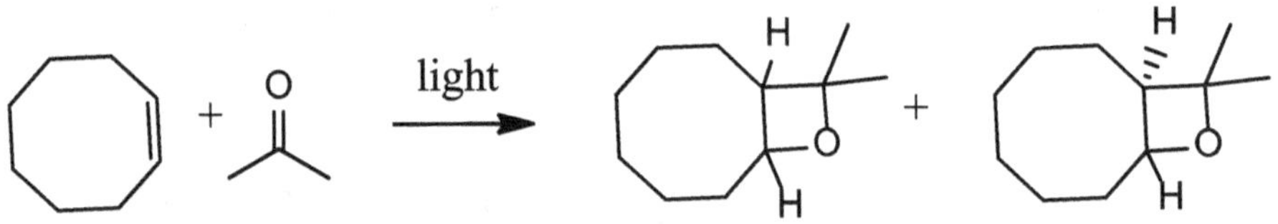

(Equation 102)

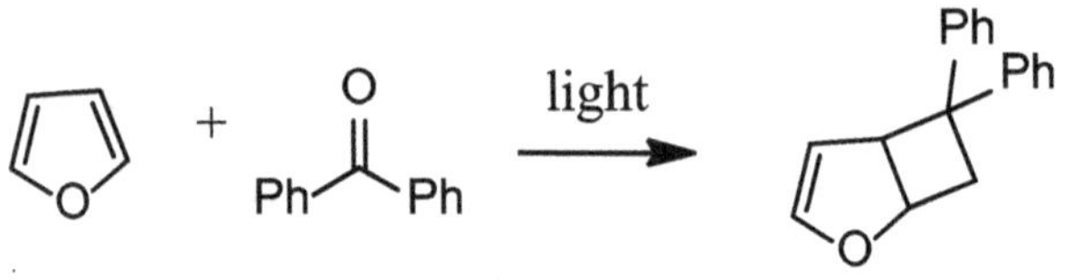

(Equation 103)

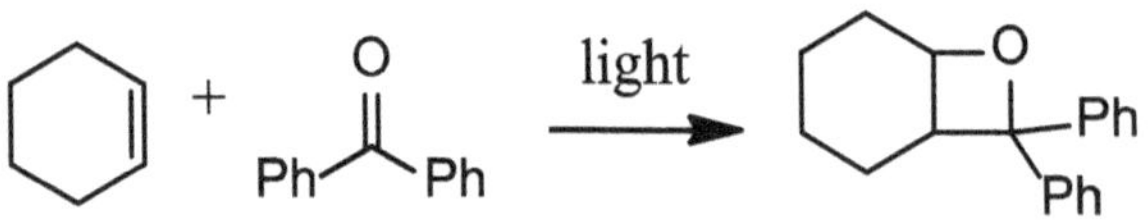

(Equation 104)

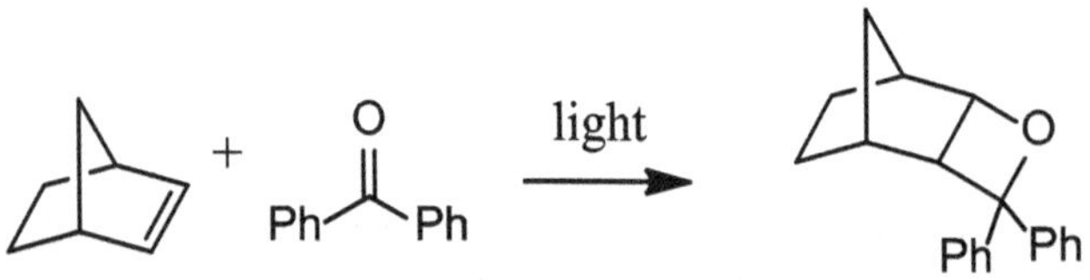

(Equation 105)

(Equation 106)

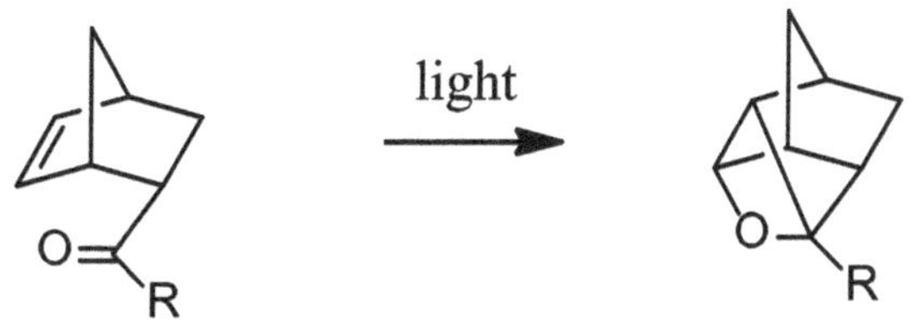

(Equation 107)

If a mixture of two different carboxylates is used, all combinations of them are generally seen as the organic product structures:

Two side reactions can limit the synthetic utility of the Patterno-Buchi reaction. If reactive hydrogen atoms (such as allylic hydrogen atoms) are present in the olefin, hydrogen abstraction by the excited ketone will compete with Patterno-Buchi reaction and complex mixture will be formed.

If the triplet energy of the ketone is comparable to or exceeds, that of the olefin, energy transfer will compete with or support addition. The problem is especially keen with aliphatic ketones because of their high triplet energies. Acetone (ET = 78 Kcals/mol), for instance, transfer of triplet energy to norbornene and thus produce dimmers whereas benzophenone (ET= 69 K Cal/mol) adds to norbornene.

Scheme 273

This reaction could be carried out with acetylene but the resulting unsaturated oxetanes are unstable and finally yield α,β-unsaturated ketones.

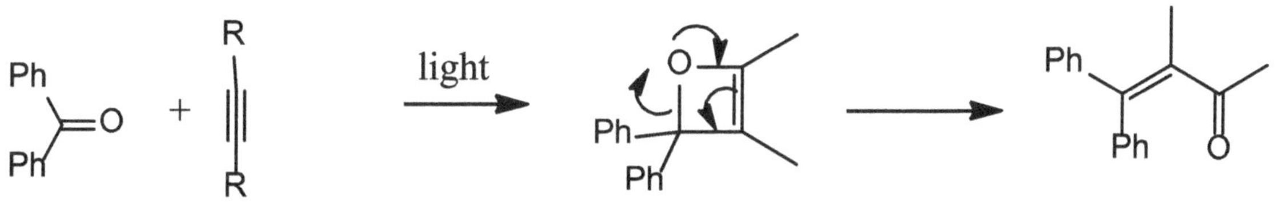

Scheme 274

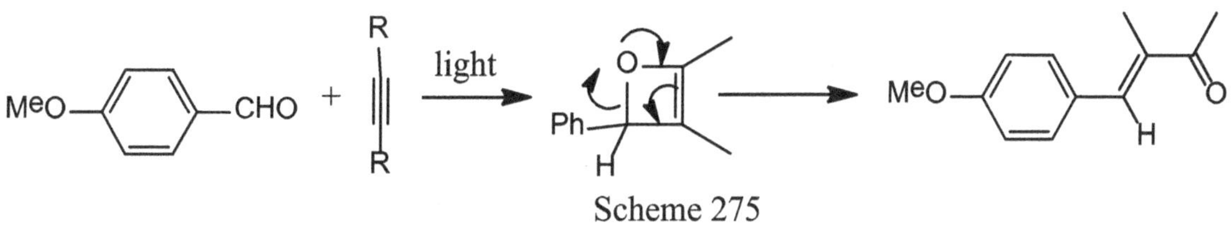

Scheme 275

Reaction of $^1(n,\pi^*)$ state ketone with electronegatively substituted olefins also give oxetanes. These reactions differ from the usual Patterno-Buchi reaction in substitution effects on the olefin and its retention of the stereochemical integrity of the olefin. The $^1(n,\pi^*)$ state thus seems to be nucleophilic and the addition probable does not involve biradical intermediate.

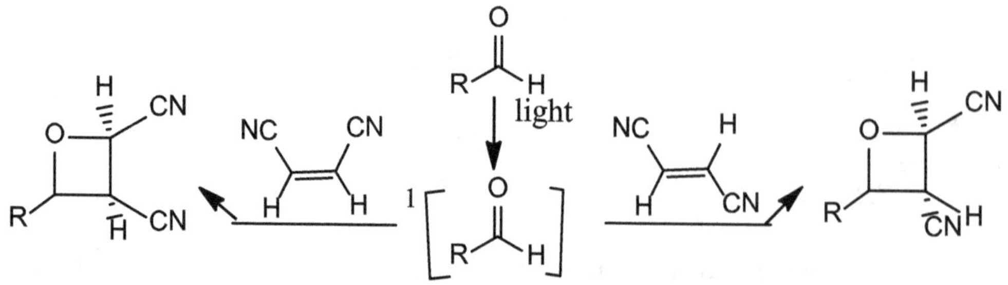

Scheme 276

6.8   PHOTOCHEMISTRY OF α,β-UNSATURATED KETONES

α,β-unsaturated ketones undergo photochemical primary processes characteristic of both CO and = groups. Therefore the photochemistry of α,β-unsaturated ketones is very rich and complicated. Only few reactions are discussed.

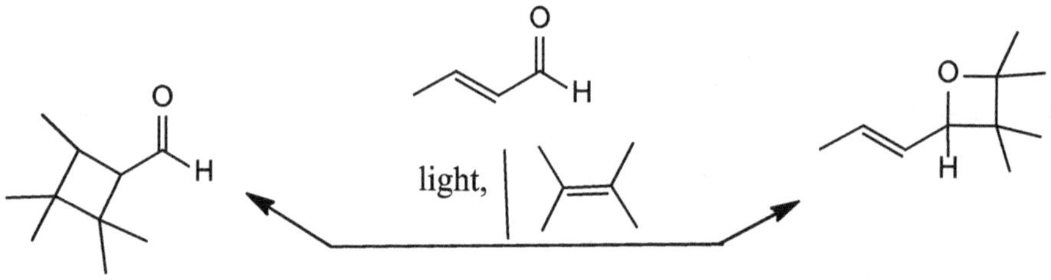

Scheme 277

Acyclic α,β-unsaturated carbonyl compounds containing γ-hydrogen atom often undergo double bond migration.

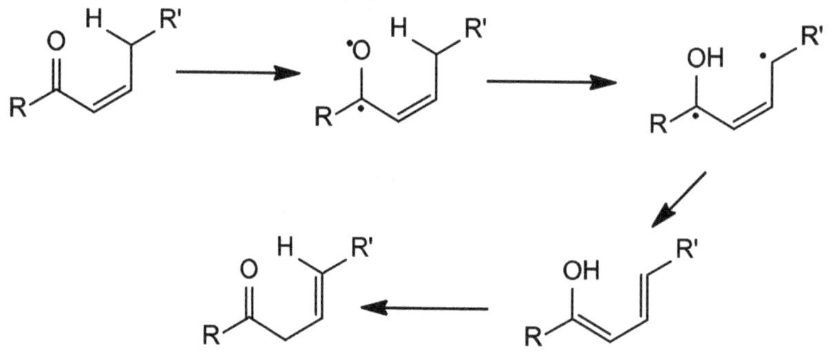

Scheme 278

Why less stable conjugated compound is formed? Because, light radiation required for conjugated system is of longer wavelength while non-conjugated ones require radiation of shorter wavelength. Therefore energy supplied is not enough and the unconjugated remains unaffected. Potochemical deconjugation of synthetically useful way of effecting isomerisation of α,β-unsaturated ketones and esters to β,γ-isomers.

Intramolecular hydrogen abstraction is also the dominant process for acyclic α,β-unsaturated ketones. The intermediate 1,4-diradical thus formed undergo cyclization to form the enol of a cyclobutyl ketone. Among the byproducts of such photolysis are cylobutanols resulting from alternative modes of cyclization of the diradical intermediate.

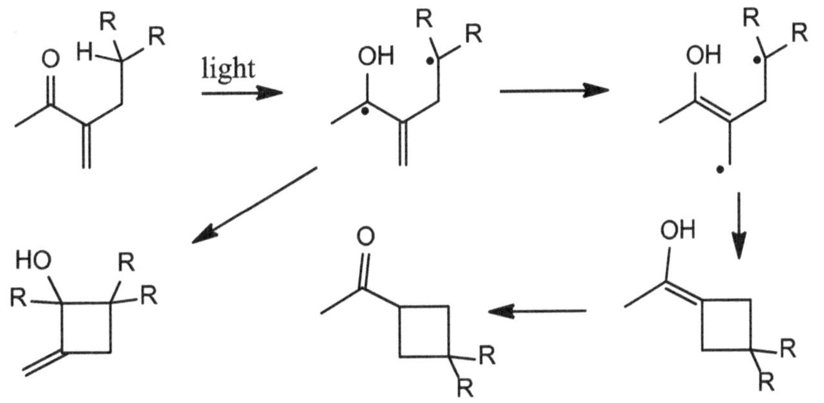

Scheme 279

If a mixture of two different carboxylates is used, all combinations of them are generally seen as the organic product structures:

On irradiation, cylcohexenone unergo cyclization via head to head or head to tail attack to form two isomers. These are formed by $3(\pi,\pi^*)$ hexenone and ground state of cyclohexenone. The ratio of the two dimers depends on the solvent used. Polar solvents favour the formation of head to head dimer.

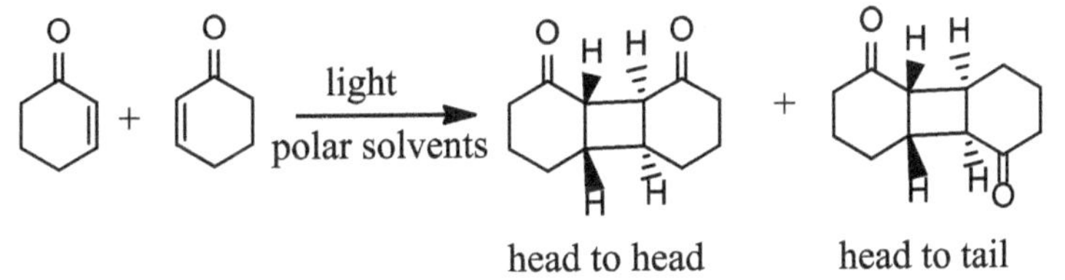

head to head   head to tail

(Equation 108)

A related reaction can be carried out using olefins in place of cyclohexenone. For instance, with 1,1-dimethoy ethylene give two products. The highly strained trans adduct is the major product.

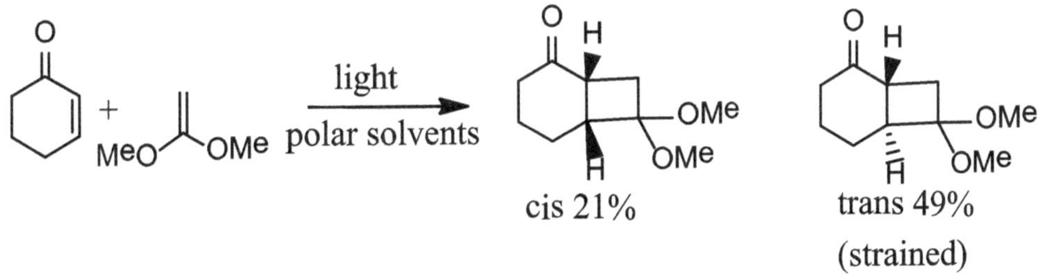

cis 21%   trans 49%
(strained)

(Equation 109)

In the case of cylcoheptenone and larger rings, the main initial photo products are the trans cycloalkenones produced by the photoisomerization. In the case of 7 or 8 membered rings, the trans bonds are sufficiently strained, that rapid reaction follows. In nucleophilic solvents, dimerization occurs whereas in nucleophilic solvents, addition occurs.

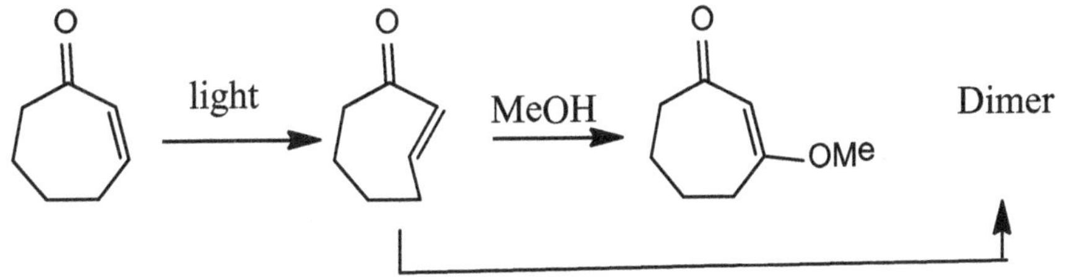

Scheme 280

4,4-Dialkylcyclohexenone undergo a photochemical rearrangement which involves a formal shift of the C-4-C-5 bond to C-3 and formation of a new bond between C-2 & C-4. This reaction is general and also proceeds in the case of 4-alkyl-4-arylcyclohexenones.

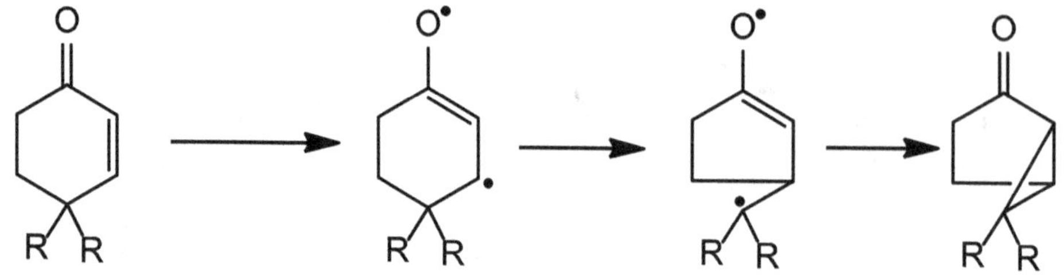

Scheme 281

The reaction is stereospecific and can be described as a [π2a+σ2a] cycloaddition. This mechanism requires that inversion of configuration occurs at C-4 as the new σ bond at the back lobe of the reacting C-4-C-5 σ bond.

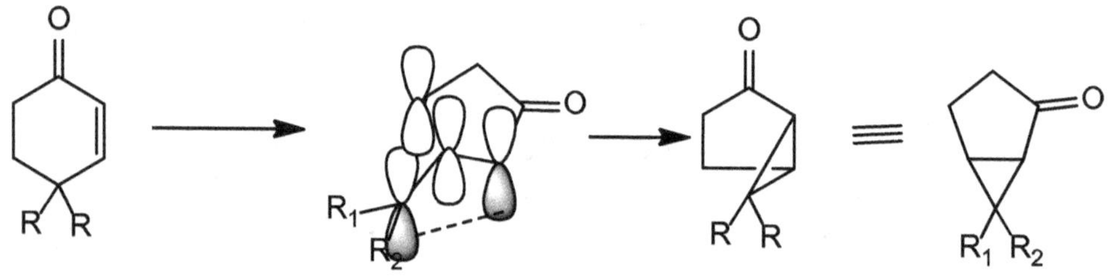

Scheme 282

It has been demonstrated in several system that the reaction is in fact stereospecific with the expected inversion occurring at C-4. The ketone (**73**) provides a specific example. The stereoisomeric product is (**74**) and (**75**), both are formed but in each product inversion has occurred at C-4.

In contrast to the rearrangement described in 4,4-dialkyl cyclohexenones, the reaction is not entirely stereospecific and a minor stereoisomer is described. This suggests the existence of competitive reaction pathway, which is not concerted. Note that the endo product is predicted by the concerted mechanism. It is the major product although it is sterically more congested than exo isomer.

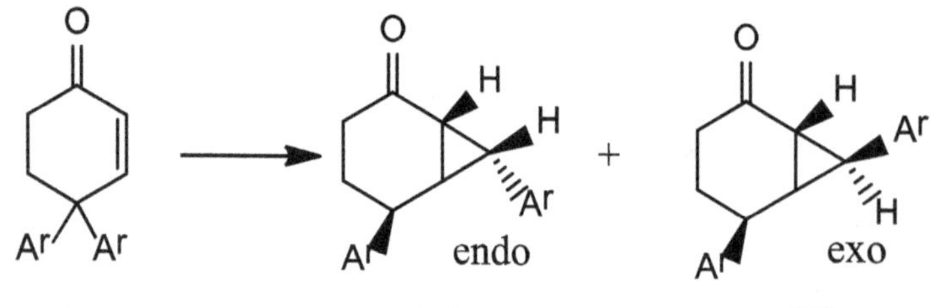

(Equation 110)

In compounds in which the two aryl groups are substituted differently, it is found that the substituents which stabilize radical character favour rearrangement. Thus the p-cyanophenyl substituent migrates in preference to the phenyl migration. This rearrangement can be considered to occur via a transition state in which 2-4 bridging is accompanied by 4-3 aryl migration.

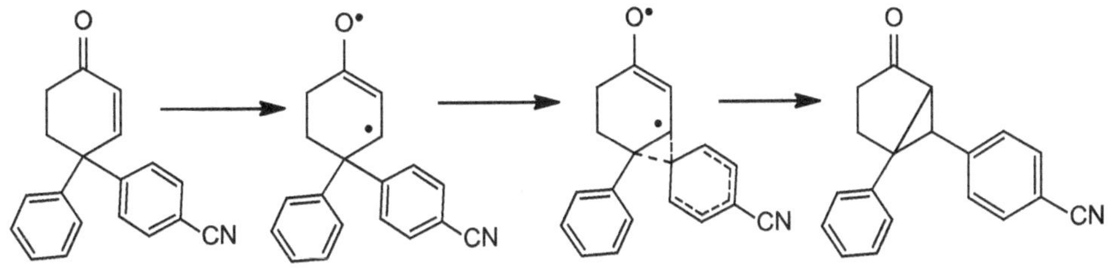

Scheme 283

With other ring sizes the photochemistry of unsaturated cyclic ketones takes different course. For cyclopentenones, the principal products resist from hydrogen abstraction process. Irradiation of cyclopentenones in cyclohexane gives a mixture of 2 and 3-cyclohexyl cyclopentanone. These products can be formed by intermolecular hydrogen abstraction followed by recombination of the product radical.

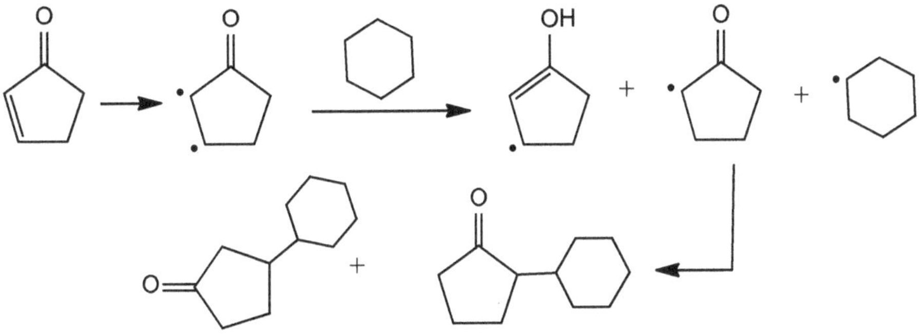

Scheme 284

6.9   DIENONE-PHENOL REARRANGEMENT

The most widely studied compounds that undergo photochemical rearrangements are the dienones. These undergo the dienone-phenol rearrangement in acid solution and also undergo this rearrangement under the influence of light, but in this case the reaction is more complex, e.g., 4,4-diphenylcyclohexa-2,5-dienone (a cross-conjugated dienonoe) in aqueous dioxan solution gives four products.

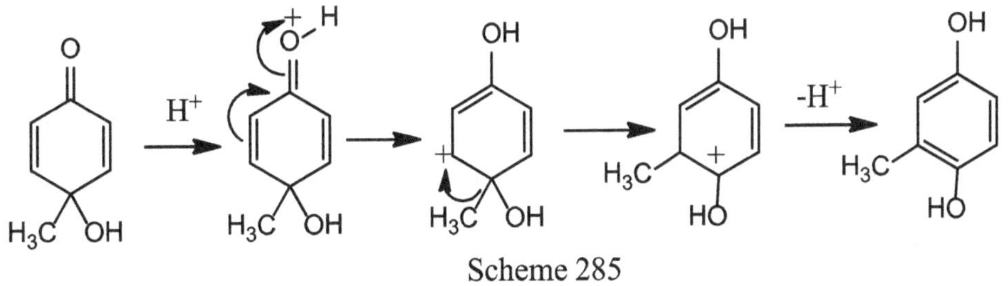

Scheme 285

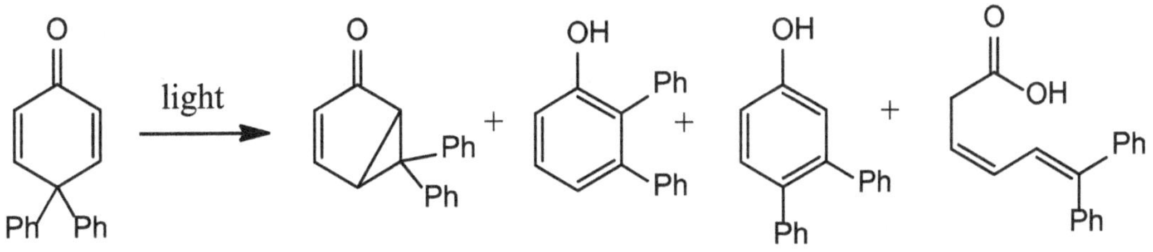

Scheme 286

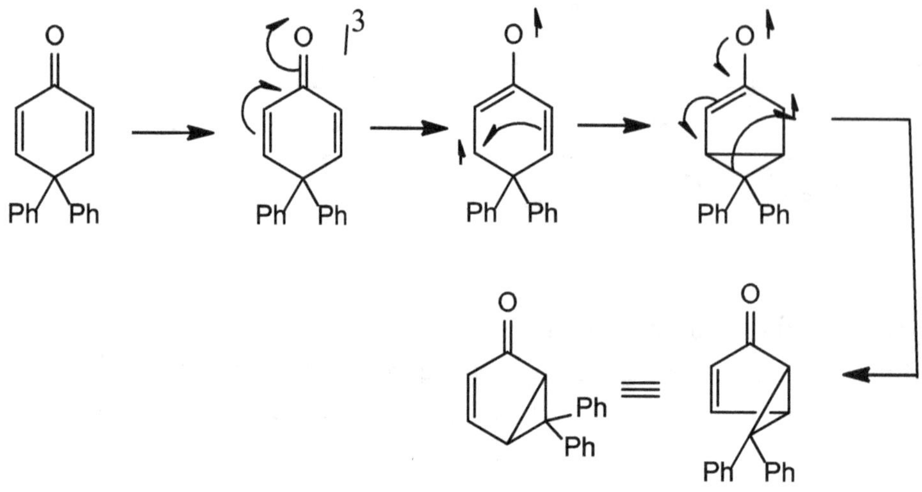

Scheme 287

On the assumption of a similar mechanism, the formation of the other products can be explained as follows:

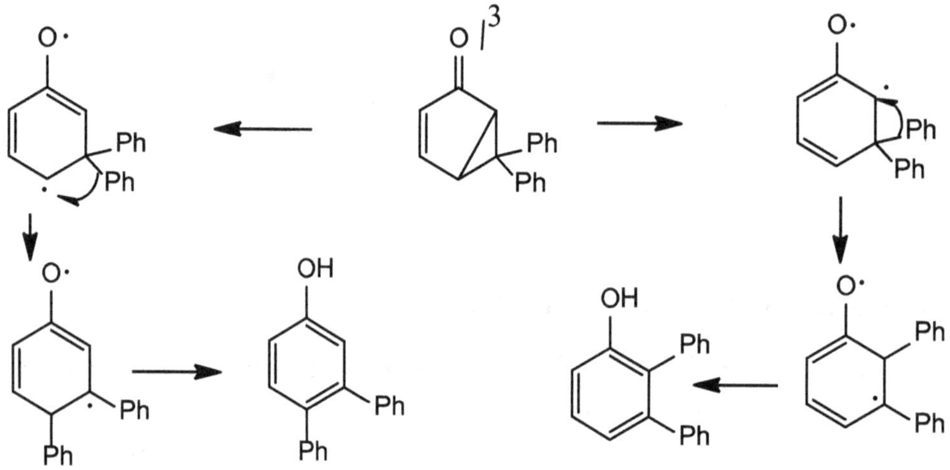

Scheme 288

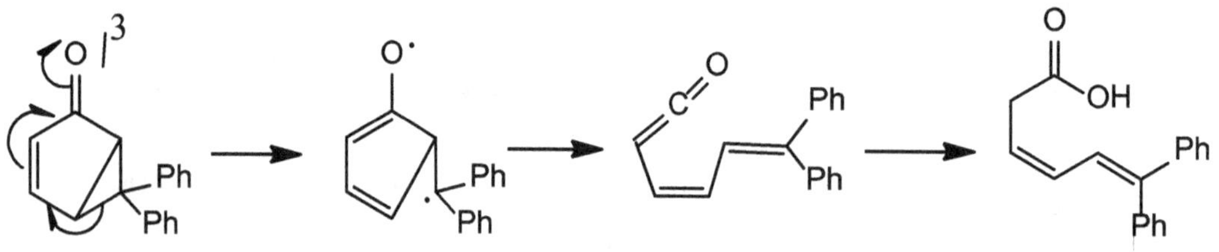

Scheme 289

Conjugated ketones undergo a variety of complex photochemical rearrangements, some of which have been studied in considerable detail. For instance, irradiation of santonin in refluxing acetic acid gives an ester of isophotosantanoic lactone with an empherical formula of santonin by the addition of molecule of water whereas irradiation in ethanol gives an isomer of santonin, lumisantonin. In both of these products, extensive skeletal rearrangement has taken place. Lumisatonin undergoes photochemical rearrangement to a ketene that cyclize thermally to mazdasantonin or reacts with water or alcohols to give photosantoanoic acid or its ester. Irradiation of mazdasantonin also gives the ketene. This process is an example of α-cleavage in an unsaturated ketone.

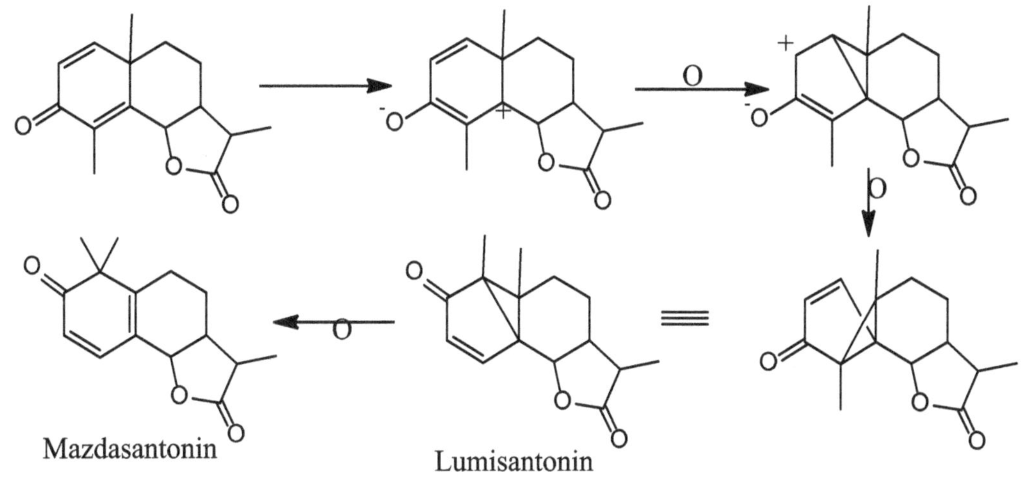

Mazdasantonin   Lumisantonin

Scheme 290

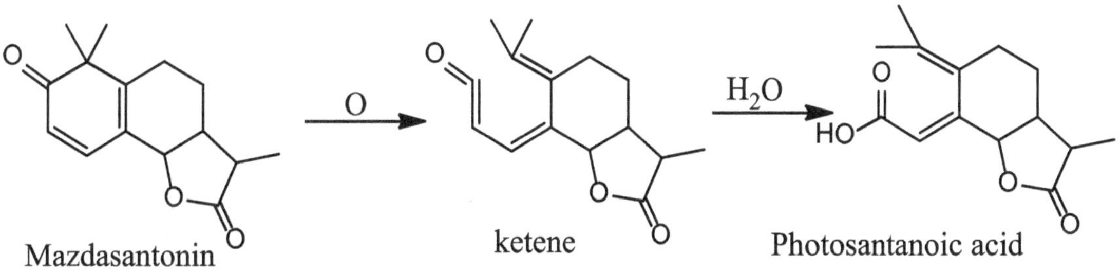

Mazdasantonin   ketene   Photosantanoic acid

Scheme 291

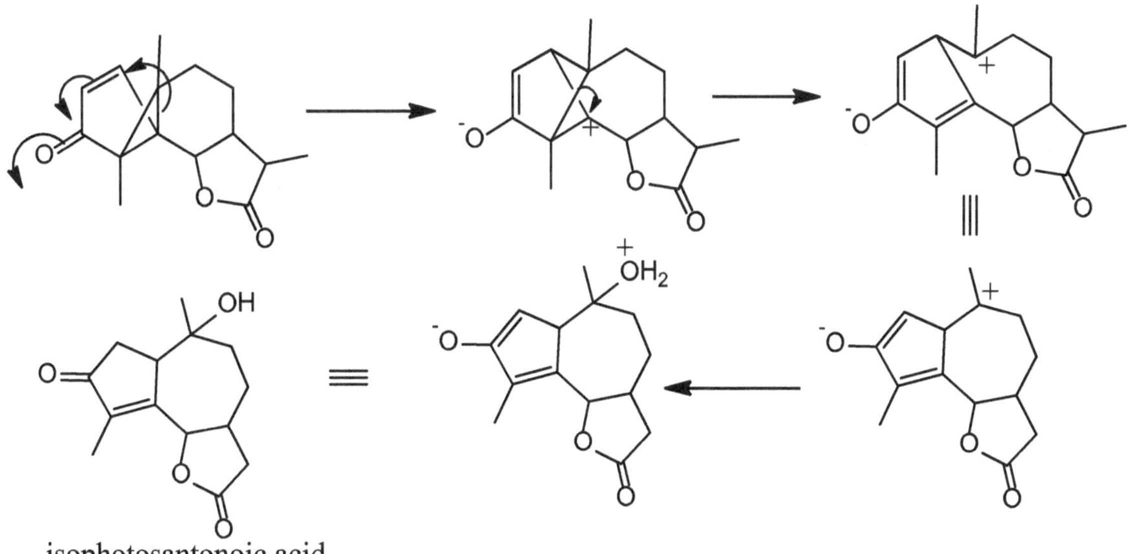

isophotosantonoic acid

Scheme 292

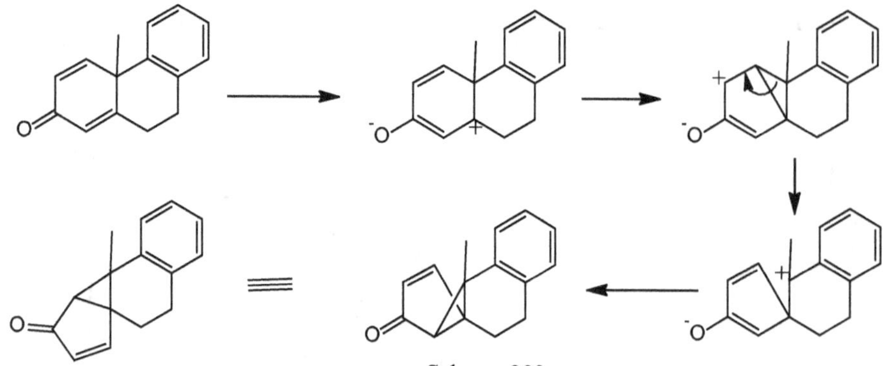

Scheme 293

The **DeMayo reaction** is a photochemical reaction in which the enol of a 1,3-diketone reacts with an alkene (or another species with a C=C bond) and the resulting cyclobutane ring undergoes a retro-aldol reaction to yield a 1,5-diketone:

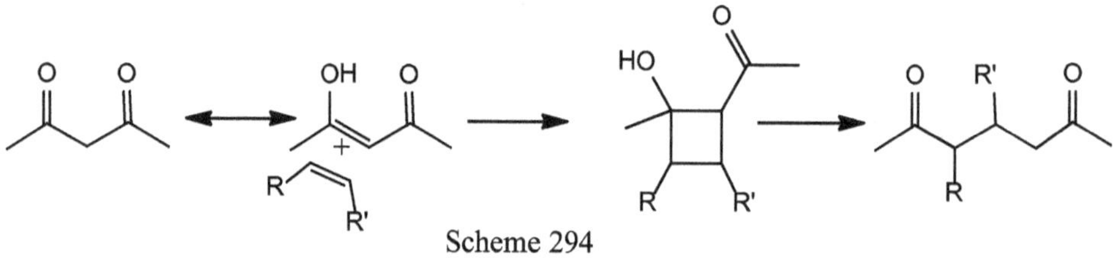

Scheme 294

The net effect is to add the two carbon atoms in the C=C double bond between the two carbonyl groups of the diketone. It is thus useful in syntheses both as a relatively selective way to join two parts of a molecule and as a way to apply the more developed chemistry of 1,3-diketone synthesis to 1,5-diketones. The first part is a [2+2] cycloaddition. The ensuing retro-aldol cleavage is favored by the relative instability of the cyclobutane ring [2009Mis173].

EXERCISE

1. Discuss Norrish type I cleavage with suitable examples.
2. What is the electronic configuration in ground and first excited state of formaldehyde molecule?
3. Write a reasonable mechanism for the following conversions:

$$CH_3COCH_3 \xrightarrow{\text{light}} H_2C=CH_2$$

4. Outline the mechanism for the following reaction.

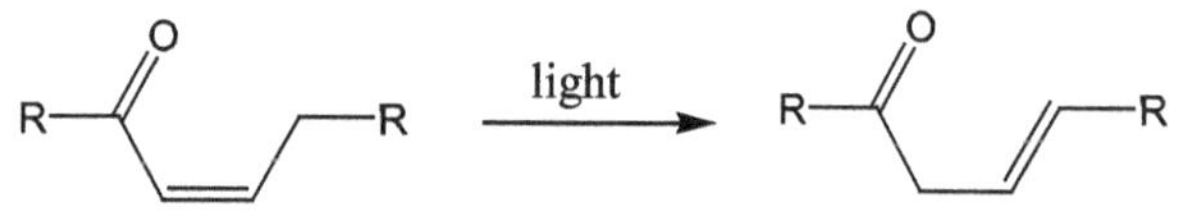

5. Outline the mechanism for the following reaction.

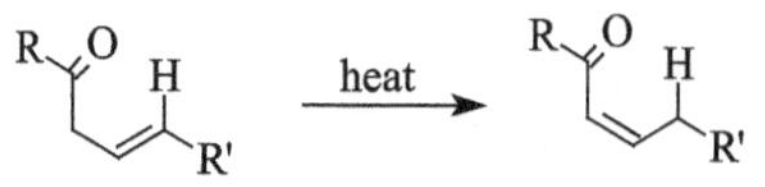

6. During Norrish II type reaction, the reaction goes via δ-hydrogen abstraction. Justify this statement with suitable example

7. Predict the product for the following:

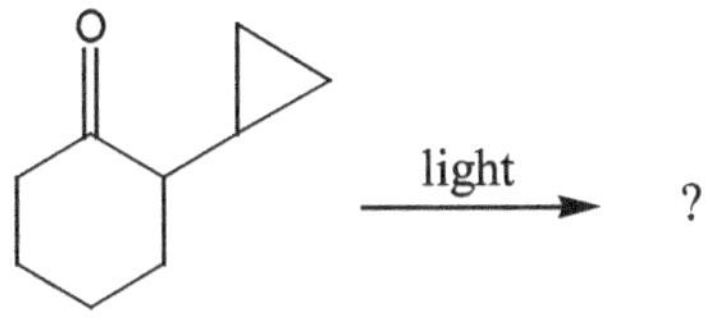

8. Predict the product for the following:

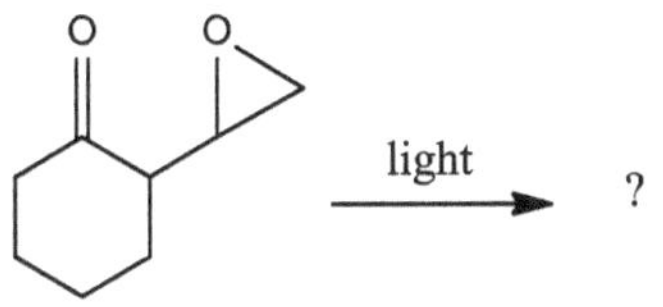

9. Benzophenone undergo photoreduction while p-methoxy benzophenone doesnot. Why?

10. Predict the product for the following:

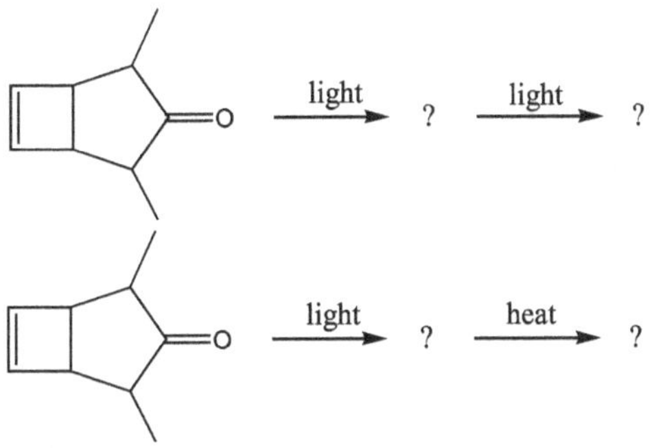

11. Explain Patterno-Buchi reaction with suitable example.

12. Write a reasonable step for the following conversions

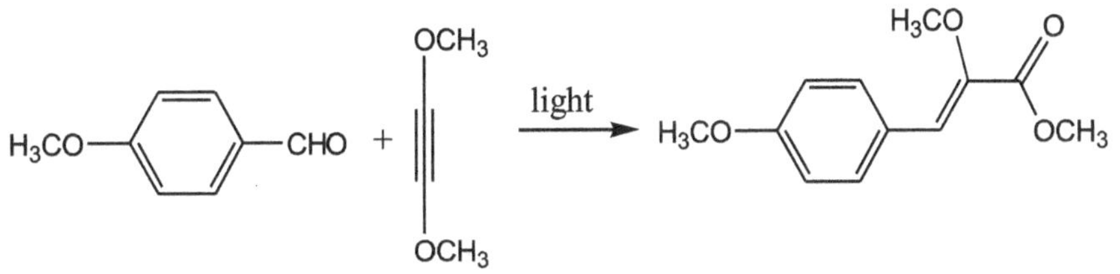

13. Predict the product for the following with suitable mechanism:

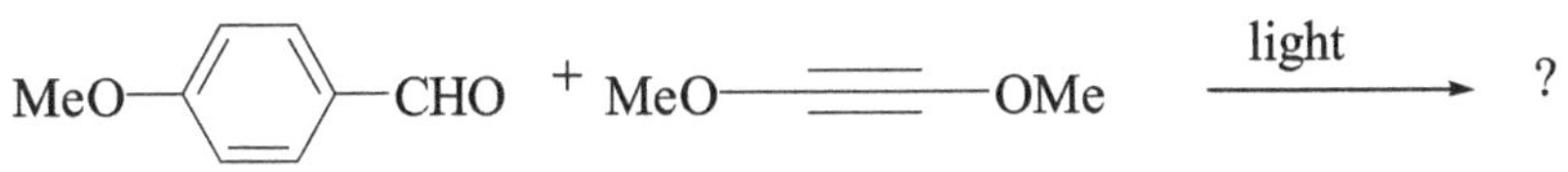

14. Predict the product(s) for the following:

ii.    PhCOPh   + $\xrightarrow{\text{light}}$?

15. Write a reasonable step for the following conversions

i

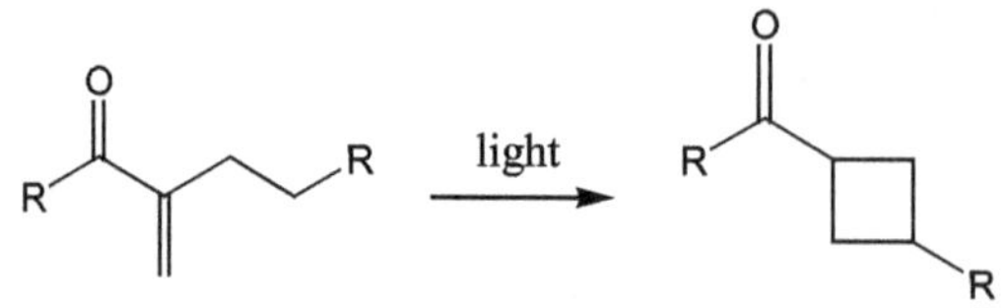

light

ii

light

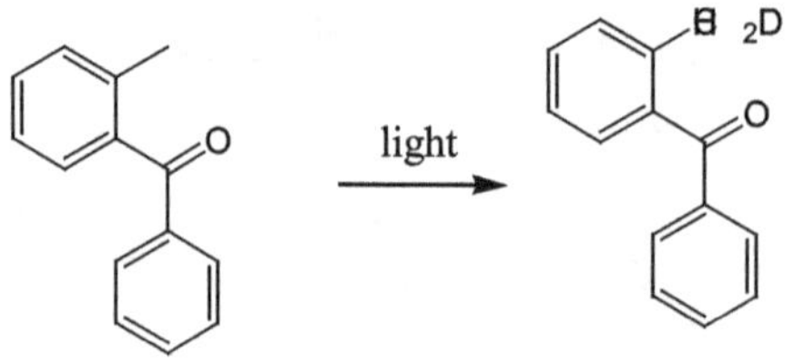

16. Predict the following with reasonable steps:

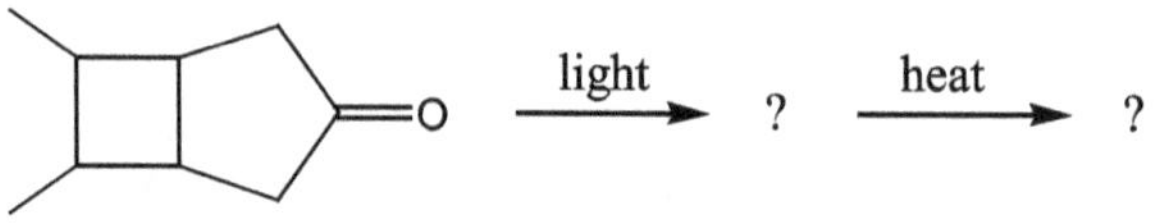

light → ? heat → ?

17. Predict the product for the following:

light → ? + ?

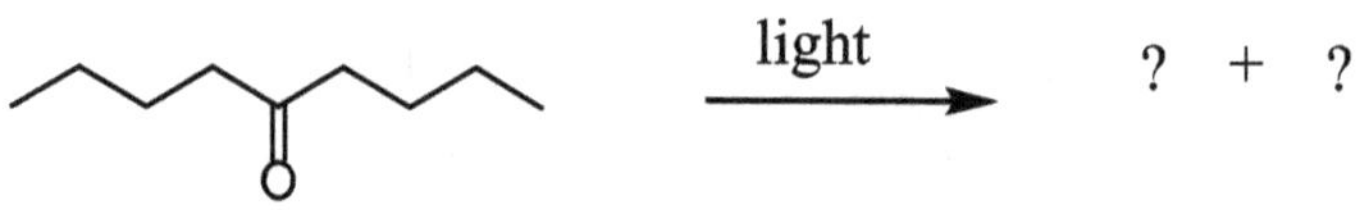

18. Outline the mechanism for the following conversion:

iii

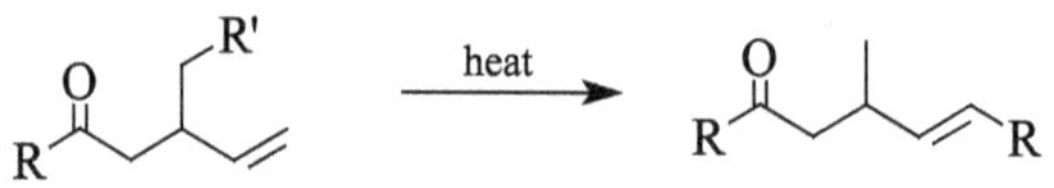

heat →

19. Predict the product for the following:

light → ?

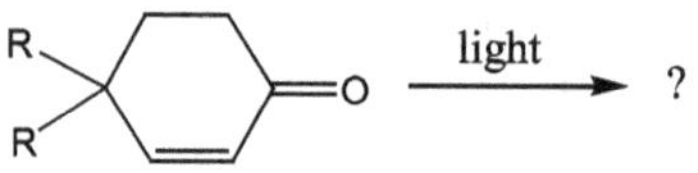

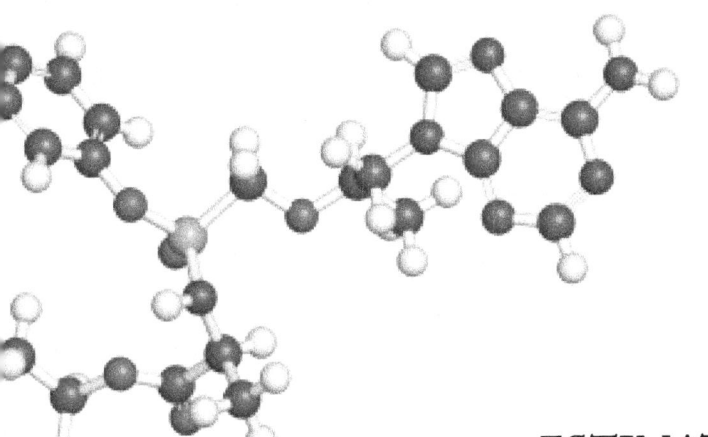

ESTIMATION OF CARBONYL COMPOUNDS

7.1 DETERMINATION OF EQUIVALENT WEIGHT OF AN ACID BY TITRATION WITH ALKALI

Reagents Required
Stock solution: If the acid is water soluble, pure acid (0.8-1g) are accurately weighed into a 100 ml standard flask and dissolved in water and make the solution up to the mark. If the acid is water insoluble, pure acid (0.8-1g) are accurately weighed into a 100 ml standard flask and dissolved in ethanol (20ml) and make up to the mark either adding EtOH or water.

Sodium hydroxide :	0.1N
Potassium hydrogen phthalate :	0.1N
Phenolphthalein :	1%
Hydrochloric acid :	0.1N

Procedure:
Pipette out 10 ml of the stock solution into a conical flask. Add few drops of Phenolphthalein indicator and titrate the solution against standard sodium hydroxide solution till a faint pink colour is obtained.

Sodium hydroxide is standardised against standard Potassium hydrogen phthalate solution using phenolphthalein as indicator.

Calculation:

$$\text{The equivalent weight of an acid} = \frac{w \times 1000}{V}$$

Where "w"   =   weight of acid taken

V   =   volume of 1N NaOH solution required.

Normality of the given acid solution is determined by using the formula $N_1V_1 = N_2V_2$

7.2   DETERMINATION OF EQUIVALENT WEIGHT OF AN ACID BY ANALYSIS OF ITS SILVER SALT

Silver salts are the most suitable because they are normal salts, are generally insoluble in water, are easily purified, contain no water of crystallization and are easily decomposed by heat. They are highly sensitive to light and should be dried and preserved in a dark place.

Preparation of silver salt:

About 2 gm. of the acid are taken in a beaker, dissolved in water (15 ml) and are added dilute ammonia solution drop wise from a pipette till there is a slight excess. The excess of ammonia is removed by boiling and the cold liquid is treated with litmus to see if it is neutral. Silver nitrate (4g in 20 ml) is added to the above solution slowly. Silver salt gets precipitated. Allowed to settle, and then add a drop of silver nitrate and see if there is any precipitation. If no precipitation forms stop adding silver nitrate solution otherwise add. Precipitated silver salt is filtered, washed twice with cold water, then with cold methanol, followed by washing with ether and dried on a porous plate and then in the vacuum desiccator.

Ignition of the silver salt:

0.5 to 1.0 gm. Of the silver salt is weighed out accurately in a tarred porcelain crucible and the salt is heated very cautiously at first so as to avoid any of the decontaminant being forced out by the violence of the reaction. When the organic matter has been completely destroyed, the crucible is maintained at a dull red heat for a few minutes and then cooled in a desiccator. Repeat the process for several times until a constant weight. The silver salt is weighed and from the weight of silver the equivalent weight of the acid is calculated.

$$\text{Equivalent weight of an acid} = \frac{w \times 109}{x} - 109 + 1$$

Where w   =   weight of silver salt taken

x   =   weight of silver obtained

7.3   DETERMINATION OF pKa VALUE OF ORGANIC ACIDS

Reagents Required:

Acetic acid, monochloroacetic acid and propionic acid :	0.1N
NaOH solution:	0.1N
Buffer solution pH =	4.0 & 9.2

Procedure:

Standardise the pH meter using standard solutions (pH = 4.0 & 9.2). Pipette out 50 ml of 0.1N acetic acid solution into a 100 ml beaker. Note down its pH. Titrate it against standard alkali solution adding at the rate of 0.1 ml from a microburette. After each addition, stir the solution and note down the pH. Continue the titration until pH no longer changes. After completion of the titration, rinse the electrode with water and check its pH. Plot a graph of pH versus volume of alkali added (or its equivalents) and determine the pKa value of acetic acid from the graph as shown below. In a similar way, determine the pKa value of monochloroacetic acid and propionic acid respectively.

7.4   DETERMINATION OF SAPONIFICATION VALUE OF OIL

The determination of this constant involves an application of the method of quantitative hydrolysis of esters. Technically, the "saponification value" is the number of milligrams of potassium hydroxide required to hydrolyze 1 gram of the oil or fat. For scientific purposes it is more appropriate to calculate the weight in grams of oil hydrolyzed by one gram equivalent of caustic potash (56.1 g). The latter is called the "saponification equivalent"

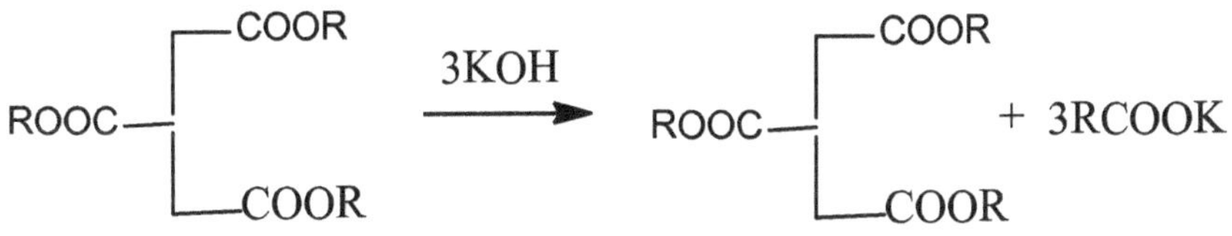

$$
\text{ROOC}\!-\!\begin{cases} -\text{COOR} \\ \\ -\text{COOR} \end{cases} \xrightarrow{\ 3\text{KOH}\ } \text{ROOC}\!-\!\begin{cases} -\text{COOR} \\ \\ -\text{COOR} \end{cases} +\ 3\text{RCOOK}
$$

$$
\begin{aligned}
\text{One mole of oil or ester, M} \ &= \ 3 \times 56.1 \text{ g of KOH} \\
&= \ 3 \times 56.1 \times 1000 \text{ mg of KOH}
\end{aligned}
$$

$$
1 \text{ gm of oil or ester} = \text{Saponification value} \ = \ \frac{3 \times 56.1 \times 1000}{M}
$$

Reagents Required

| Alcoholic potassium hydroxide | : | 0.5 N |

Oils ; coconut oil, castor oil, olive oil, gingelly oil, groundnut oil etc.,

| Hydrochloric acid | : | 0.5N |

Phenolphthalein indicator

| Potassium hydrogen phthalate | : | 0.1N |

Procedure:

Weigh out accurately 1-2 g of the oil into a RB flask fitted with a reflux condenser. Add 25 ml of 0.5N alcoholi1c KOH solution by means of a pipette. Reflux the mixture on a water bath for about 30 minutes till the liquid becomes quite clear. Run the blank simultaneously with the same quantity of 0.5N HCl acid solution using phenolphthalien as indicator. Calculate the saponification value of an oil by using the formula

$$
\text{Saponification value} \ = \ \frac{(V_2 - V_1) \ \text{x N x} \ 56.1}{w}
$$

where V_2 = volume of HCl required for the blank,
V_1 = volume of HCl required for the sample,
w = weight of the oil taken and
N = normality of HCl acid used.

The saponification value of the following oils may be determined: coconut oil, castor oil, olive oil, gingelly oil, groundnut oil and palm oil.

Under identical condition, the equivalent weight of an ester can be calculated using the formula:

$$\text{The equivalent weight of an ester} = \frac{W \times 1000}{V}$$

Where "w"   =   weight of acid taken

V   = $(V_2 - V_1)$ volume of 1N NaOH solution consumed for hydrolysis of ester.

7.5   ESTIMATION OF ESTER

Reagents Required

Stock solution: Ester (2g) is dissolved in absolute alcohol in a 100ml stand flask.

On the assumption of a similar mechanism, the formation of the other products can be explained as follows:

Alcoholic potassium hydroxide (0.5 N) prepared by dissolving potassium hydroxide pallets (15g) in an equal weight of water and diluting to 500 ml by the addition of absolute alcohol. The solution is agitated with anhydrous sodium sulphate (10 g) until it clarifies, after which the clear solution is decanted.

Hydrochloric acid   :   0.5N
Phenolphthalein indicator

Potassium hydrogen phthalate   :   0.1N
Sodium hydroxide   :   0.1N

Procedure: Pipette out 20 ml of the stock solution into a 250 ml RB flask fitted with reflux condenser followed by the addition of alcoholic potassium hydroxide (30ml) and is boiled gently on a water bath for about 1.5 to 2hours. After cooling, the condenser is washed by little water and the excess KOH is titrated with 0.5N hydrochloric acid using phenolphthalein as indicator. The equivalent weight of an ester can be calculated by using the equation:

$$\text{Equivalent weight of an ester} = \frac{m \times 1000}{(V_2 - V_1)}$$

where V_2 =   volume of 0.5N HCl solution required for the blank
V_1 =   volume of 0.5N HCl solution required for the actual experiment
m   =   weight ester present in 20 ml of stock solution.

7.6   ESTIMATION OF AN ACID AND AN ESTER IN A MIXTURE

Reagents Required

Stock solution: Mixture of ester (1g) and acid (1g) dissolved in acid free ethanol in a 50 ml standard flask. [Mixture: Ethyl acetate and acetic acid, ethyl cinnamate and cinnamic acid, ethyl benzoate and benzoic acid etc.].

Alcoholic potassium hydroxide   :   0.5N
Hydrochloric acid   :   0.5N

Phenolphthalein indicator

Potassium hydrogen phthalate	:	0.1N
Sodium hydroxide	:	0.1N

Procedure: Pipette out 20 ml of the stock solution into a conical flask and the free acid is titrated against standard NaOH solution. Let "v"ml of normal NaOH be required to react with the free acid present in 20 ml of the mixture.

Pipette out 20 ml of the stock solution into a 250 ml RB flask fitted with reflux condenser followed by the addition of alcoholic potassium hydroxide (30ml) and is boiled gently on a water bath for about 1.5 to 2hours. After cooling, the condenser is washed by little water and the excess KOH is titrated with 0.5N hydrochloric acid using phenolphthalein as indicator. Let the volume of alkali required for both the free acid and ester present in 20 ml of the stock solution be "y".

If E_1 and E_2 are the equivalents of the acid and ester respectively in the mixture, then

$$\text{The percentage of free acid} = \frac{x \times E_1}{1000} \times 100$$

$$\therefore \quad \text{ester} = \frac{(y-x) \times E_2}{1000} \times 100$$

7.7 DETERMINATION OF IODINE VALUE OF AN OIL

Definition: Iodine value is usually expressed as the number of parts by weight of iodine absorbed by 100 parts by weight of an oil or fat.

The determination iodine value of the oil is of great help in characterizing an oil and also in finding the proportion of an adulterant in a sample of the oil. The drying power of an oil is generally proportional to its iodine value. Linseed oil which is a drying oil has a high iodine value while the non-drying coconut oil has a very low iodine value.

7.7.1 Wijs' method

Reagents required

Oils: coconut oil, castor oil, olive oil, gingelly oil, groundnut oil etc.,

Iodine monochloride: Finely powdered iodine (6.5 g) dissolved in glacial acetic acid (50 ml) contained in a litre round – bottomed flask by warming on a water bath. After iodine has dissolved, 50 ml of the solution are transferred when cold into another flask and pure dry chlorine is passed in till the colour changes from dark brown to a clear orange tint. The remaining iodine solution is now added, the colour of the solution becomes light brown. The excess of iodine prevents the formation of iodine trichloride. The solution is next heated on a water bath for twenty minutes. This treatment makes the solution much more stable and the solution maintain its strength over long periods when preserved in a stoppered bottle in the dark. In its interaction, the molecule of iodine monochloride is equivalent to one molecule of iodine.

Potassium iodide	:	10%
Potassium dichromate	:	0.1N
Sodium thiosulphate	:	0.1N

Sulphuric acid   :   2N
Starch   :   1%

Procedure: One gm of the oil are weighed into a stoppered bottle and dissolved in carbon tetrachloride (10ml) followed by the addition of iodine monochloride solution (25ml). The bottle is kept aside for an hour, after which potassium iodide solution (20ml) are added and then water (20ml). The mixture is titrated with standard thiosulphate solution using starch as indicator. A blank determination is carried out without the oil using exactly the same quantity of carbon tetrachloride and the Wijs'solution.

1ml of 0.1N sodium thiosulphate solution   =   12.69 mg of iodine.

$$\text{Iodine number} = \frac{(V_2\text{-}V_1) \times 0.01269 \times 100 \times N}{m}$$

where V_2 =   volume of thiosulphate solution required for the blank
V_1 =   volume of thiosulphate solution required for the actual experiment
N   =   normality of thiosulphate solution and
m   =   mass of an oil.

7.7.2   Iodine value by Hanus method

The reagent used is iodine monobromide, prepared by dissolving powdered iodine crystals (13.2 g) in 1 litre of acetic acid followed by the addition of liquid bromine (3 ml). This solution is quite stable. The time of reaction is 40minutes and the rest of the procedure is the same as in Wijs' method.

7.7.3   Iodine value by chloramine-T method

The method reported here by Rai et al [1995Anal] makes use of the fact that, in the presence of ethanoic acid, chloramine-T adds on to the alkene to give mainly acetoxy(chloro)alkane or tosylamino(chloro)alkanes involving one mole of chloramine-T per alkene bond.

1 mole of CAT per C=C unit = 1 mole of iodine = 2000ml of 1N sodium thiosulphate
1ml of 0.1N sodium thiosulphate solution = 12.69 mg. of iodine

$$\text{Iodine number} = \frac{(V_2\text{-}V_1) \times M \times 12.69 \times 100}{m \times 0.1}$$

where V_2 =   volume of thiosulphate solution required for the blank
V_1 =   volume of thiosulphate solution required for the actual experiment
M   =   normality of thiosulphate solution and
m   =   mass of an oil in mg.

Reagents required:
Chloramine-T.$2H_2O$: 0.1N prepared by dissolving 2.81 g of chloramine-T.$2H_2O$ in 100 ml glacial acetic acid.

Potassium iodide   :   10%
Potassium dichromate   :   0.1N
Sodium thiosulphate   :   0.1N
Starch   :   1%
Sulfuric acid   :   2N

Procedure:

An accurately weighed (50-100 mg) sample of an oil was placed in a clean dry iodine flask. A solution of 0.1N chloramine-T in glacial acetic acid was added to it and kept at room temperature for 1-2h. A blank was prepared b transferring an equal volume of Chloramine-T solution into an iodine flask which was allowed to stand for 2h under identical conditions. After the reaction, 10% potassium iodide (10 ml), water (10 ml) and 2N Sulfuric acid (10 ml) were added to each of the flask, and the liberated iodine was titrated with standard sodium thiosulfate solution (0.1N) using starch as indicator. From the difference in the volume of thiosulfate solutions consumed, the iodine number was calculated using the above formula.

Table:

Oil	Iodine value	Saponification value
Tung oil	163 – 173	
Cod liver oil	145 – 180	
Grape seed oil	124 – 144	180 – 200
Palm oil	44 – 51	190 – 205
Groundnut oil	84-100	
Olive oil	80 – 88	184 – 196
Castor oil	82 – 90	176 – 186
Coconut oil	7 – 10	250 – 265
Mustard oil	103	172
Gingelly oil	103-117	
Jojoba oil	80 ~82	86 – 96
Poppy seed oil	133 ~133	196
Cotton seed oil	100 – 117	192
Corn oil	109 – 133	190
Safflower oil	145	186 – 198
Rape seed oil	94 – 120	175
Wheat germ oil	115 – 134	180 – 195
Sunflower oil	118 – 144	188 – 194
Linseed oil	136 – 178	
Soybean oil	120 – 136	180 – 200
Peanut oil	84 – 106	190
Rice bran oil	95 – 108	185 – 195
Walnut oil	120 – 155	189 – 197
Kapok seed oil	85 – 100	

7.8 ESTIMATION OF THE AMOUNT OF KETONE PRESENT BY HALOFORM METHOD [MIS355]

Ref: Laboratory manual of organic chemistry by Dey and Seetharaman, P. 355.

Iodine in alkaline solution with acetone to form iodoform according to the following equations

$$CH_3COCH_3 + 3I_2 + 4KOH \longrightarrow CH_3COOK + 3KI + 3H_2O$$

1 molecule of acetone (58g)   =   3 molecules of iodine

                                            =   6 litres of 1N of thiosulphate solution

Hence 1 ml of N/10N of thiosulphate solution   =   58 × 60000 g acetone

                                            =   0.00097 g of acetone

$$\text{Hence w g of acetone} = (V_2 - V_1)\, N/10 \text{ thio} = \frac{(V_2 - V_1) \times 0.00097 \times 10}{w \times 100}\ g$$

Stock solutions:

Acetone solution : 0.25 g of distilled acetone dissolved in water and make up to 100 ml in a standard flask.

Iodine solution	:	0.1N
Potassium iodide	:	10%
Potassium hydroxide	:	1N
Potassium dichromate	:	0.1N
Sodium thiosulphate	:	0.1N
Starch	:	1%
Sulfuric acid	:	2N

Procedure:

Pipette out 10 ml of ketone solution into an iodine flask. Dilute to about 50 ml with water. Add 1N sodium hydroxide solution (25ml), mix well and allow to stand at room temperature for 10 minutes. Introduce 0.1N iodine solution (100ml) using burette. Allow the mixture to stand for 20 minutes at room temperature. Then add 2N sulphuric acid (25ml) and titrate immediately with standard sodium thiosulphate solution. Run a blank under similar condition. The amount of iodine which has interacted is determined by difference.

$$\% \text{ purity of acetone} = \frac{(V_2 - V_1) \times 0.00097 \times 10 \times 100}{w \times 100}$$

V_2	=	Volume of 0.1N sodium thiosulphate consumed for blank
V_1	=	Volume of 0.1N sodium thiosulphate consumed for test solution
N	=	Normality of sodium thiosulphate
W	=	Weight of the sample
M	=	Molecular weight of the sample

112. Derive an expression for the determination of molecular weight by this method to estimate the following compounds?

113. Give the mechanism of haloform reaction.

114. Could the reaction be applied to estimate the following compounds?

7.9   DETERMINATION OF NUMBER OF KETO GROUPS

7.9.1   Chloramine-T method [2001CA269]

The method reported here by Rai et al [2001CA269] makes use of the fact that aldoximes are known to undergo an oxidative dehydrogenation by chloramine-T, yielding the respective nitrile oxides (Scheme 56,

P66 [3], while ketoximes give pale blue a-chloronitroso compounds (Scheme 62, p69) [3] by consuming one molecule of chloramine-T per one molecule of the oxime.

Generally known volume of chloramine-T is added to known mass of oxime, after the completion of the reaction the unreacted chloramine-T is determined iodometrically. By carrying out a parallel blank experiment, the amount of chloramine-T consumed is determined. As the overall reaction requires one molecule of chloramine-T per mole of the oxime, which is equivalent to one mole of iodine, the molecular mass M_{oxime} of the oxime is determined by using the following equation.

$$Moxime = \frac{w \times 2000}{(V_1 - V_2) \times N}$$

Where V_1 = Volume of 0.1N sodium thiosulphate consumed for blank
V_2 = Volume of 0.1N sodium thiosulphate consumed for test solution
N = Normality of sodium thiosulphate
w = mass of the oxime dissolved in 5 ml of 95% ethanol (in gms)

Since the molecular mass of each aldehyde and ketone containing one CO group is 15, the number of oxo groups present in the investigated carbonyl compound is given by the equation:

$$n = \frac{M}{Moxime - 15}$$

where n is number of the oxo groups, and M is molecular mass of the carbonyl compound investigated.

Reagents required:

Oxime   :   Prepared by the usual method.
Chloramine-T.2H_2O :   0.01N prepared by dissolving 0.281 g of chloramine-T.2H_2O in 100 ml glacial acetic acid.
Potassium iodide   :   10%
Potassium dichromate   :   0.01N
Sodium thiosulphate   :   0.01N
Starch   :   1%
Sulfuric acid   :   1N

Procedure

An accurately weighed (10-30mg) sample of an oxime was dissolved in 5 ml of 95% ethanol taken in an Erlenmeyer flask, and 20 ml of 0.0 1 N chloramine-T solution was added, and this was kept aside at room temperature for varying time intervals. A blank solution was prepared by transferring an equal volume of chloramine-T into the flask containing 5 ml of 95% ethanol. 1N sulfuric acid (1 ml), 10% aqueous potassium iodide (1 ml), and water (2 ml) were added, and the iodine evolved was titrated with a standard sodium thiosulfate solution (0.01 N), using starch as an indicator. From the difference in the volumes of the sodium thiosulfate solution consumed, the searched molecular mass was established experimentally from the equation shown above.

7.9.2 Oxime method

The method reported here makes use of the fact that aldoximes are known to undergo hydrolysis by acid, liberated hydroxylamine gets oxidized to N_2O by added ferric alum at the same time gets reduced to ferrous sulphate. Amount of ferrous sulphate liberated was determined experimentally by titrating against standard potassium permanganate.

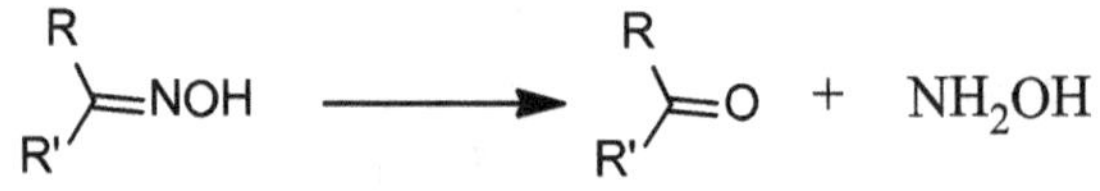

$$R\underset{R'}{\diagup}\!\!=\!\!NOH \longrightarrow R\underset{R'}{\diagup}\!\!=\!\!O \ + \ NH_2OH$$

$$2NH_2OH \longrightarrow N_2O + H_2O + 2H_2$$

$$2H_2 + 2Fe_2(SO_4)_3 \longrightarrow 4FeSO_4 + 2H_2SO_4$$

$$\therefore \quad NH_2OH \equiv 24FeSO_4 \equiv O \equiv 2 \text{ litres of } 1N \text{ KMmO}_4$$

Hence the weight of oxime taken and V ml of 1N KMn)4 are required to react with the ferrous salt formed, then the weight of oxime containing one keto group

$$\frac{w \times 2000}{V \times N} = M_{oxime}$$

Where V =   Volume of 0.1N $KMnO_4$ consumed
N   =   Normality of $KMnO_4$ solution
w   =   mass of the oxime dissolved

Since the molecular mass of each aldehyde and ketone containing one CO group is 15, the number of oxo groups present in the investigated carbonyl compound is given by the equation:

$$n = \frac{M}{M_{oxime} - 15}$$

where n is number of the oxo groups, and M is molecular mass of the carbonyl compound investigated.

7.10   ESTIMATE THE AMOUNT OF ENOLIC COMPOUND PRESENT IN THE GIVEN SOLUTION

7.10.1 Determination of percentage of enol by K. H. Meyer's method

The method reported here makes use of the fact that the enol form reacts with bromine very rapidly compared with the keto form. The excess of bromine can be easily removed by adding alcoholic β-naphthol solution. For instance,

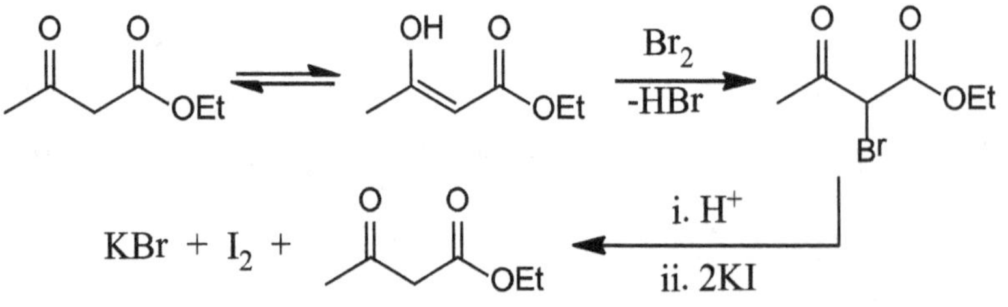

Scheme 295

1 mole of ethyl acetoacetate (130g) = 1 mole of enol = 1 mole of iodine = 2000ml of 1N thiosulphate

2000ml of 1N thiosulphate = 130 gm of ethyl acetoacetate

$\therefore$ 1 ml of 1N thiosulphate = 130/2000 gm of ethyl acetoacetate

$\therefore$ 1 ml of 1/10N thiosulphate = 13/2000 gm (= 0.0065 gm) of ethyl acetoacetate

If "V" be the volume of 0.1N thiosulphate used and "x" the weight of ester taken, then

$$\% \text{ of enol} = \frac{V \times 0.0065 \times 100}{x}$$

Reagents required:

Stock solution : Weigh exactly 0.35 gm of ethyl acetoacetate into a 100 ml standard flask and dissolved in absolute ethanol.

Potassium iodide	:	10% solution
Alcoholic bromine solution	:	0.1N solution
Alcoholic β-naphthol	:	5% solution
Potassium dichromate	:	0.01N solution
Sodium thiosulphate	:	0.01N solution
Starch	:	1% solution

Procedure:

Pipette out 10 mL of the stock solution into a conical flask. Cool the solution to –5°C to –10°C in freezing mixture. Add with swirling, ice-cold alcoholic bromine solution till a permanent faint yellow colour persists (about 20 mL). Immediately, in a single portion add alcoholic β-napthol solution to remove excess of bromine. Quickly add potassium iodide solution (5cc) and titrate the liberated iodine against standard thiosulphate using starch as an indicator. Calculate % enol by using the above formula.

7.10.2 Determination of percentage of enol by chloramine-T method

The method reported here by Rai et al [2004OC831] makes use of the fact that chloramine-T would quickly add to the enolic C-C double bond to form transient intermediate () as shown in the reaction mechanism.

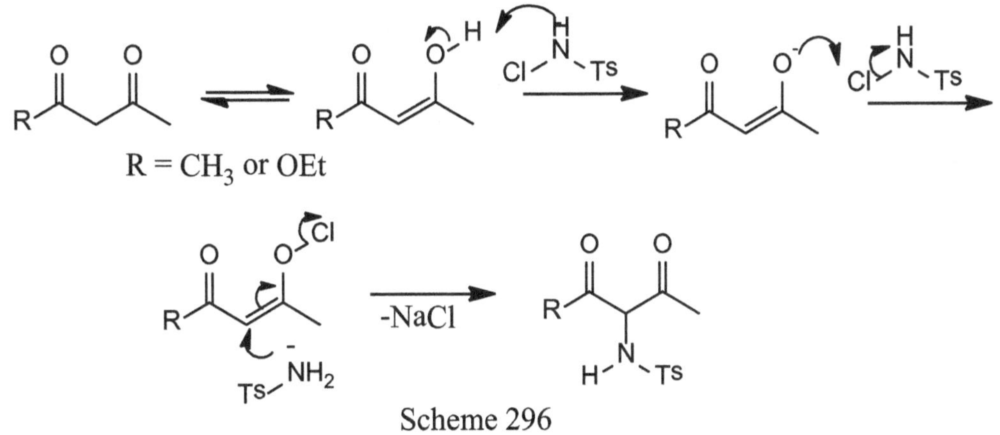

R = CH$_3$ or OEt

Scheme 296

It can be supposed that the enolization of () is somewhat slow due to the presence of electron donating –HNTs group. So further attack of CAT is restricted. In case of earlier method, after addition of one mole of bromine over enol releases new species similar to that of (), wherin enolization is too fast and the addition of β-naphthol is required to arrest the reaction.

1 mole of enol (M) = 1 mole of CAT = 1 mole of iodine = 2000ml of 1N sodium thiosulphate

2000ml of 1N sodium thiosulphate = M gm of enol

∴ 1 ml of 1N sodium thiosulphate = M/2000 gm of enol

∴ 1 ml of 1/10N sodium thiosulphate = M/2000 × 10 gm of enol

∴ 1 ml of 1/100N sodium thiosulphate = M/2000 × 100 gm of enol

If "V" be the volume of 0.01N sodium thiosulphate used and "x" the weight of ester taken, then

$$\text{\% of enol} = \frac{(V2\text{-}V1) \times M \times 10 \times 100}{x \times 0.01 \times 100 \times 2000}$$

Where M is the molecular mass of β-keto carbonyl compounds

V_2	=	Volume of 0.1N sodium thiosulphate consumed for blank
V_1	=	Volume of 0.1N sodium thiosulphate consumed for test solution
N	=	Normality of sodium thiosulphate
x	=	weight (in mg) of β-keto carbonyl compounds dissolved in 50 ml of absolute alcohol.

Reagents required:

Stock solution : Weigh exactly 100-200 mg. of β-keto carbonyl compound into a 50 ml standard flask and dissolved in absolute ethanol.

Chloramine-T	:	0.01N
Potassium iodide	:	10%
Potassium dichromate	:	0.01N
Sodium thiosulphate	:	0.01N
Starch	:	1% solution

Procedure:

Pipette out 10 mL of the stock solution into a conical flask. Cool the solution to –5°C to –10°C in freezing mixture. Add with swirling, ice-cold CAT solution (0.01N) and kept aside for varying time intervals 2 min for ethyl acetoacetate and 4 min for acetyl acetone). A blank solution was prepared by transfering an equal volume of CAT into the flask containing 5 ml absolute ethanol. 2 M sulphuric acid (1 ml), 10% KI (1 ml) and water (2 ml) were added and the iodine evolved was titrated against standard thiosulphate using starch as an indicator. Calculate % enol by using the above formula

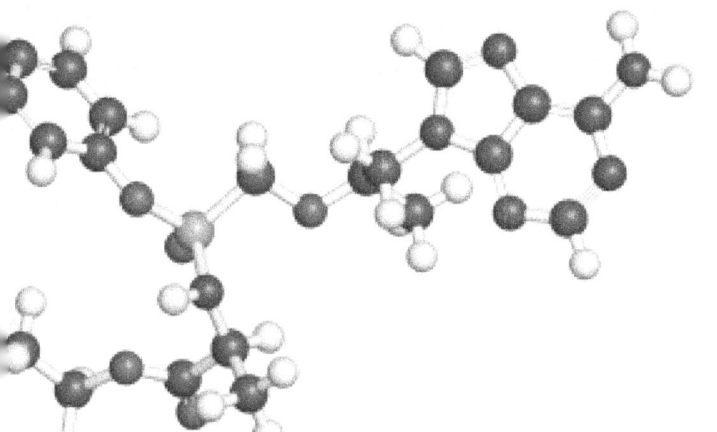

8

PROBABLE NET QUESTIONS

1. Consider the following statements: Synthetic transformation of –CO- to –CH$_2$- can be achieved by a. Clemenson reduction and b. Wolf-Kishner reduction. Which of the statements given above is/are correct?

    i.   a only
    ii. b only
    iii. Both a and b
    iv. Neither 1 nor 2

2. Among the following compounds, the formyl anion equivalent is

    i.   1,3-dithiane
    ii. 1,3-dioxalane
    iii. Acetylene
    iv. 1,4-dithiane

3. The conversion shown below is an example of
    $$RCON_3 \longrightarrow RNH_2$$

    i.   Beckmann rearrangement
    ii. Curtius rearrangement
    iii. Lossen rearrangement
    iv. Hoffmann rearrangement.

4. The following reaction is an example of

    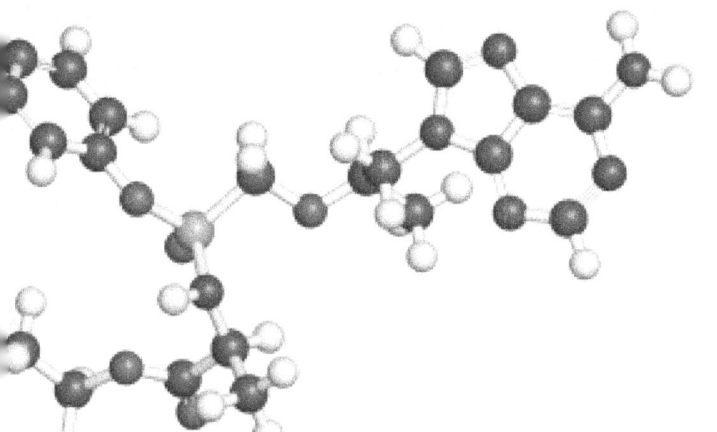

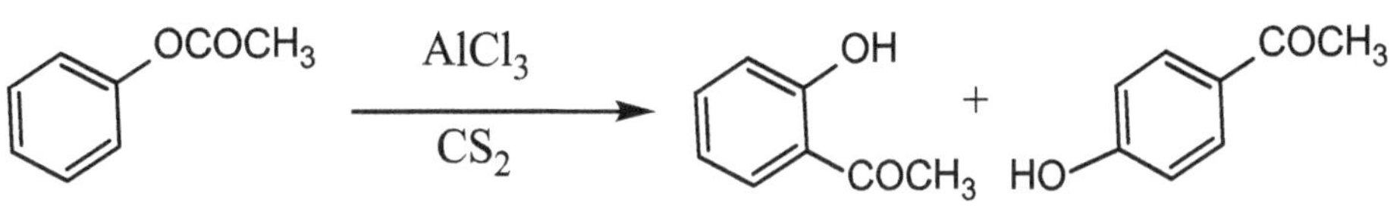

    i.   Claissen rearrangement
    ii. Reimer-Iieman reaction
    iii. Friedel-Craft's acylation
    iv. Fries rearrangement

5. The reaction given below proceeds through

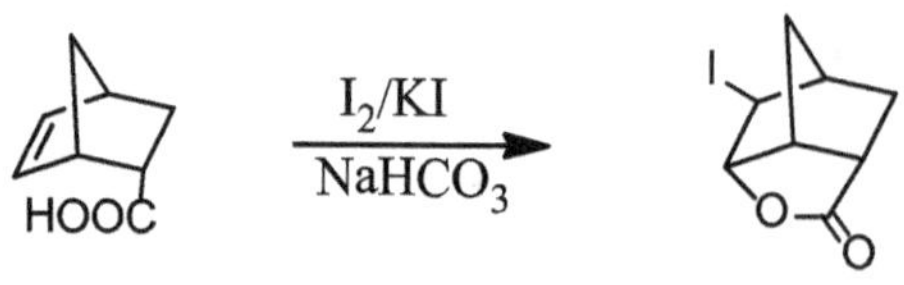

i 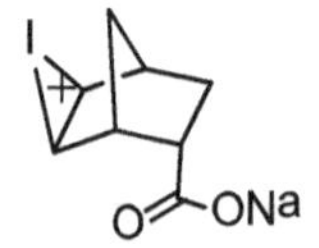   ii

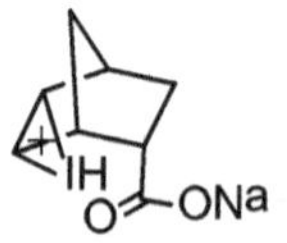

iii 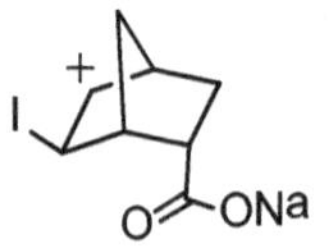   iv

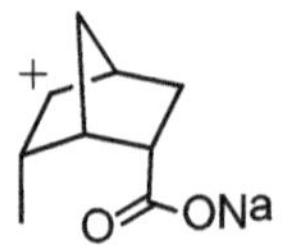

6. Hofmann rearrangement of acetamide yields

   i.   ethylamine
   ii.   methylamine
   iii. acetanilide
   iv.   acetonitrile

7. The following reaction is an example of:

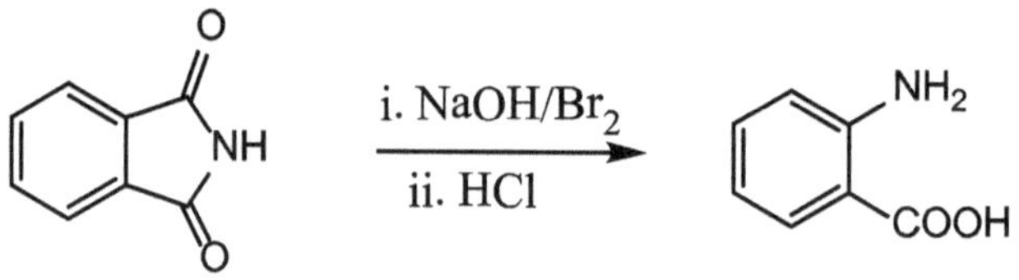

   i.   Beckmann rearrangement
   ii.   Curtius rearrangement
   iii. Lossen rearrangement
   iv.   Hoffmann rearrangement.

8. The conversion of acetophenone to acetanilide is best accomplished by using:

   i.   Beckmann rearrangement
   ii.   Curtius rearrangement
   iii. Lossen rearrangement
   iv.   Hoffmann rearrangement.

9. Identify the products A and B in the following reaction sequence

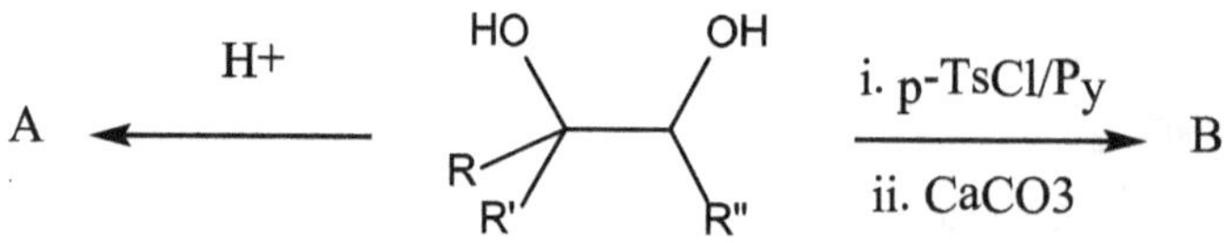

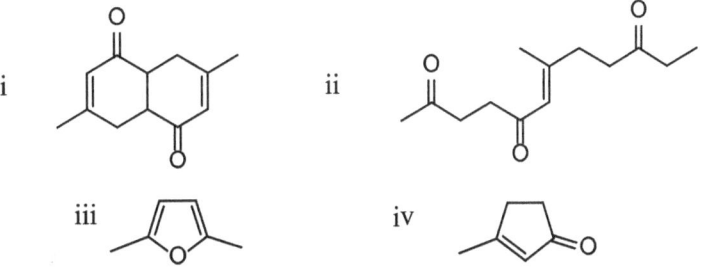

i. A is ⟨structure⟩   B is ⟨structure⟩

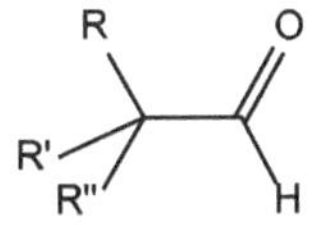

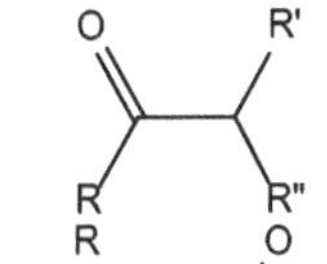

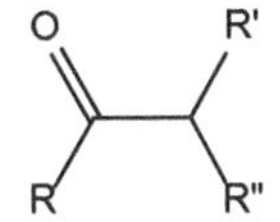

ii. A is ⟨structure⟩   B is ⟨structure⟩

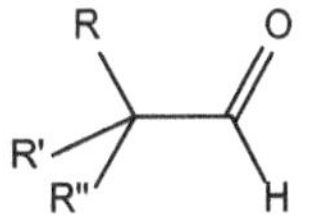

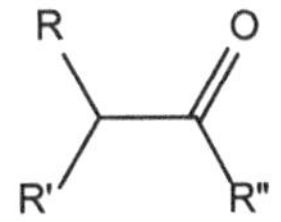

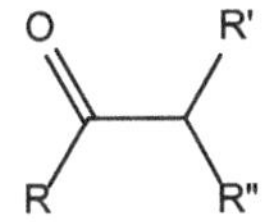

iii. A is ⟨structure⟩   B is ⟨structure⟩

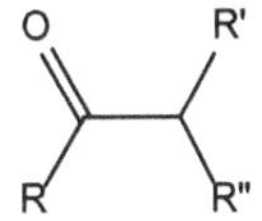

iv. A is ⟨structure⟩   B is ⟨structure⟩

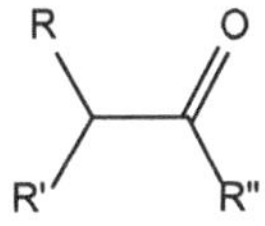

10. The conversion of acetophenone to phenyl acetate is best accomplished by using:

  i.   Beckmann rearrangement
  ii. Curtius rearrangement
  iii. Bayer-Villeger oxidation
  iv. Lossen rearrangement

11. Treatment of furfuraldehyde with NaOH yields furfuryl alcohol and 2-furoic acid. This reaction is called

  i.   Perkin reaction
  ii. Dickmann reaction
  iii. Reimer-Tieman reaction
  iv. Cannizarro reaction

12. The major product formed in the reaction of 2,5-hexanedione with P_2O_5 is

i ⟨structure⟩   ii ⟨structure⟩

iii ⟨structure⟩   iv ⟨structure⟩

13. Predict the product for the following:

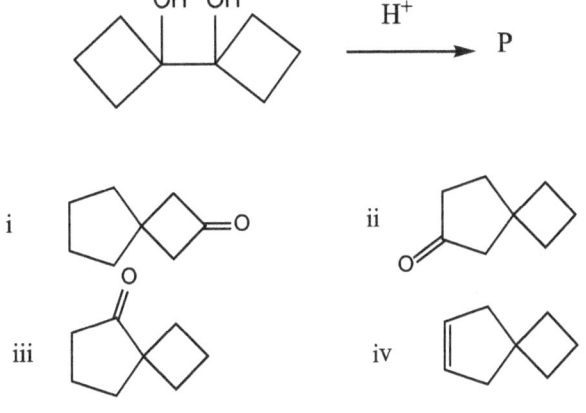

i ⟨structure⟩   ii ⟨structure⟩

iii ⟨structure⟩   iv ⟨structure⟩

14. The major products formed in the reaction are

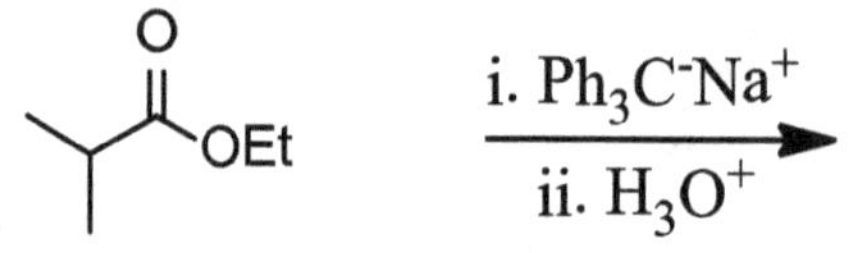

i. $Ph_3C^-Na^+$

ii. H_3O^+

i 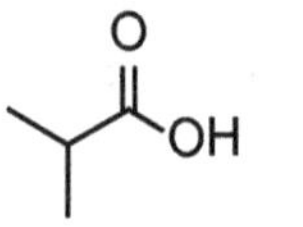   + $Ph_3CCH_2CH_3$

ii 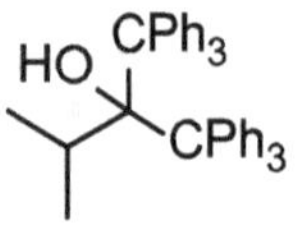   + EtOH

iii 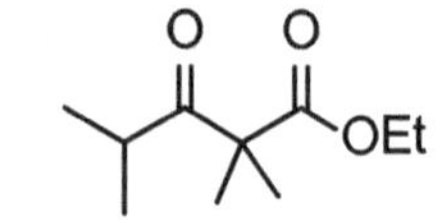   + EtOH

iv 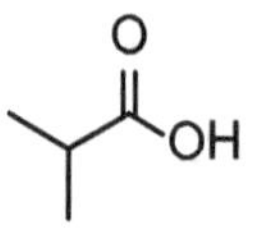   + $CH_2=CH_2$

15. Predict the product for the following:

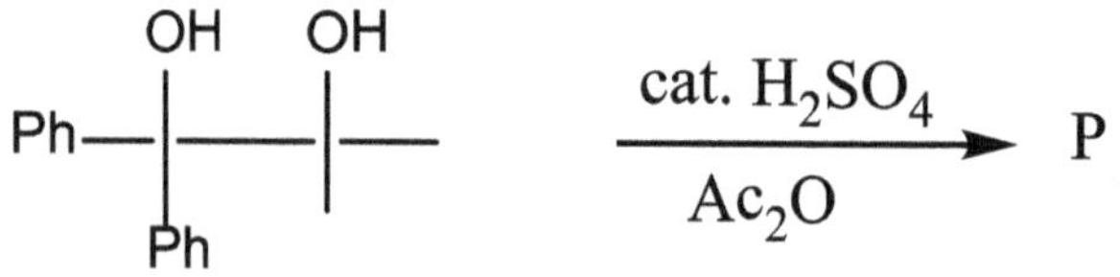

cat. H_2SO_4

Ac_2O P

i

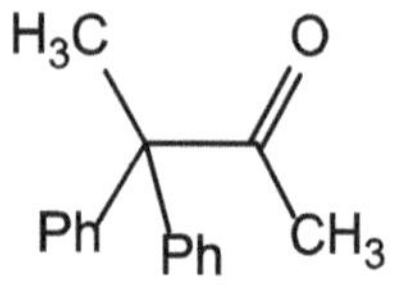

i

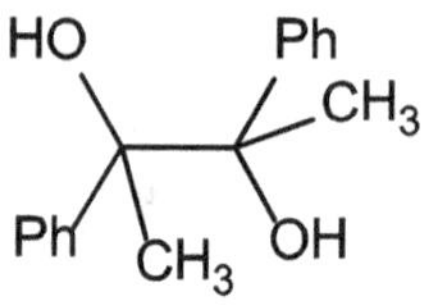

iii

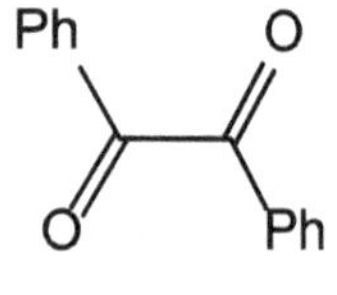

iv

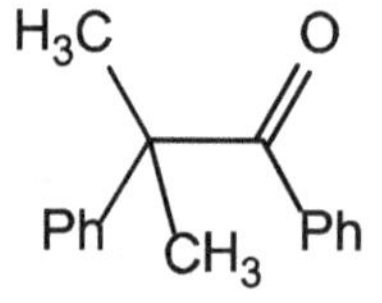

16. The major product in the Beckmann rearrangement of $CH_3C(=NOH)CH(CH_3)_2$ is

   i. $CH_3NHCOCH(CH_3)_2$
   ii. $CH_3CONHCH_3$
   iii. $(CH_3)_2CHNHCOCH_3$
   iv. $(CH_3)_2CHNHCOCH(CH_3)_2$

17. A compound 'X' on treatment with $SOCl_2$ produce 'Y'. 'Y' on hydrolysis yields a mixture of benzoic acid and methyl amine. The compound 'X' is

i $\underset{\underset{HO}{\big|}}{\overset{C_6H_5}{\diagdown}}\underset{N}{\overset{CH_3}{\diagup}}$ ii $\underset{\underset{N}{\big|}}{\overset{C_6H_5}{\diagdown}}\underset{OH}{\overset{CH_3}{\diagup}}$

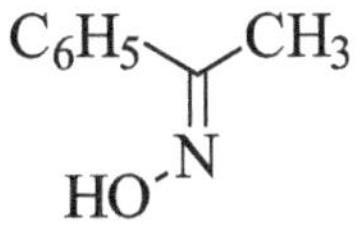

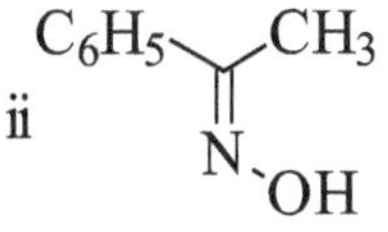

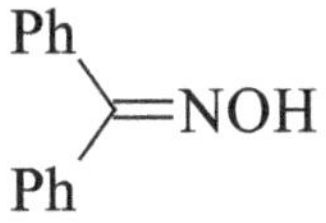

iii $\underset{Ph}{\overset{Ph}{\diagdown}}{=}NOH$ iv $CH_3{-}\bigcirc{-}CH{=}NOH$

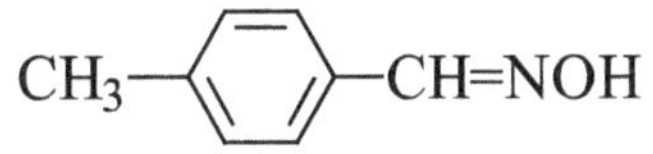

18. Reaction of CH_3CHO with excess of HCHO in presence of NaOH yields

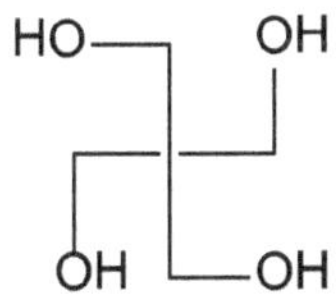

   i.   Cross Cannizarro reaction followed by Aldol condensation
   ii. Consecutive Aldol condensation followed by cross Cannizarro reaction
   iii. Aldol condensation
   iv. Cannizarro reaction

19. Which of the compound listed below would be the product(s) of the following reaction?

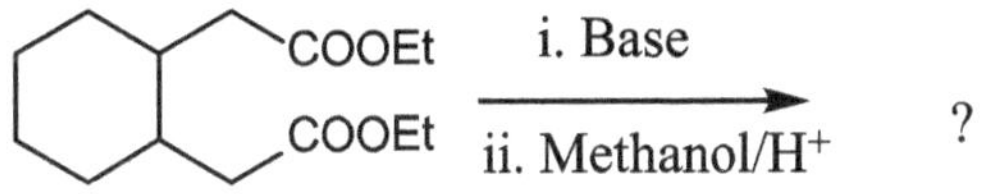

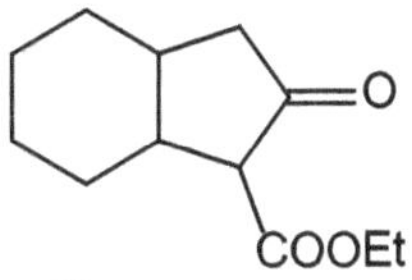

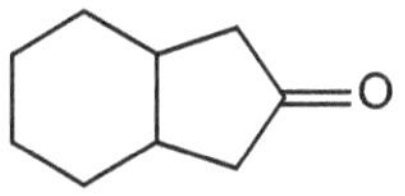

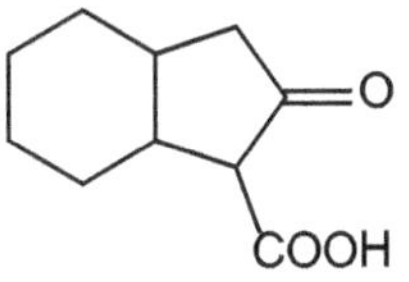

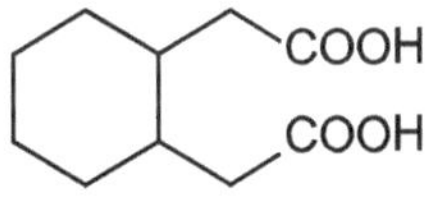

20. Which of the compound listed below would be the products (s) of the following reaction?

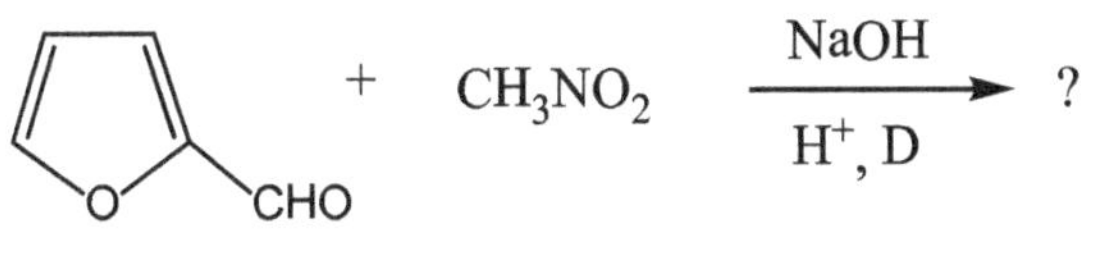

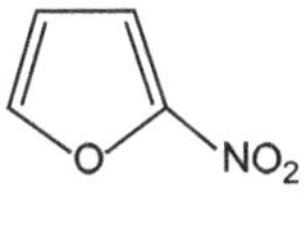

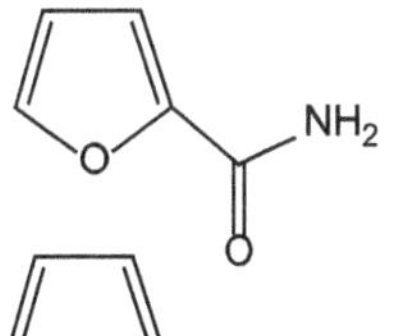

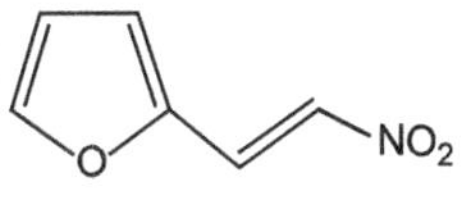

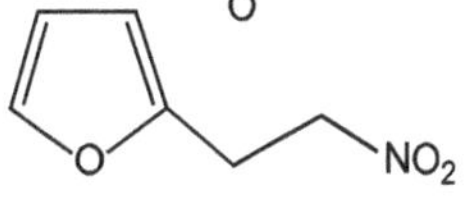

21. Aliphatic aldehydes do undergo

    i.   Cannizarro reaction

    ii.   Perkin reaction

    iii.   Claisen reaction

    iv.   Aldol condensation

22. β-Ketoesters can be synthesized by

    i.   Aldol condensation

    ii.   Perkin reaction

    iii.   Claisen condensation

    iv.   Benzoin condensation

23. Predict the product for the following reaction:

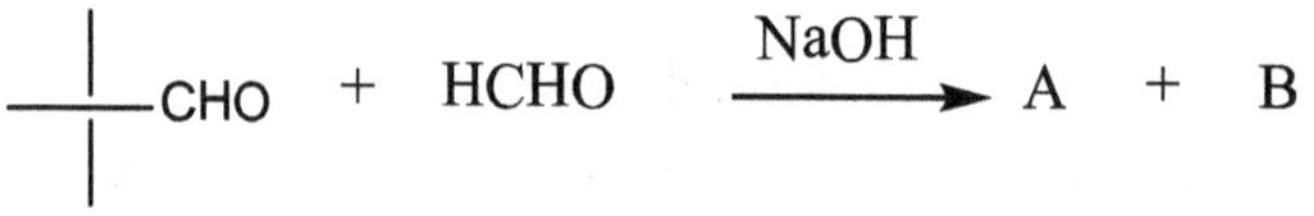

$$\text{--CHO} \ + \ \text{HCHO} \ \xrightarrow{\text{NaOH}} \ A \ + \ B$$

    i   A = —COOH   B = CH_3OH

    ii   A = —CH_2OH B = HCOOH

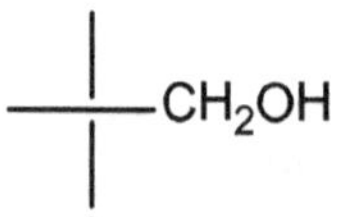

    iii   A = —COOH   B = HCOOH

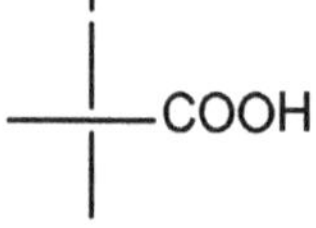

    iv   A = —CH_2OH B = CH_3OH

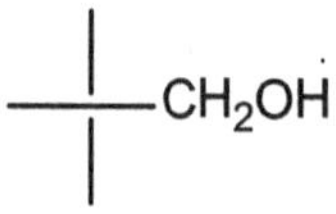

24. Predict the product(s) for the following reaction:

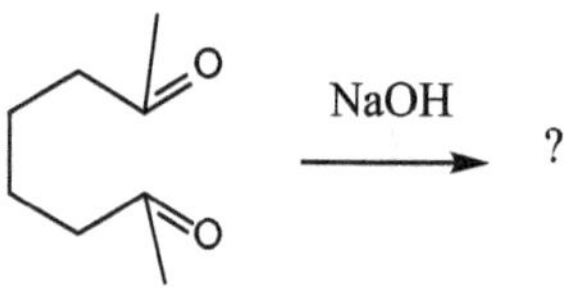

i a 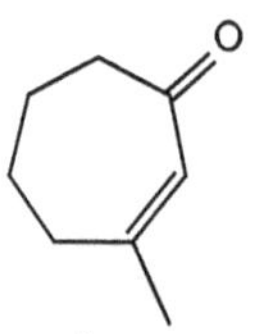   ii b

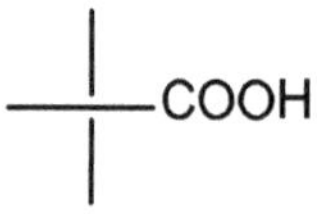

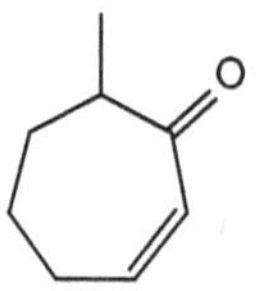

iii c 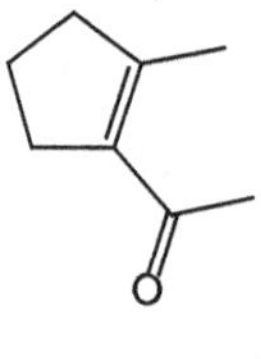   iv d

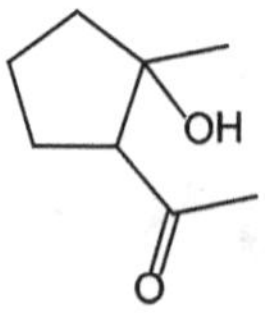

25. The products A and B in the following sequence of reactions are

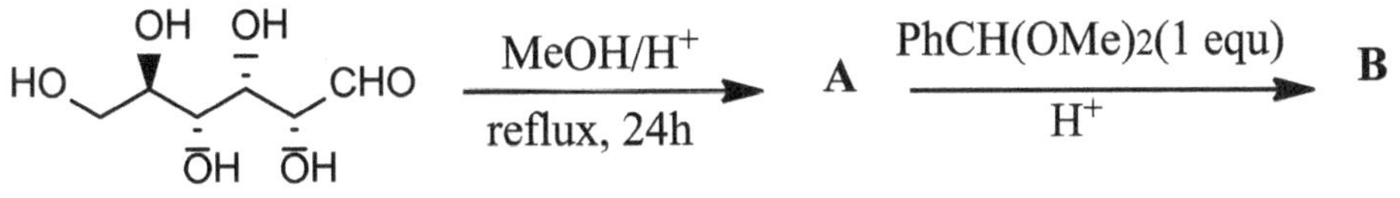

i   A = 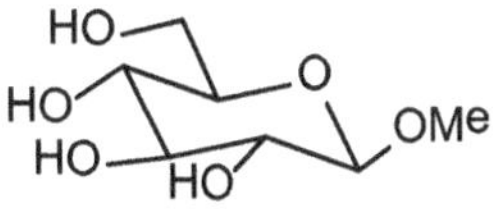   B =

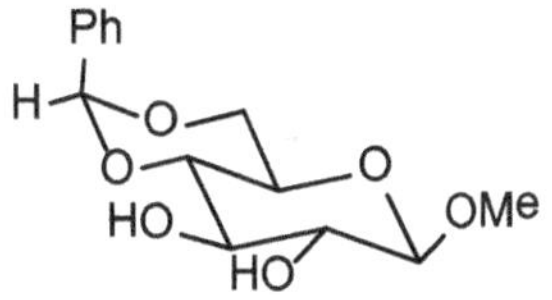

ii   A = 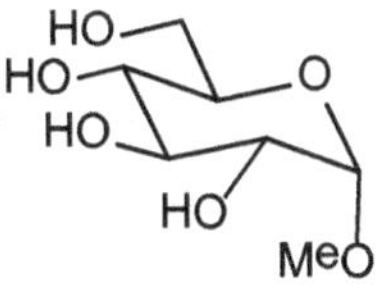   B =

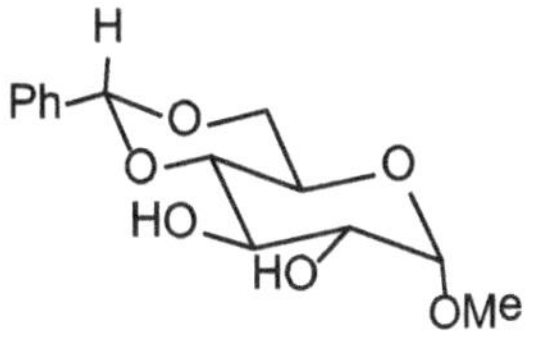

iii   A = 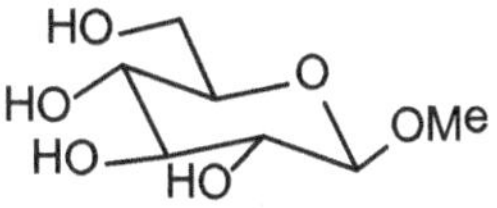   B =

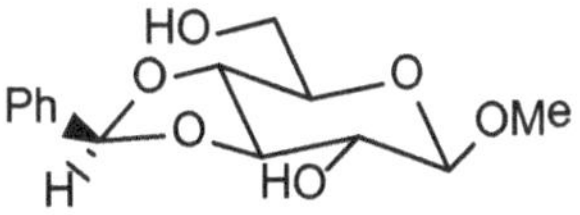

iv   A = 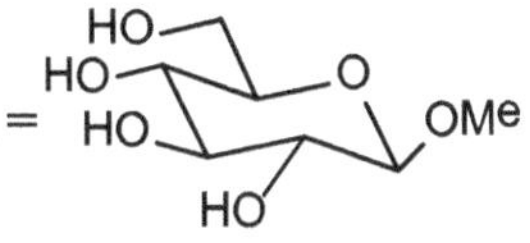   B =

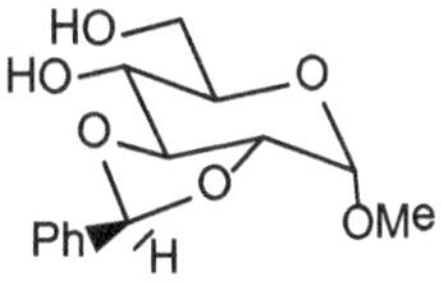

26. The major product in the reaction of glucose with benzaldehyde and pTsA is

i 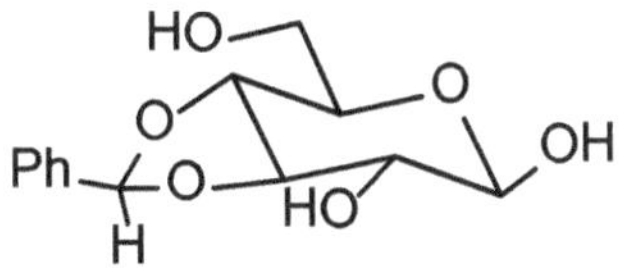   ii

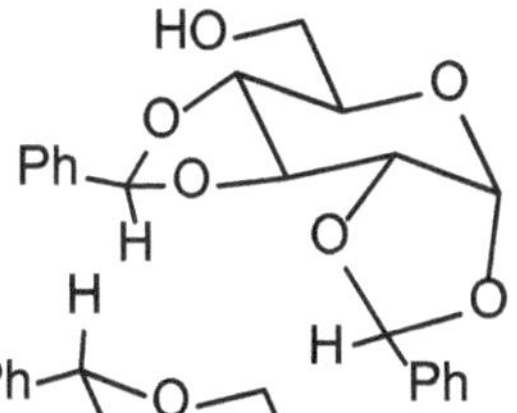

iii 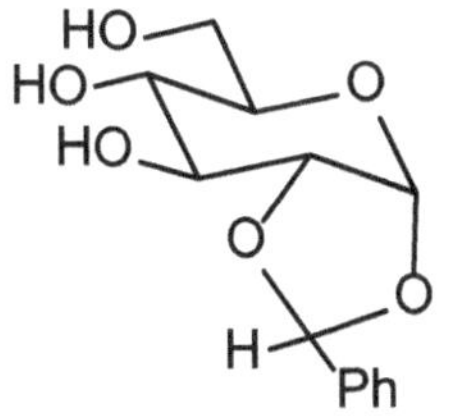   iv

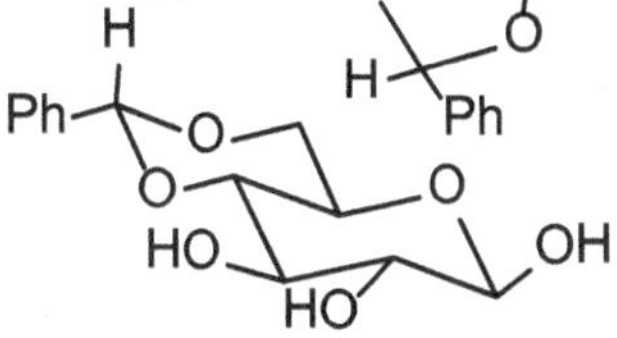

27. Only two products are obtained in the following reaction sequence. The structures of the products from the list I-IV are

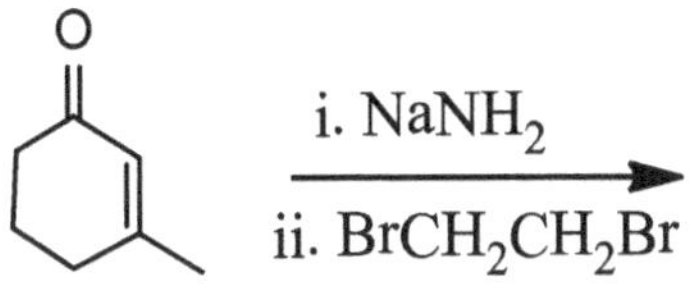

i. NaNH$_2$

ii. BrCH$_2$CH$_2$Br

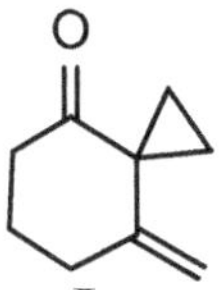

I    II 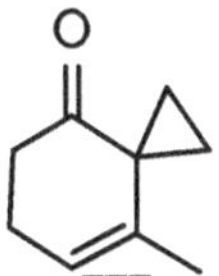   III 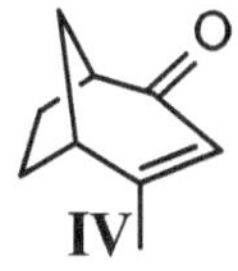   IV

    i. I and III,

    ii. I and II

    iii. II and IV

    iv. III and IV

28. Treatment of benzyl acetate with excess of methylmagnesium bromide followed by hydrolysis yileds

    i. benzyl alcohol and acetic acid

    ii. a,a-dimethylbenzoic acid and acetic acid

    iii. t-butanol and benzyl alcohol

    iv. t-butanol and toluene

29. The increasing order of acidity of the following compounds A-D is:

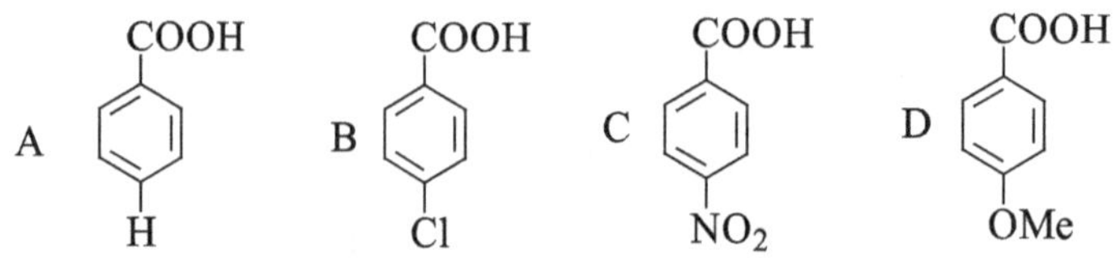

    i. $A < B < C < D$

    ii. $B < A < D < C$

    iii. $D < A < B < C$

    iv. $C < B < D < A$

30. The conversion of an amide into an amine can be achieved by:

    i. Hoffmann rearrangement

    ii. Claisen rearrangement

    iii. Beckmann rearrangement

    iv. S_N^2 replacement

31. Benzaldehyde can be prepared by reacting phenyl magnesium bromide with:

    i. HCHO

    ii. $COCl_2$

    iii. HCOOH

    iv. HCO_2Et

32. Conversion of acetophenone to ethyl benzene can be achieved by reacting with

    i. lithium aluminium hydride

    ii. Zinc amalgum and HCl

    iii. Lithium, liquid ammonia and t-butanol

    iv. Potassium permanganate

33. Which one of the following would readily give Tollen's test?

i ii OCH$_3$ iii OH   iv CH$_3$COCH$_3$

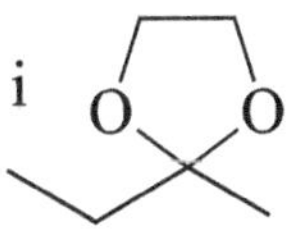

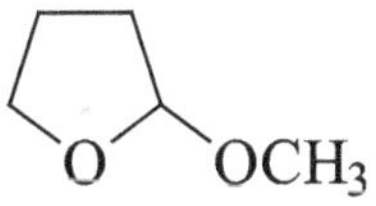

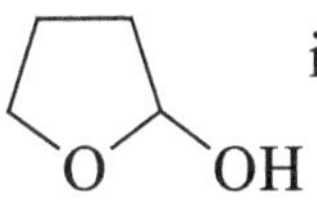

34. The major products A and B in the following reaction sequence are

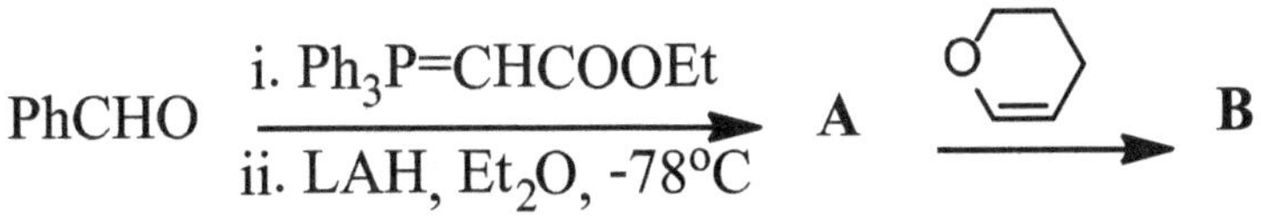

PhCHO $\xrightarrow[\text{ii. LAH, Et}_2\text{O, -78°C}]{\text{i. Ph}_3\text{P=CHCOOEt}}$ A $\xrightarrow{}$ B

i A = Ph$\diagup$=$\diagdown$OH   B = Ph$\diagup$=$\diagdown$O-

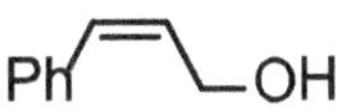

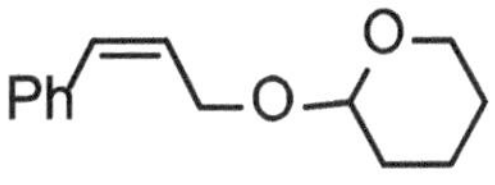

ii A = Ph$\diagup$=$\diagdown$OH   B = Ph$\diagup$=$\diagdown$O-

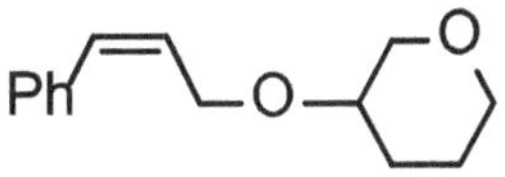

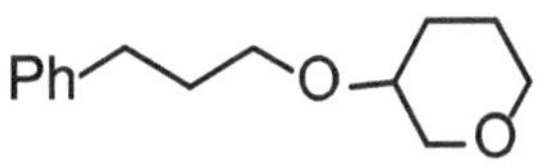

iii A = Ph$\diagdown$OH   B = Ph$\diagdown$O-

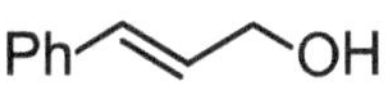

iv A = Ph$\diagup$OH   B = Ph$\diagup$O-

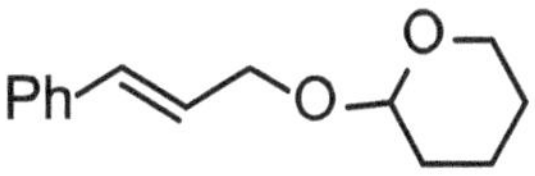

35. The major product formed in the following reaction sequence is

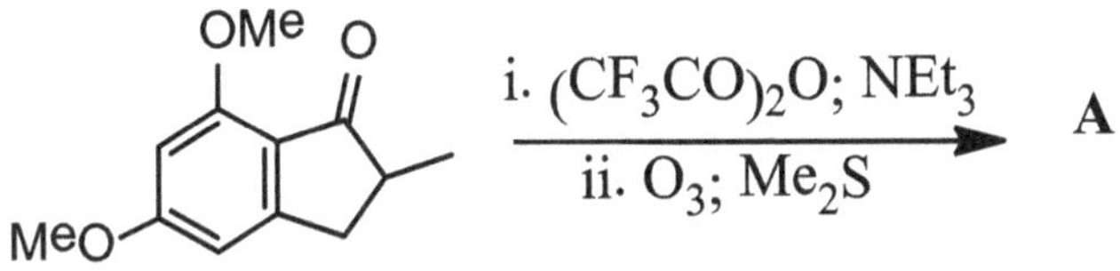

$\xrightarrow[\text{ii. O}_3; \text{Me}_2\text{S}]{\text{i. (CF}_3\text{CO)}_2\text{O; NEt}_3}$ A

i 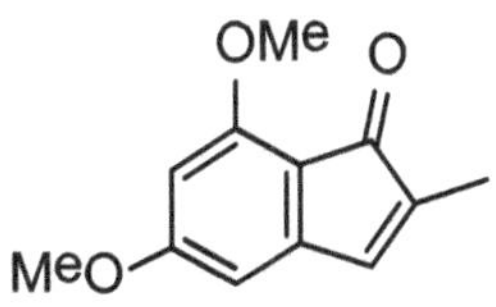   ii

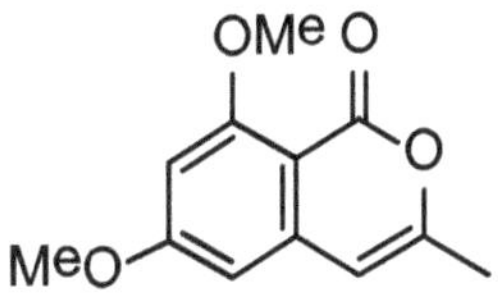

iii 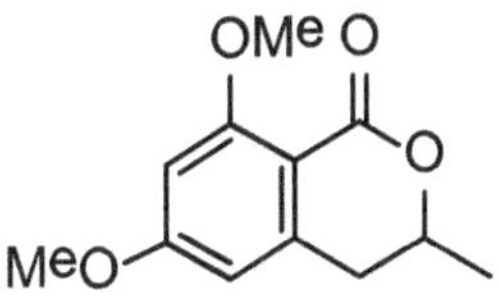   iv

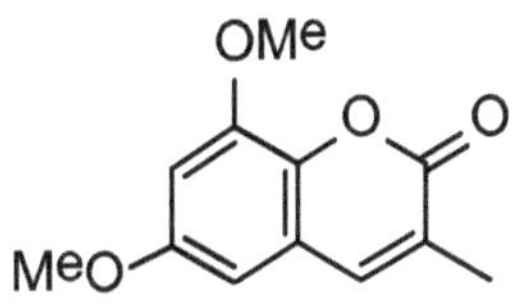

36. The major products A and B in the following reaction sequence are

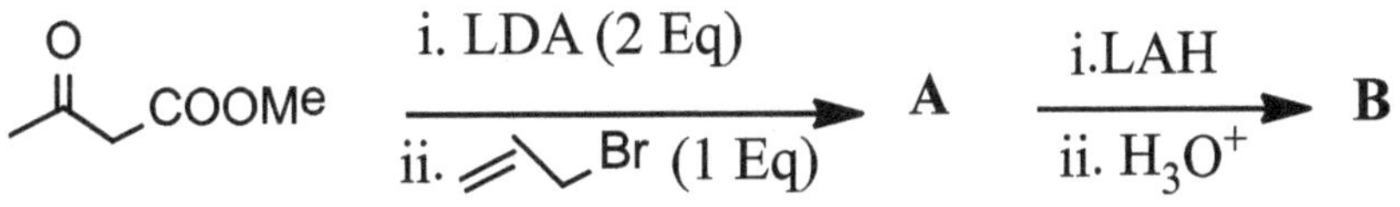

$\xrightarrow[\text{ii. } \diagup\diagdown\text{Br (1 Eq)}]{\text{i. LDA (2 Eq)}}$ A $\xrightarrow[\text{ii. H}_3\text{O}^+]{\text{i.LAH}}$ B

i   A =   B =

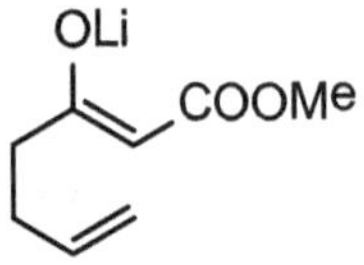

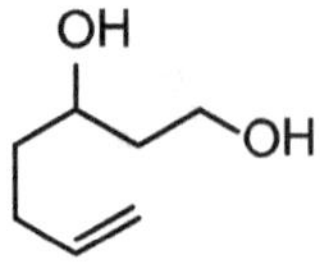

ii   A 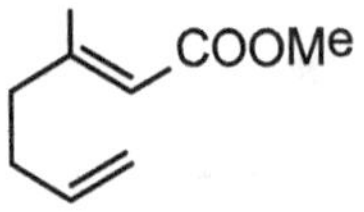 =   B =

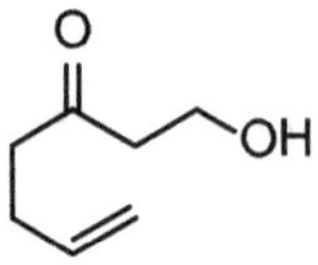

iii   A 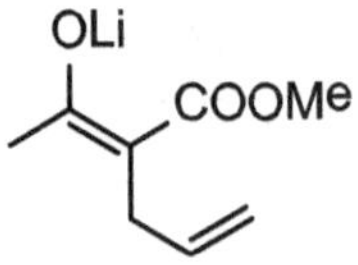 =   B =

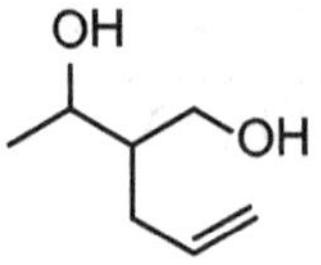

iv   A 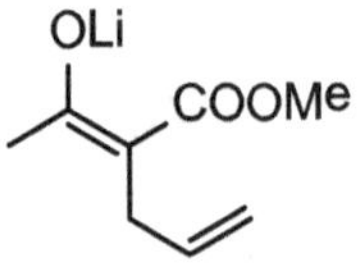 =   B =

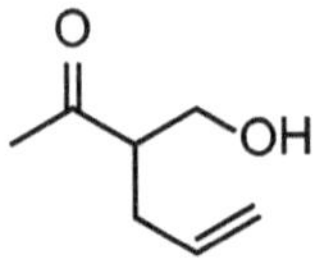

37. The major products A and B in the following reaction sequence are

A 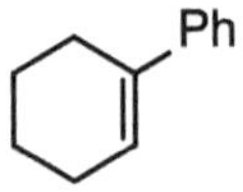 =   B =

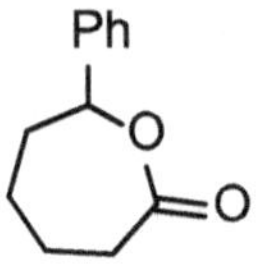

A 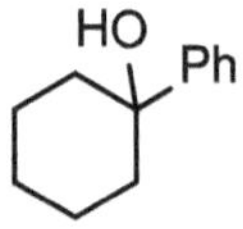 =   B =

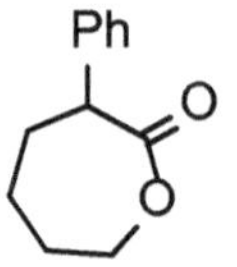

A 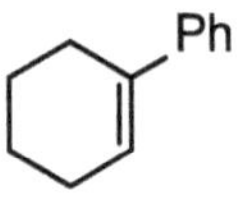 =   B =

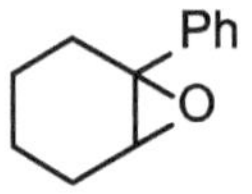

A 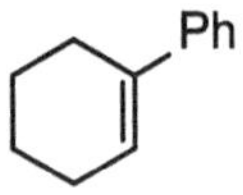 =   B =

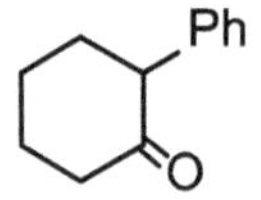

38. In the following reaction sequence, structures of the major products A and B are

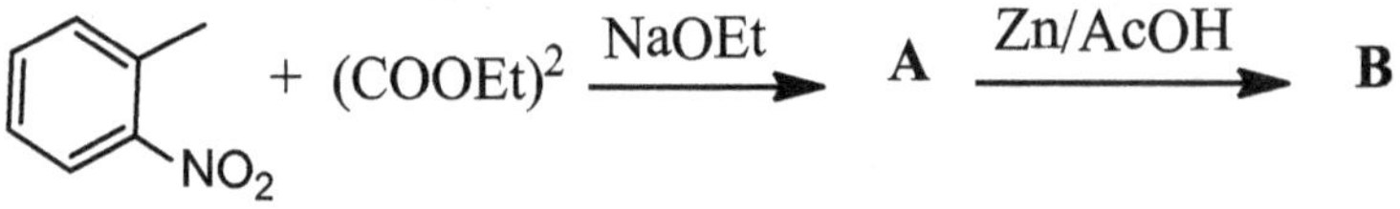

i   **A =** **B =**

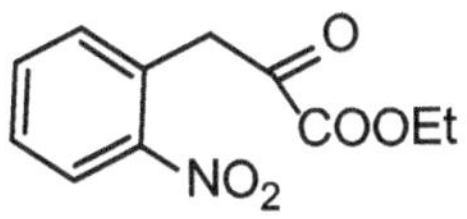

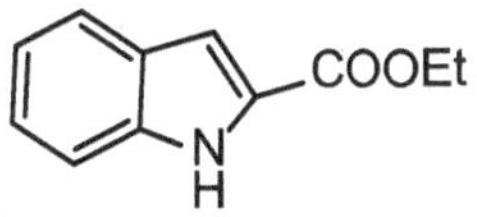

ii   **A 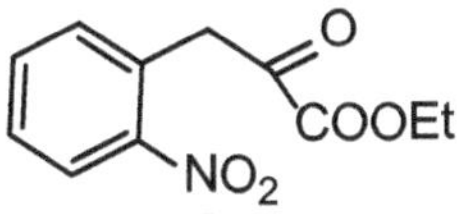 =** **B =**

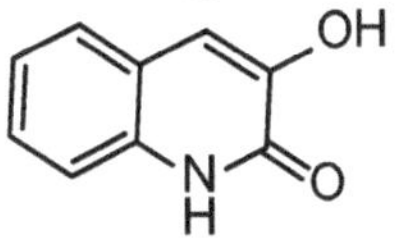

iii   **A 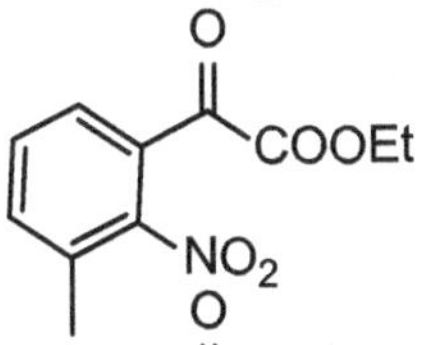 =** **B =**

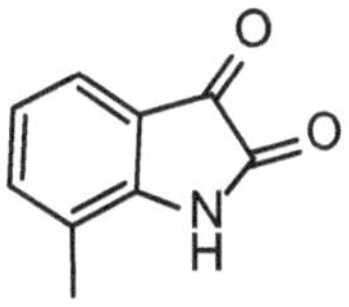

iv   **A 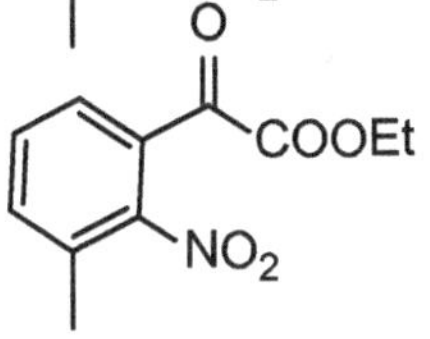 =** **B =**

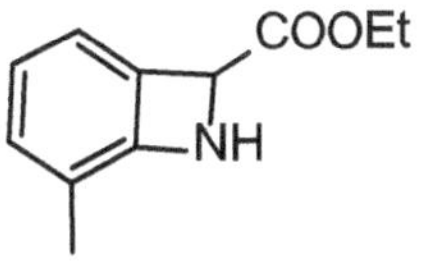

39.   Conversion of acetophenone to benzoic acid can be achieved by reacting with

    i.   lithium aluminium hydride
    ii.   Zinc amalgum and HCl
    iii.   Lithium, liquid ammonia and t-butanol
    iv.   Potassium permanganate

40.   Michael reaction is an example of

    i.   a reaction proceeding via a cyclic transition state, in the presence of light
    ii.   an electrophilic addition to an α,β-unsaturated ketone, under acid catalysis.
    iii.   an electrophilic addition to an olefinic bond, in a trans fashion
    iv.   a nucleophilic addition to an α,β-unsaturated ketone, under base catalysis.

41.   In the conversion of a Grignard reagent into an aldehyde, the other component used is:

    i.   ethyl formate
    ii.   ethyl acetate
    iii.   ethyl cyanide
    iv.   Formic acid

42.   An organic compound 'X' with molecular formula C_4H_8O on oxidation with perbenzoic acid yields ethyl acetate. When X is treated with iodine and alkali, the product will be

    i.   Ethanoic acid
    ii.   Propanoic acid
    iii.   Butanoic acid
    iv.   Butene

43. Among the following which one undergo haloform reaction?

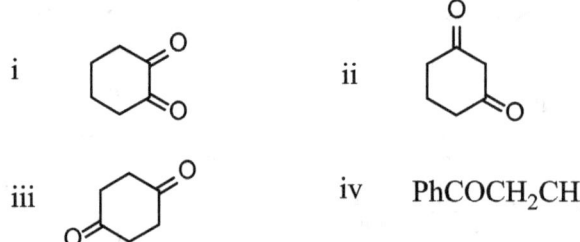

i

ii

iii

iv   PhCOCH₂CH₃

44. Reaction of ethyl formate with an excess of phenyl magnesium bromide generates

  i.   benzoic acid
  ii.   benzaldehyde
  iii. diphenylmethanol
  iv.   ethyl benzoate

45. The major products A and B in the following reaction sequence are:

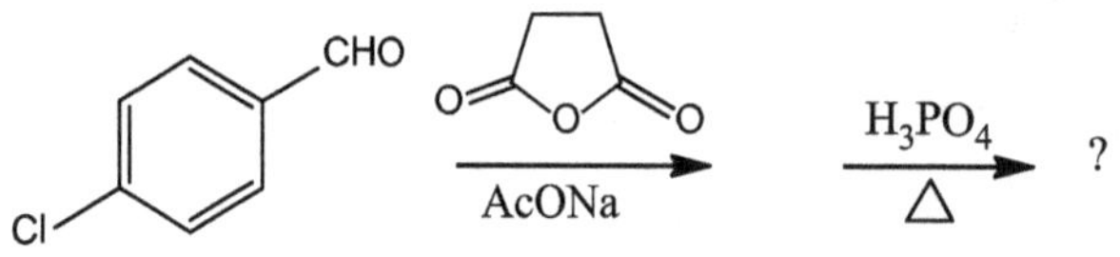

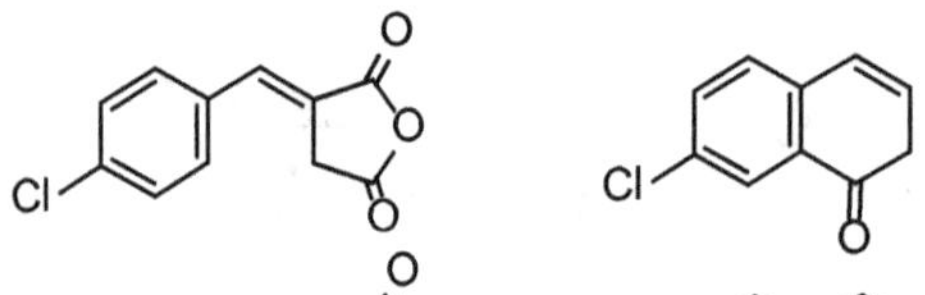

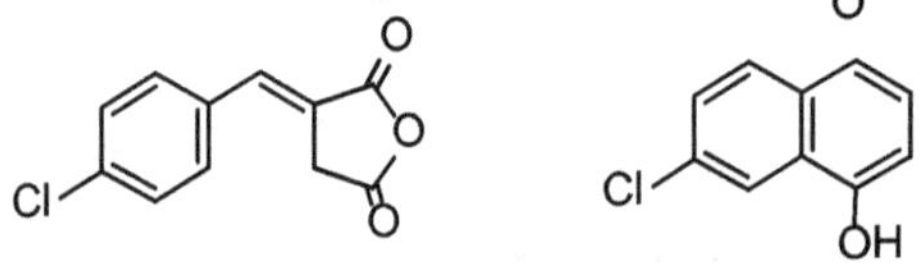

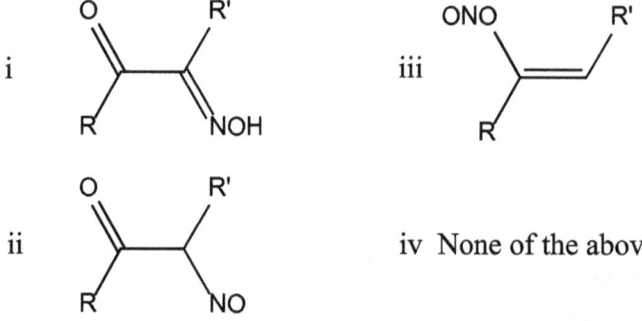

46. Reaction of RCOCH₂R' with nitrous acid yields

i

ii

iii

iv None of the above

47. The silver salt of a fatty acid on refluxing with an alkyl halide gives an,

    i.   Ether
    ii. Amine
    iii. Acid
    iv. ester

48. Among formaldehyde, acetaldehyde and benzaldehyde, the aldehydes which can undergo Cannizaro reaction on treatment with sodium hydroxide are

    i.   acetaldehyde
    ii. formaldehyde and acetaldehyde
    iii. acetaldehyde and benzaldehyde
    iv. formaldehyde and benzaldehyde

49. The commonly used reagent for the reduction of an ester to primary alcohol is:

    i. H_2/Pd
    ii. LiOH
    iii. LiAlH$_4$
    iv. NaBH$_4$

50. Conversion of the following reaction involves:

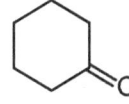 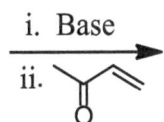 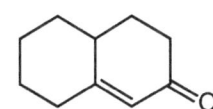

    i.   Aldol condensation followed by Michael addition
    ii. Michael addition followed by Aldol condensation
    iii. Consecutive aldol condnsatuon
    iv. Consecutive Michael addition

51. The two reactions involved in the Robinson annulation are:

    i.   hydroboration and oxidation
    ii. Michael reaction and aldol condensation
    iii. Perkin reaction and Michael reaction
    iv. Oppenauer oxidation and Friedel-Craft's reaction

52. p-Nitrobenzoic acid on treatment with diborane gives

    i.   p-aminobenzoic acid
    ii. P-nitrobenzyl alcohol
    iii. p-aminobenzyl alcohol
    iv. P-nitrobenzaldehyde

53. The suitable reagent for the following conversion is

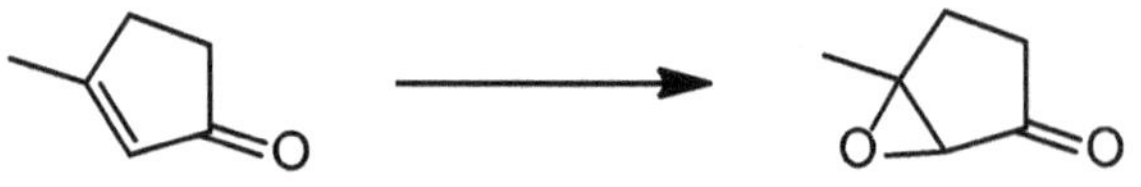

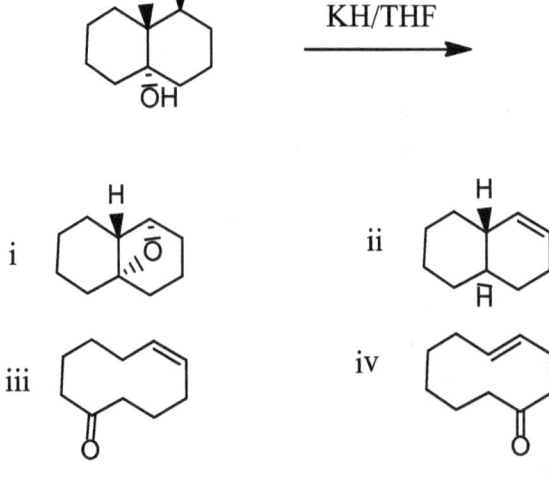

   i.   m-CPBA
   ii. H_2O_2/ AcOH
   iii. ^tBuOOH/HCl
   iv. H_2O_2/NaOH

54. Conversion of acetophenone to benzoic acid can be achieved by reaction with

   i.   iodine and sodium hydroxide
   ii. sodium hydroxide alone
   iii. hydroxylamine followed by reaction with H_2SO_4
   iv. meta-chloroperbenzoic acid

55. The major product formed in the following reaction sequence is

   i. m-CPBA
   ii. $BF_3.Et_2O$

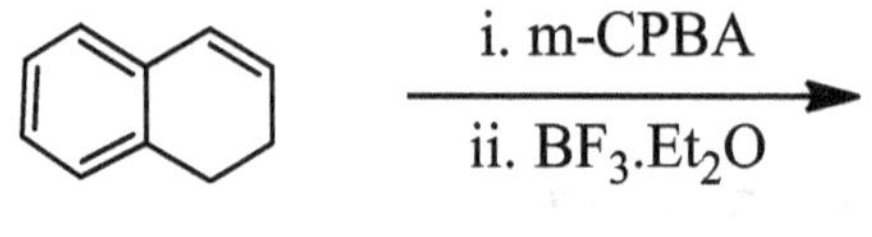

   i 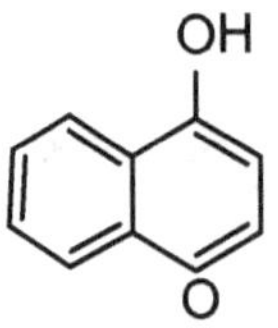   ii

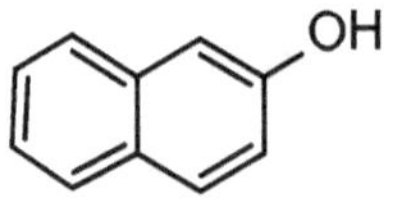

   iii 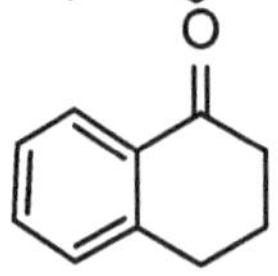   iv

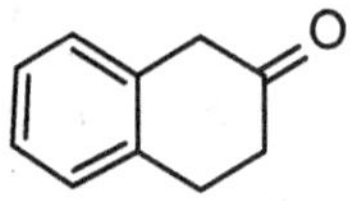

56. The major product formed in the following reaction sequence is

   KH/THF

   i   ii

   iii   iv

57. What is the major product of the following reaction?

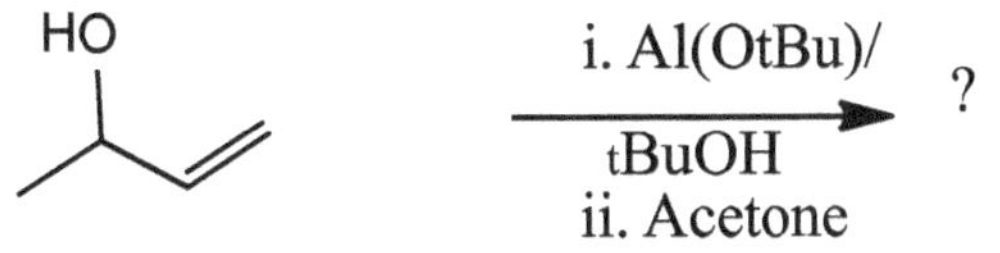

i 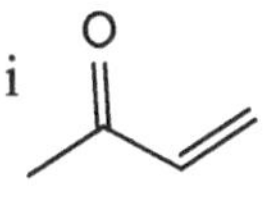
ii 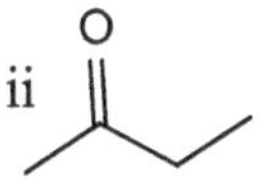
iii 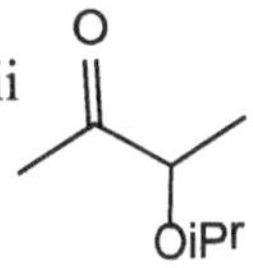

iv 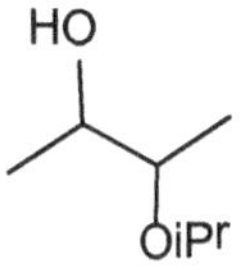

58. The products A and B in the following reaction sequence are

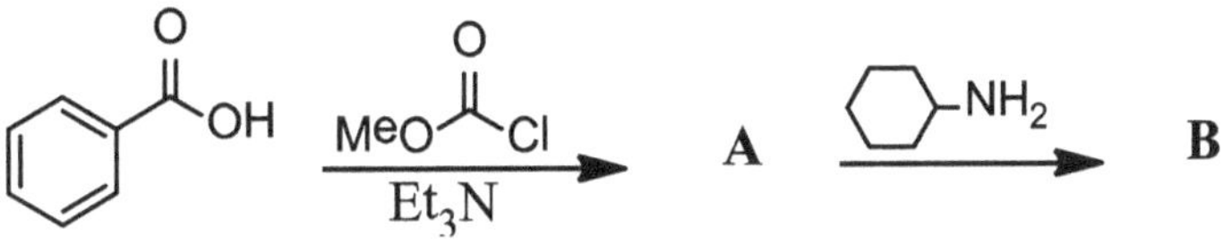

i A = 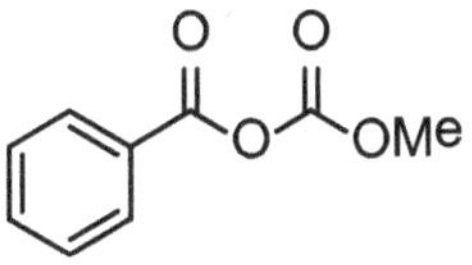

B = 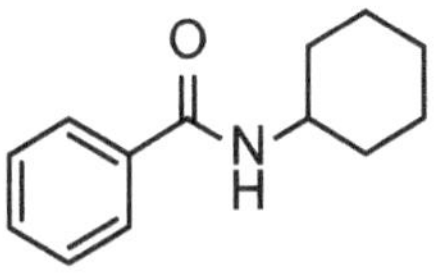

ii A = 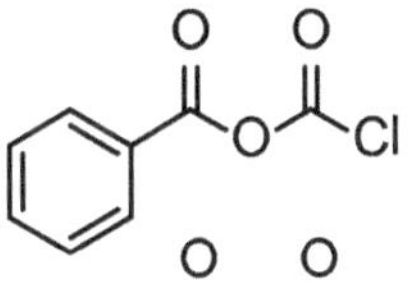

B = 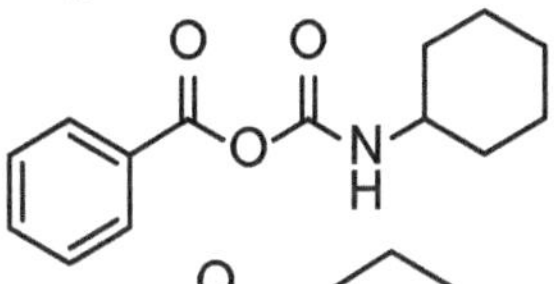

iii A = 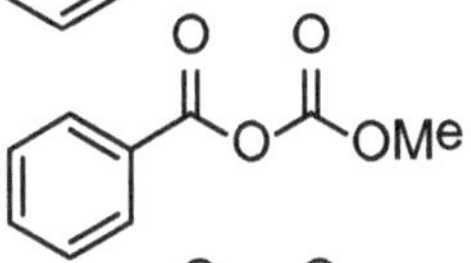

B = 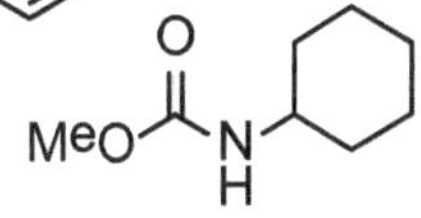

iv A = 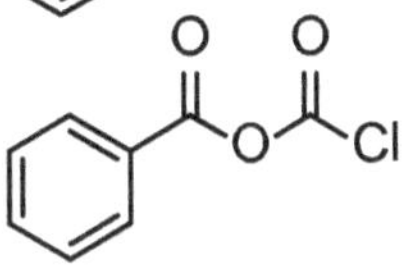

B = 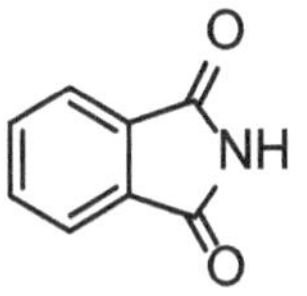

59. Predict the product for the following reaction is

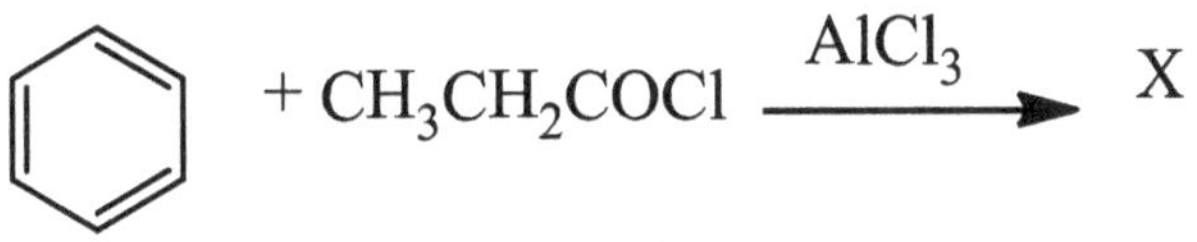

i 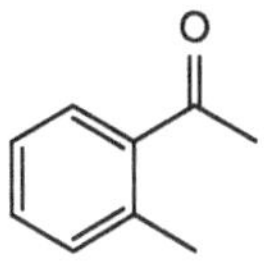
ii 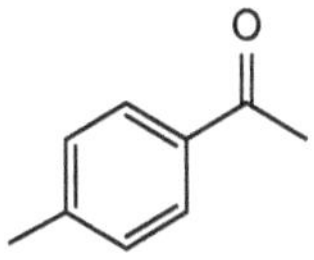
iii 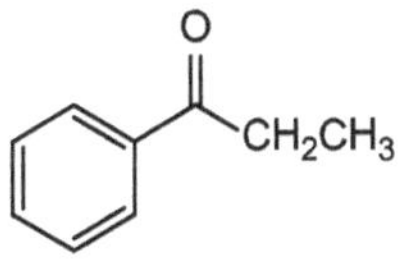

iv 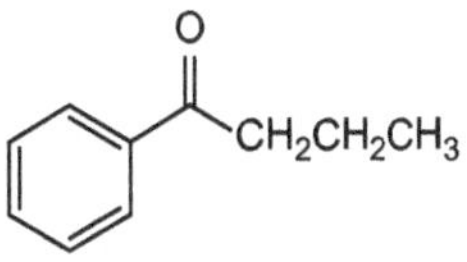

60. The conversion of $-COCH_2-$ to CH_2CH_2- using Zn-amalgum and HCl is known as Clemenson reduction. An alternative method for the same conversion is known as:

   i.   Birch reduction
   ii. Wolf-Kishner reduction
   iii. Meerwin-Pondroff-Verley reduction
   iv. Rosenmund reduction

61. Predict the product for the following reaction

$$RCHO \xrightarrow[\text{ii. } H^+]{\text{i. } Ph_3P=CHOR'} ?$$

   i.   RCH=CHOR'

   ii.   RCH$_2$CHO

   iii. No product formation

   iv.   RCH=CHR'

62. What is the product B in the following sequence of reaction?

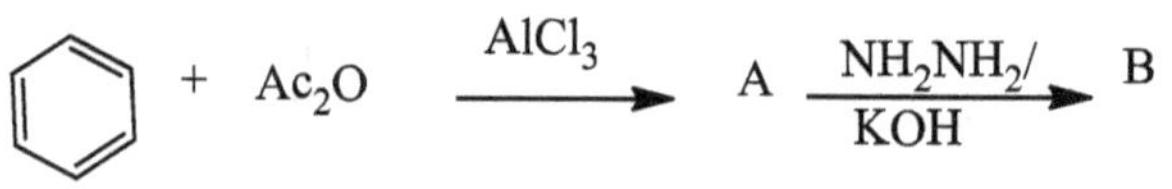

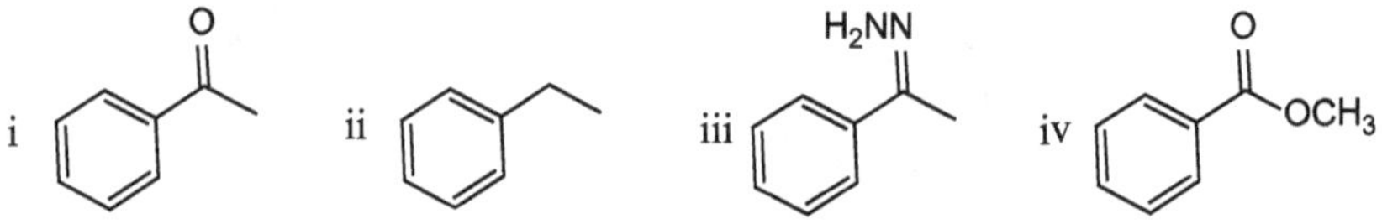

63. The following reaction show Umpolung activity

   i.   Grignard reaction
   ii. Benzoin condensation
   iii. Aldol condensation
   iv. Both (i) and (ii)

64. Among the following molecule which one undergo intersystem crossing efficiently.

   i.   Butadiene
   ii. Benzil
   iii. Benzopheone
   iv. acetophenone

65. In the concerted process, the reaction goes via –

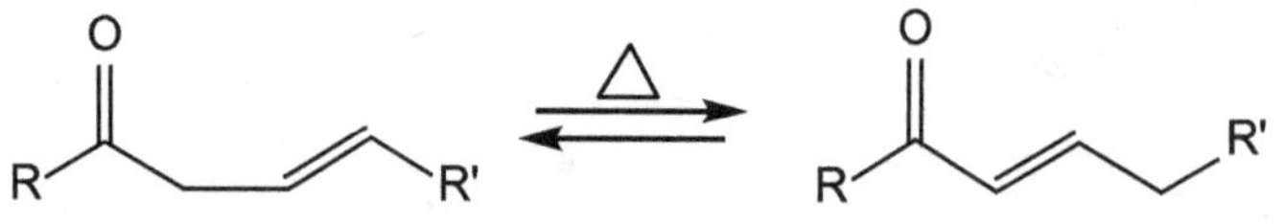

i.   1,3-sigmatropic rearrangement
ii.   1,5- sigmatropic rearrangement
iii. 2,3- sigmatropic rearrangement
iv.   3,2- sigmatropic rearrangement

66.   In the concerted process, the reaction goes via –

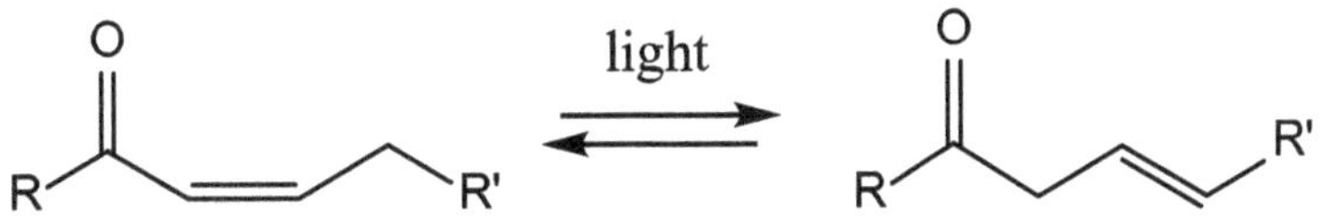

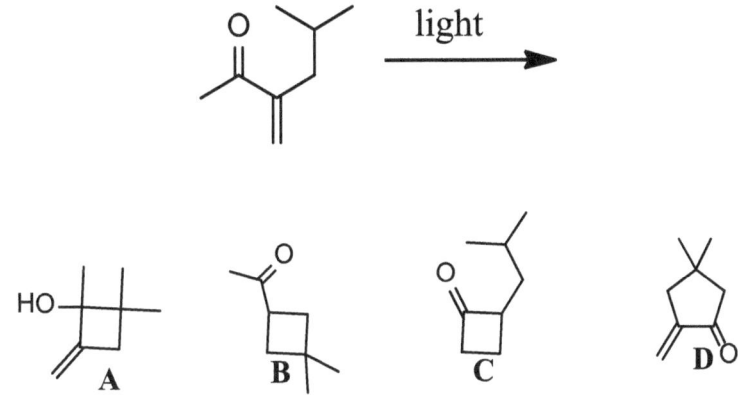

i.   1,3-sigmatropic rearrangement
ii.   1,5- sigmatropic rearrangement
iii. 2,3- sigmatropic rearrangement
iv.   3,2- sigmatropic rearrangement

67.   Amongst the following, the major products formed in the following photochemical reaction are

i.   A and C
ii.   B and C
iii. A and D
iv.   A and B

68.   Identify the final product for the following reaction.

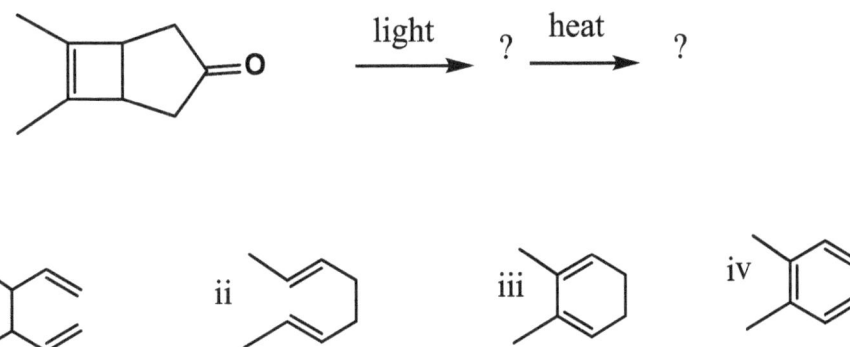

69.   Identify the final product for the following reaction.

light ? heat ?

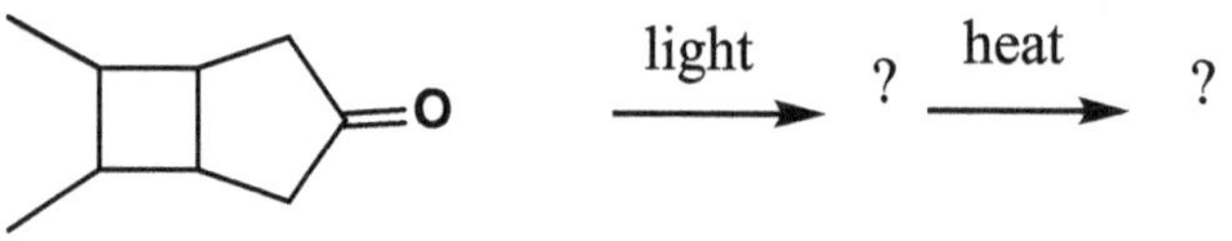

i 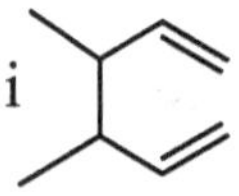   ii 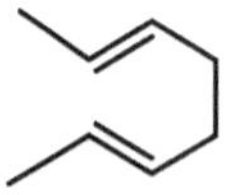   iii 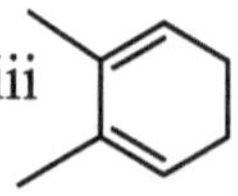   iv

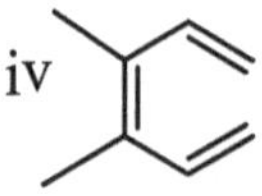

70.   The following photochemical conversion proceeds through

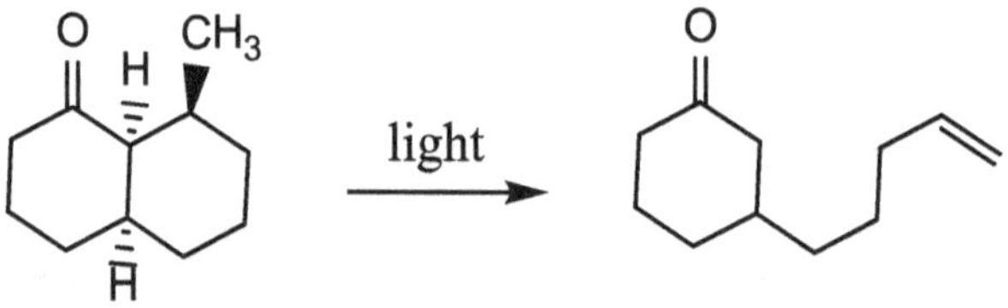

light

   i.   Barton reaction
   ii.   Paterno-Buchi reaction
   iii. Norrish-I type reaction
   iv.   Norrish-II type reaction

71.   The following photochemical conversion proceeds through

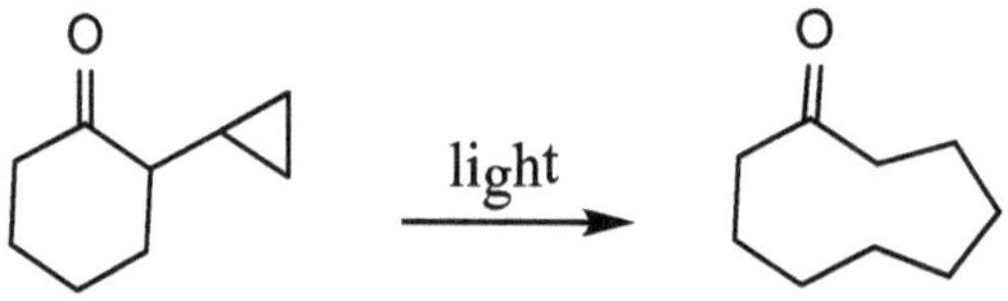

light

   i.   Barton reaction
   ii.   Paterno-Buchi reaction
   iii. Norrish-I type reaction
   iv.   Norrish-II type reaction

72.   Predict the product in the following reaction formed via consecutive 1,5-sigmatropic rearrangement is -

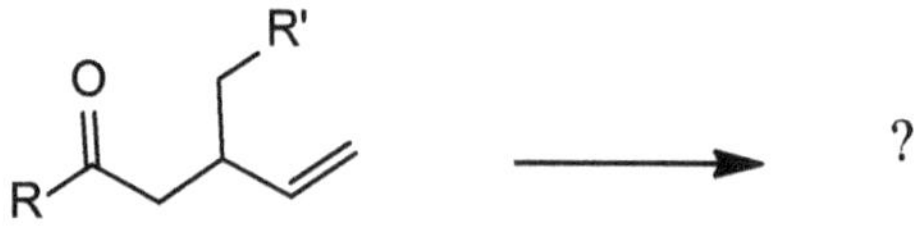

?

i 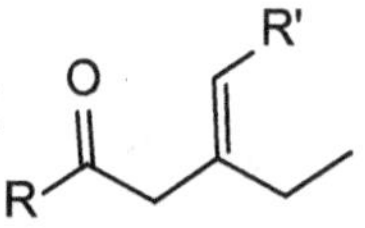   ii

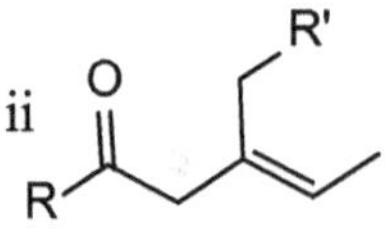

iii 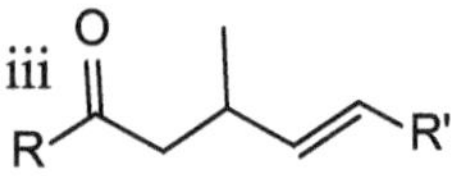   iv

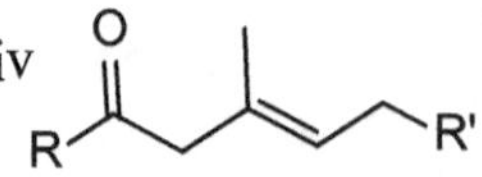

73. Which one of the following is the mechanism of hydrolysis of t-butyl acetate under acidic condition?

    i.   Acyl oxygen bond cleavage, unimolecular
    ii. Acyl oxygen bond cleavage, bimolecular
    iii. Alkyl oxygen bond cleavage, unimolecular
    iv. Alkyl oxygen bond cleavage, bimolecular

74. Hydrolysis ot t-butyl acetate undergo via $A_{AL}1$ mechanism. This was determined by

    i.   Cross over experiment
    ii. Trapping the intermediate
    iii. Isotopic labeling experiment
    iv. Isolation of the product

75. Which one of the following is the mechanism of hydrolysis of ethyl benzoate by refluxing with dilute aqueous NaOH solution?

    i.   Acyl oxygen bond cleavage, unimolecular
    ii. Acyl oxygen bond cleavage, bimolecular
    iii. Alkyl oxygen bond cleavage, unimolecular
    iv. Alkyl oxygen bond cleavage, bimolecular

76. Glyceryl trioleate on complete hydrogenation gives

    i.   Glyceroyl tristearate
    ii. Glyceroyl tripalmitate
    iii. Glyceroly trilinoleate
    iv. Glyceroly trilinolinate

77. The compound given below can be obtained by

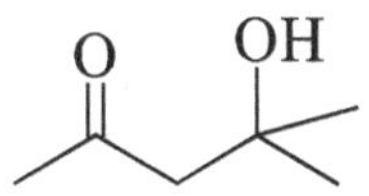

    i.   Claisen condensation
    ii. Aldol condensation
    iii. Dehydration
    iv. esterification

78. The IUPAC name of the compound given below, is

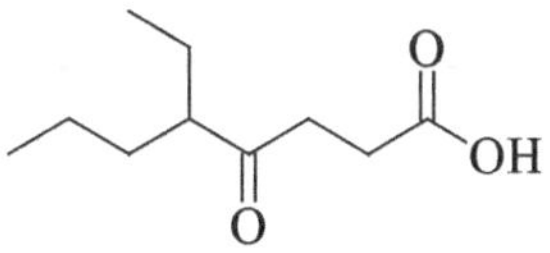

    i.   4-oxo-5-propylheptanoic acid
    ii. 2-carboxyethyl 3-hexylketone

    iii.   3-(3-hexylcarbonyl)propanoic acid

    iv.   5-ethyl-4-oxooctanoic acid

79.   Which of the following ketone absorbs IR near 1750 cm^{-1}?

      i CH_3COCH_3 ii

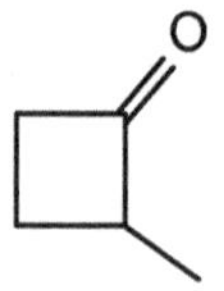

      iii 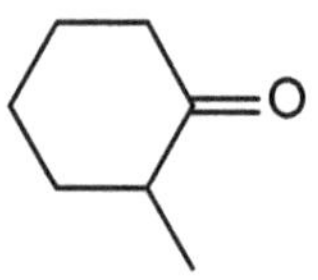   iv

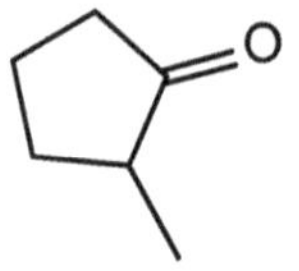

80.   The UV spectrum of acetone in hexane (solvent) shows two peaks: i. λ_{max} = 179nm, λ_{max} = 15 and ii. λ_{max} = 189nm, λ_{max} = 900. Identify the transition associated with λ_{max} =189nm.

    i. $\sigma \rightarrow \pi^*$

    ii. $n \rightarrow \pi^*$

    iii. $\pi \rightarrow \pi^*$

    iv. $\pi \rightarrow \sigma^*$

81.   Which of the following has $n \rightarrow \pi$ electron transition?

    i.   Acetic acid

    ii.   Ethylene

    iii. Amylonitrile

    iv. Vinyl acetylene

82.   In the IR spectrum, carbonyl absorption band for the following compound appears at

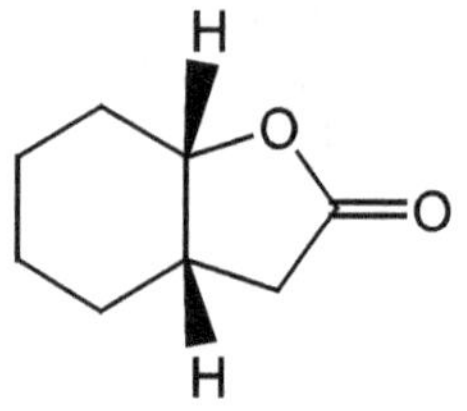

    i.   1810 cm^{-1}

    ii.   1770 cm^{-1}

    iii. 1740 cm^{-1}

    iv. 1690 cm^{-1}

83.   The correct ^{13}C NMR chemical (δ) shift values of carbon labeled **a-e** in the following ester are

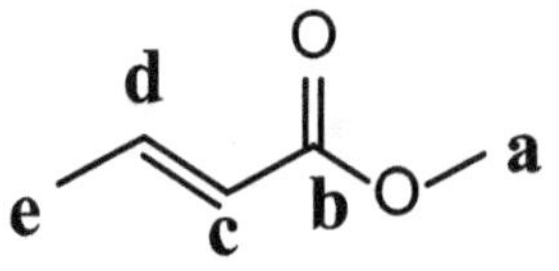

    i.   a: 19; b: 143; c: 167; d: 125; e: 52

    ii.   a: 52; b: 143; c: 167; d: 125; e: 19

    iii.   a: 52; b: 167; c: 143; d: 125; e: 19

    iv.   a: 52; b: 167; c: 125; d: 143; e: 19

84.   An organic compound (MF: $C_{15}H_{14}O$) exhibited the following spectral data. ^{1}H NMRδ 2.4 (3H, s), 7.2 (2H, d, J = 8Hz), 7.7 (2H, d, J = 8Hz) and ^{13}C NMR: δ 21, 129, 130, 136, 141, 190 ppm. The compound is

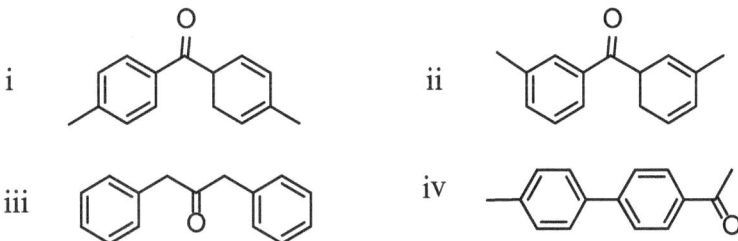

85.   How many ^{1}HNMR signals would be given by $CH_3COCH_2COCH_3$ at room temperature?

    i.   2

    ii.   3

    iii.   4

    iv.   5

86.   How many ^{1}HNMR signals would be given by $CH_3COCH_2COCH_3$ at -78°C?

    i.   2

    ii.   3

    iii.   4

    iv.   5

87.   Among the following, the correct statement for the following reaction is

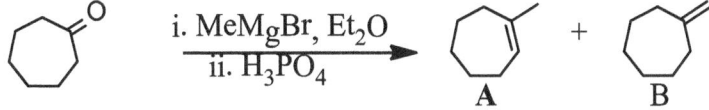

    i.   A is the major and it will have five signals in the proton decoupled ^{13}C NMR spectrum

    ii.   A is the minor and it will have eight signals in the proton decoupled ^{13}C NMR spectrum

    iii.   B is the major and it will have five signals in the proton decoupled ^{13}C NMR spectrum

    iv.   B is the minor and it will have five signals in the proton decoupled ^{13}C NMR spectrum

88.   Which one of the following is the mechanism of hydrolysis of ethyl acetate under acidic condition?

    i.   Acyl oxygen bond cleavage, unimolecular

    ii.   Acyl oxygen bond cleavage, bimolecular

    iii.   Alkyl oxygen bond cleavage, unimolecular

    iv.   Alkyl oxygen bond cleavage, bimolecular

Answer:

Q.No	Ans	Q.No	Ans	Q.No	Ans	Q.No	Ans
1	iii	2	i	3	ii	4	iv
5	i	6	ii	7	iv	8	i
9	iii	10	iii	11	iv	12	iii
13	iii	14	iii	15	i	16	i
17	i	18	ii	19	ii	20	iii
21	iv	22	iii	23	ii	24	iii
25	ii	26	ii	27	i	28	iii
29	iii	30	i	31	iii	32	ii
33	iii	34	iv	35	ii	36	i
37	i	38	i	39	iv	40	iv
41	iv	42	ii	43	ii	44	iii
45	ii	46	i	47	iv	48	iv
49	iii	50	ii	51	ii	52	i
53	i	54	i	55	iv	56	iii
57	i	58	i	59	iii	60	ii
61	ii	62	ii	63	iv	64	iii
65	ii	66	i	67	iv	68	iii
69	ii	70	iv	71	iii	72	iii
73	iii	74	iii	75	i	76	i
77	ii	78	iv	79	iv	80	ii
81	i	82	iii	83	iv	84	i
85	ii	86	iii	87	iv	88	ii

REFERENCES

1849Misc106: H. Fehling, Annalen der Chemie und Pharmacie, 1849. **72**, 106–113.

1882Misc1635: Tollens, B., Berichte der Deutschen Chemischen Gesellschaft (in German). 1882, **15**: 1635–1639.

1887Ber651: Claisen L., Lowman O., Ber 1887, 20, 651.

1891Ber2962: Biginelli P, Ber 1891, 24, 2962.

1894Ber102: Dieckmann W., Ber, 1994, 27, 102.

1897Ber1622: Gattermann C., Koch J. A., Ber, 1897, 30, 1622.

1904BSC1306: *Bouveault, L. Bull. Soc. Chim. Fr.* (in French). 1904, ***31***: 1322–1327.

1906RPCS355: Tischenko W. J., Russ. Phy. Chem. Soc., 1906, 38, 355]

1912ADP647: *Mannich, C.; Krösche, W., Archiv der Pharmazie.* **250**, *1912, 647–667*

1913CL1011: Mukaiyama, T.; Narasaka, K.; Banno, K. Chem. Lett.1913, 1011-1014. 4

1925JCST1874: *Stephen, H, J. Chem. Soc., Trans.,* 1925, ***127***, 1874–1877

1929BSC37: C. D. Ninitzescu, Bull. Soc, Chim. Roumania, 1927, 37, 1929.

1935JCS1285: *Rapson, William Sage; Robinson, Robert, J. Chem. Soc., 1935, 1***285**

1937Nat195: Norrish R. G. W.; Bamford C. H.; Nature, 1937, 140, 195.

1937RTCPB137: Oppenauer, R. V. , Recl. Trav. Chim. Pays-Bas, 1937, 56, 137–144.

1942OR0001: Blicke, F. F. Org. React. **1942**, 1, 01

1942OR155: Martin, E. L.. Org. React. **1942**, 1, 155.

1946JBC549: Hollander V.P.; Gallagher T. F. ; J. Biol. Chem., 1946, 162, 549.

1946OR307: Wolff, H. Org. React. **1946**, 3, 307

1948OR378: Todd, D. Org. React. **1948**, 4, 378

1948JACS1187: Alexander, E; Ruth B. W.,. J. Am. Chem. Soc.,. 1948, ***70***: 1187–1189

1948Misc362: Mosettig, E.; Mozingo, R., Organic Reactions. 1948, **4**: 362–377.

1948OR229: Berlinear E., Organic Reactions, 1949, 5, 229-69.

1951JOC661: Pollard, C.B.; David C. Young, J. Org. Chem., 1951, ***16***, 661–672

1952JACS5828: Donald J. Cram, Fathy Ahmed Abd Elhafez J. Am. Chem. Soc.; 1952; 74; 5828–5835

1952CR505: E. Arundale, L. A. Mikeska Chem. Rev.; 1952; 51, 505–555

1955JACS3199: *Mandell, L. (1955) [. "The Mechanism of the Wettstein-Oppenauer Oxidation". J. Am. Chem. Soc.* **78**,: *3199–3201*

1955JACS3272: Leonard, N. J.; Gelfand, S. , *J. Am. Chem. Soc.* **78**, *3272*

1955OS38: yron R. Byron; Moffett, R. B.; McIntosh A. V., syntheses. 1955, ***24***: 38.; Collective Volume, ***3***, p. 234

1957JACS6562:   *Kornblum, N.; Powers, J. W.; Anderson, G. J.; Jones, W. J.; Larson, H. O.; Levand, O.; Weaver, W. M., J. Am. Chem. Soc., 1957, 79, 6562*

1958Ber61:   Leopold Horner; Hoffmann, H. M. R.; Wippel, H. G. *Ber.* 1958, 91, 61–6.

1959JACS4113:   *Kornblum, N.; Jones, W. J.; Anderson, G. J., J. Am. Chem. Soc., 1959, 81, 4113–4114*

1959JACS2748:   D. J. Cram and K. R. Kopecky, J. Am. Chem. Soc., 1959, 81, 2748

1956JACS5129:   G. Stork, and H. K. Landesman, J. Am. Chem. Soc., 1956, 78, 5126.

1958CI979:   R. E. Ireland, Chem. Ind. (London), 1958, 979

1963JOC3134:   K. C. Beannock, R. D. Burprit, V. W. Goodlett and J. G. Thweatt, J. Org. Chem., 1963, 28, 3134.

1963JACS1245:   D. J. Cram. and D. R. Wilson, J. Am. Chem. Soc., 1963, 85, 1245.

1963JACS3027:   Pfitzner, K. E.; Moffatt, J. G., *J. Am. Chem. Soc.,* 1963, **85**: 3027

1963JOC1933:   Burgert B. E.;Kruger J. I., J. Org. Chem., 1963, 28, 1933.

1964JACS2909:   Szmant, H. H.; Harmuth, C. M., J. Am. Chem. Soc., 1964, . **86** (14): 2909

1964CR81:   O'Brien C., Chem. Rev., 1964, 64, 81.

1966JACS1830:   Denny D. Z. and Denny D. R., 1966, J. Am. Chem. Soc., 1966, 88, 1830.,

1967JACS5734:   Shapiro, Robert H.; Heath, Marsha J. *J. Am. Chem. Soc.,* **1967**, 89, 5734– 5735

1967JACS5505:   *J. R. Parikh; W. v. E. Doering., J. Am. Chem. Soc., 1967, 89, 5505–5507*

1967JOC780:   Peterson, D. J. J. Org. Chem. 1967, 32, 780-784

1967OR204:   Jones, G. Org. React. 1967,15, 204-599

1968ACSP900:   Fétizon, Marcel; Golfier, Michel. C. R. Acad. Sc. Paris (C). 1968, *267*: 900

1969JCS1118:   Fétizon, Marcel; Golfier, Michel; Louis, Jean-Marie. J. Chem. Soc. D (19): 1969, 1118– 1119

1969JCSCS1102:   Fétizon, M.; Golfier, M.; Louis, J. M., J. Chem. Soc. Chem. Commun., 1969, 1102

1970ACS1191:   Sandstorm J.; Wennerbeck I, Acta Chem. Scand., 1970, 24, 1191.

1971JOC1339:   Fetizon, M.; Balogh, V.; Golfier, J. Org. Chem., 1971, **36** (10): 1339.

1971TL4995:   *Heathcock, C, H.; Ellis, J. E.; McMurry, J. E.; Coppolino, A., Tett. Lett,, 1971, 12, 4995– 96*

1972Misc266:   House H. O., Modern Synthetic Reactions (The Organic Chemistry Monograph Series), 2nd Ed 1972, 856.

1972JACS7586:   *E. J. Corey; C. U. Kim, J. Am. Chem. Soc., 1972, 94, 7586–7587.*

1972MIsc303:   Taylor, R. In **Comprehensive Chemical Kinetics** Bamford, C.H.; Tipper, .F.H. Eds.; Elsevier Publishing Co; New **York**, *1972;* Vol. *13, 303-316.*

1973TL4833:   Julia, M.; Paris, J.-M. *Tett. Lett.* 1973, 14, 4833–4836

1974JACS4708:   John E. McMurry; Michael P. Fleming, J. Am. Chem. Soc., 1974, **96**, 4708– 4709

1974P109:   Evangelidou-Tsolis E, Ramirez F and Pilot J. P., Phosphorous, 1974, 4, 109.

1975TL1811:   Lipton, M.F.; Kolonko, K.J.; Buswell, R.L.; Capuano, L.A. *Tetrahedron Lett.,* **1975**, 1811

1974CL32:   Mukaiyama, T.; Izawa, T.; Saigo, K. Chem. Lew. 1974, 323-326

1973JACS4287:   Johnson, C. R., Kirchoff, R. A., Reischer, R. J., Katekar, G. F., J. Am. Chem. Soc., 1973, 95, 4287.

1975Syn236:   Larcheveque M.; Valette G.; Cuvigny T.; Normant H., Synthesis, 1975, 256.

1976OR405:   Shapiro, R. H. *Org. React.,* **1976**, 23, 405. (Review)

1976Angew639:   Stetter H., Angew. Chem., Int. Ed. Engl., 1976, 15, 639.

1976BCS3597:   Yamasihi M., Watanabe Y., Mitsudo T. –A., Bull. Chem. Soc., Japan, 1976, 49, 3597.

1977OR73:   Wadsworth, W. *Org. React.* 1977, 25, 73.

1977JOC3114:   Van Leusen, Daan; Oldenziel, Otto; Van Leusen, Albert. J. Org. Chem. , 1977, **42**, 3114– 3118

1978JCSPT829:   Kocienski, P. J.; Lythgoe, B.; Ruston, S. J. Chem. Soc., Perkin Trans. 1 1978, 829

1978JACS3611:   F. N. Tebbe, G. W. Parshall and G. S. Reddy,. *J. Am. Chem. Soc.* , 1978, ***100***, 3611–3613.

1979JOC4148:   Mancuso, A. J.; Brownfain, D. S.; *Swern, D.* , *J. Org. Chem.*, 1979, **44**, 4148– 4150

1979CL785:   Isobe, K.; Fuse, M.; Kosugi, H.; Hagiwara, H.; Uda, H. Chem. Lett. **1979**, 785 and reference therin..

1980TL1031:   W. C. Still and J. H. McDonald,III, Tett. Lett., 1980, 21, 1031.

1980CC1026:   Terashima S.; Tanno N.; Koga K., J. Chem. Soc., Chem. Commun., 1980, 1026.

1980JACS1577:   Soderquist, J. A.; Hassner, A. J. Am. Chem. Soc. **1980**, 102, 1577.

1981Misc129:   Green T. W., Protective groups in organic synthesis, Wiley, New York, 1981,129.

1981JACS870:   Hevesi, L.; Piquard, J.-L.; Wautier, H. J. Am. Chem. Soc. **1981**, 103, 870 and reference therin.

1981Syn145:   Manueso A. J.; Swern D., Synthesis 1981, 165-85.

1976BCS3597:   Hickmott P. W., Tetrahedron, 1982, 38, 1975 and 3363.

1981SynMeth26: Eyley S. C., Rainey D. R., Synthetic methods, 1981, 4, 26-86.

1983CS62:   Murthy C. A.; Lokanatha Rai K. M., Curr. Sci., **1983**, 52, 62.

1983SCR55:   Adlington, R.M.; Barret, A.G.M. *Acc. Chem. Res.*, **1983**, 16, 55. (Review)

1983ACR399:   *Huffman, J. W. Accounts of Chemical Research. 1983,* **16** *, 399*

1983T3207:   Albright J. D., Tetrahedron, 1983, 39, 3207.

1984S369:   Tsuji J., Synthesis, 1984, 369.

1985IJCB502:   Murthy C. A.; Lokanatha Rai K. M., J. Indi. Chem. Sce B., 1985, 24B, 502.

1985TL6377:   Pelter A, Ward R S, Pritchchard, Kay I. T, Tett. Lett., 1985, 26, 6377.

1985Misc: March, Jerry (1985), Advanced Organic Chemistry: Reactions, Mechanisms, and Structure (3rd ed.), New York: Wiley,

1986SC1343:   *Dave, Paritosh; Byun, Hoe-Sup; Engel, Robert, " Synth.Comm., 1986,* **16***,*

1986JOC1135:   *Molander, G. A.; Hahn, G., J. Org. Chem., 1986,. 51, 1135.*

1987CL2101:   Matsukawa M.; Tabuchi T.; Inanaga J.; Yamaguchi M., Chem. Lett., 1987, 2101

1987SC901:   *Molander, G. A.; Etter, J. B., Synth. Commun., 1987, 17, 901.*

1988JCSP1603:   Pelter A, Ward R S, Pritchchard, Kay I. T, J. Chem. Soc. I, 1988, 1603.

1988JOC5215:   Fauve A.; Veschambre H., J. Org. Chem., 1988, 53, 5215.

1989JA1418:   Chen, R.H.; Kafafi, S.A.; Stein, S.E. ***J. Am. Chem. SOC. 1989,*** I l l *, **1418***

1989Syn57:   Hassner A., Rai K. M. L., Synthesis, 1989, 57.

1989TL1173:   McMurry, Tett. Lett., **1989**, 30, 1173

1990JACS6447:   D. A. Evans, A. H. Hoveyda, J. Am. Chem. Soc. 1990, 112, 6447

1990Syn857:   Manueso A. J.; Swern D., Synthesis 1990, 857-70.

1991JACS7277:   *Dess, D. B.; Martin, J. C., J. Am. Chem. Soc., 1991,* **113***: 7277.*

1991Mis777:   Meth-Cohn O., Comprehensive Organic Synthesis; Ed. Trost, B. M., Pergamon: Elmsford, NY, 1991, Vol. 2, 777.

1991Misc341:   Tietze, L. F.; Beifuss, U., Comprehensive Organic Synthesis; Ed. Trost, B. M., Pergamon: Elmsford, NY, 1991, Vol. 2, 341-394

1991Misc541:   Hassner A.; Rai K. M. L., Comprehensive Organic Synthesis, Ed. by. B. M. Trost, Pergamon Press, 1991, 1, 541

1991Misc321:   Rosini, G. In Comprehensive organic Synthesis, B. Trost, I. Flemming, Eds. Pergamon, 1991, 2,321-340.

1991Misc327:   Hutchins, R. O., In Comprehensive organic Synthesis, B. Trost, I. Flemming, Eds. Pergamon, 1991, 1, 327

1991Misc795:   Davis D. R.; Garratt P., In Comprehensive organic Synthesis, B. Trost, I. Flemming, Eds. Pergamon, 1991,2,795-863

1991CR49:   Csuk R.; Glanzer B. I., Chem. Rev., 1991, 91, 49.

1991PAC307:   Brown H, C.; Ramachandran P. V., Pure and Appl. Chem., 1991, 63, 307.

1992JOC5979:   Thompson, S. K.; Heathcock, C. H., *J. Org. Chem.*, 1992, **57**, 5979–5989

992JOC4732:   Meyers A. J.; Elworthy T. R., J.Org. Chem., 1992, 57, 4732.

1993TL583:   Petasis, N. A.; Akritopoulou, I., *Tetrahedron Lett.*, 1993, **34,** 583–586

1995IJHC63:   K. M. Lokanatha Rai and C.A. Murthy, Indian J of Heterocycl. Chem., **1995**, 63

1996EF776:   Manion, **J. A,;** McMillen, D. F.; Malhotra, R. ***Energy and Fuels 1996,10, 776***

1984SC1669:   Hassner A, Rai K. M. L., and Wim Dehaen, Synth. Commun., **1994**, 24, 1669

1996JSST71:   Rai K. M. L.; Linganna N.; Hassner A. and C.A. Murthy, J. Sci. Soc.Thailand, 1996, 22, 71.

1997JACS445:   Petasis, N. A.; Zavialov, I. A., J. Am. Chem. Soc., 1997, **119,** 445–446

1998JACS11798:   Petasis, N. A.; Zavialov, I. A., J. Am. Chem. Soc., 1998, **120,**11798–11799.

1998OrgCat513:   Monfiler E.; Mortreux A., Organometallic Catalysis, 1998, 513.

1998JOC2918:   Covamubias-Zuniga A., Cantu F.; Maldanado L. A., J. Org. Chem., 1998, 63, 2918.

1999Misc426:   Williamson, K. L. Macroscale and Microscale Organic Experiments, 3rd ed. Boston: Houghton-Mifflin, 1999, 426–7.

1999IJC1126:   Linganna N, Lokanatha Rai K. M., and Shashikanth S., Indian J of Chem., Sect B, 1999, 38B, 1126.

2000IF389:   Lokanatha Rai K. M.; Linganna N., Il Formica 2000, 55, 389.,

2000OL577:   Rangan G.; Bhisma K. P., Org. Lett., 2000, 2, 577.

2002JACS1866:   *Córdova, A.; Watanabe, S.; Tanaka, F.; Notz, W.; Barbas Cf, 3. J. Am. Chem. Soc.., 2002,* ***124****, 1866–1867.*

2009Mis173:   Li, Jie Jack. Named Reactions: A Collection of Detailed Reaction Mechanisms. Springer: 2009, p. 173

2003OL3615:   D. Parrish, D. R. Shelton, R. D. Little, Org. Lett., **2003**, 5, 3615-3617

2003CC2644:   *Aggarwal, V. K.; Richardson, J.,* Chem. Commun., *: 2003, 2644*

2003COC1713:   Mahrwald R. Curr. Org. Chem., 2003, 3, 1713.

2004OrgRe01:   *Kappe C. O., Stadler A., Organic Reactions, 2004, 63, 1-116.*

2004ACR611:   *Aggarwal, V. K.; Winn, C. L., Acc. Chem. l Res., 2004,* **37***: 611*

2004TL7969:   M. Sureshbabu and K. M. Lokanatha Rai, Tett. Lett., **2004**, 45, 7969-70

2004Misc609:   Hayashi T.; Takahashi M.; Takaya T.; Ogasawara M., Organic Synthesis, 2004, 10, 609.

2005OL3183:   G., Vijaya; G. Alan; D., Stevan, Org. Lett., 2005, **7**, 3183–3186

2005JHC877:   Gaonkar S. L.; Rai K. M. L., J. Het. Chem., 2005, 42, 877.

2005TL5969:   Gaonkar S. L.; Rai K. M. l., Tett. Lett., 2005, 46, 5969.

2005TL5915:   Mori, N.; Togo, H., Tetrahedron 2005, 61, 5915 –5925.

2006Ark119:   Katritzky A. R.; Kirichenko K., Arkivoc, 2006, 119-151.

2006TL6289:   Cyrous O. K, David E. K, Billy W. D, Tett. Lett., **2006**, 47, 6289-6292

2006CSR146:   *Lang, S.; Murphy, J. A. , Chem. Soc. Rev. 2006, 35 (2): 146–156.*

2006CC1218:   Zhang, W.; Shi, M. Chem. Commun. **2006**: 1218–1220.

2006JACS15980:   McGilvray K. L.; Decan M. R.; Wang D.; Scaiano J. C.; J. Am. Chem. Soc.; **2006**; 128, 15980

2006BCC93:   Ajaykumar K and Lokanatha Rai K M., Bulg. Chem. Commun., 2006, 38, 93.

2006JMS1391:   Aparna E.; Lokanatha Rai K. M.; Sureshbabu M.; Jagadish R. L.; Gaonkar S. L.; and Byrappa K., J. Mater. Sci., 2006, 41, 1391.

2007Misc712:   Sachin K. Ghosh (2007), Advanced general organic chemistry, a modern approach, New central book agency (p) Ltd, 712.

2007Misc1359:   *Smith, Michael B.; March, Jerry; March's Advanced Organic Chemistry: Reactions, Mechanisms, and Structure (6th ed.). Hoboken, New Jersey: John Wiley & Sons, Inc. 2007, 1359–1360*

2007Misc: *Piancatelli, G.; Luzzio, F. A. (2007). e-EROS Encyclopedia of Reagents for Organic Synthesis. John Wiley & Sons.*

2007OL3429:   Ekoue-Kovi, K.; Wolf C., Org. Lett. 2007, 9, 3429 –3432.

2007JCSCC3714:   Lee, Sung-Chan; Seung Bum Park, Chem. Commun., 2007, 36, 3714- 3716.

2008EJC6302:   Ekoue-Kovi, K.; Wolf C., Eur. J. Chem., 2008, 14, 6302

2008Mic69:   Lokanatha Rai K. M., in "Synthesis of heterocycles via cycloadditions II" Ed. By A. Hassner, Springer-Verlag, Berlin Heidelberg, Germany, **2008**, 112,1-69 and references therin.

2009Misc493:   Wang, Z., Comprehensive Organic Name Reactions and Reagents, 2009, 493–496

2010TL3486:   Ebraheem M. A.; Rai K. M. L.; Kudva N. U.; Bahjat A. S.; Tett. Lett., 2010, 51, 3486.

2010SC3569:   Ebraheem M. A.; Rai K. M. L., 2010, 40, Synth. Commun., 2010, 40, 3569.

2011JOC0001:   Wrobleski, A.; Coombs, T. C.; Huh, C. W.; Li, S.-W.; Aubé, J. Org. Chem., **2011**, 78, 1

2012SC1804:   Chakrabarthy B. Sharma P. K., 2012, 42, 1804.

2013Sci 59:   Renata, H.; Zhou, Q.; Baran, P. S.. Science. **339** (6115): 59–63

2014Synl2923:   M. Attoui, J.-M. Vatèle, Synlett, **2014**, 25, 2923-2927.

2014OrgMet429:   *N. Ogiso, Eiko; C. C.Hsing; P. Maren; M., Daniel J., Organometallics.2014, 33, 429–432.*

2016EJIC1798:   Hörner, T. G.; Klüfers, P., Eur. J. Inorg. Chem., 2016, 2016, 1798-1807.

2019JCS 131:   Sumana Y Kotian, P M Abishad, K Byrappa and Lokanatha Rai K. M., J. Chem. Sci. (2019) 131:46

FURTHER READING

1. I.L. Finar, Organic Chemistry, ELBS Longmann, Vol. I & II, 1984.
2. S.K. Ghosh, Advanced General Organic Chemistry, Book and Alleied (P) Ltd, 1998.
3. Manual of Organic Chemistry - Dey and Seetharaman.
4. Encyclopedia of Chemical technology – Kirck-Othmer series
5. Basic Principles of Organic Chemistry – Roberts & Caserio
6. Organic chemistry, Ed.G. Mark Loudan, Addison-Wesley Publishig Company, London, 1983.
7. Modern Carbonyl Chemistry, Otera, J. Ed. Wiley-VCH, Weinheim, 2000.
8. March, Jerry (1985), Advanced Organic Chemistry: Reactions, Mechanisms, and Structure (3rd ed.), New York: Wiley,